Computational Methods in Financial Engineering

Erricos J. Kontoghiorghes · Berç Rustem
Peter Winker
Editors

Computational Methods in Financial Engineering

Essays in Honour of Manfred Gilli

 Springer

Prof. Erricos J. Kontoghiorghes
University of Cyprus
Department of Public
and Business Administration
75 Kallipoleos St.
P.O.Box 20537
CY-1678 Nicosia
Cyprus
erricos@ucy.ac.cy

Prof. Berç Rustem
Department of Computing
180 Queen's Gate
South Kensington Campus
Imperial College London
London SW7 2AZ
United Kingdom
br@doc.ic.ac.uk

Prof. Peter Winker
University of Gießen
Department of Economics
Licher Straße 64
35390 Gießen
Germany
peter.winker@wirtschaft.uni-giessen.de

ISBN 978-3-540-77957-5   e-ISBN 978-3-540-77958-2

DOI 10.1007/978-3-540-77958-2

Library of Congress Control Number: 2008921042

Production: le-tex Jelonek, Schmidt & Vöckler GbR, Leipzig
Cover design: WMX Design GmbH, Heidelberg

Printed on acid-free paper

9 8 7 6 5 4 3 2 1

springer.com

This book is dedicated to Manfred Gilli

Preface

Manfred is a well known researcher in computational economics and finance. He is the only known member of this community who has also been a rally racer. His many interests include fine wines, running Marathons, sailing and hill climbing. Despite this latter interest, Manfred has not particularly favoured hill climbing as a computational tool and has preferred extensive research in heuristic algorithms for computationally hard problems. His interests were initially focused on econometrics and subsequently moved to computational finance. He has contributed extensively to the areas of model solution algorithms and optimization in finance. He has forged important links with the finance sector to ensure solid research collaboration, he has helped the academic research community by the successful conferences he has organised and he has helped numerous researchers with their computational problems. The conferences Manfred has organised have contributed to the establishment of two communities. The first such conference was the Computational Economics and Finance Conference in Geneva, 1996, which helped establish the Society for Computational Economics. The most recent such conference was the Computational Management Science Meeting in Geneva, 2007.

Manfred is the author of many research papers and edited volumes in computational economics and finance. He is currently the President of the Society for Computational Economics. This volume is dedicated to Manfred in recognition of his research activities and the generous help and support he has provided to young researchers.

The chapters of this volume are selected to celebrate Manfred's contributions, thus far, in three distinct areas. The first, portfolio optimisation and derivatives, includes contributions in portfolio optimisation with four chapters by Kuhn, Parpas, Rustem; Maringer; Specht, Winker; AitSahlia, Sheu, Pardalos, one chapter focusing on value at risk models by Alentorn, Markose, and two chapters on derivatives by Dalakouras, Kwon, Pardalos; Chiarella, El-Hassan, Kucera. The second area is estimation and classification with two chapters on interest and exchange rates by La Rocca, Perna; Krishnakumar, Neto, three chapters on prediction, classification and decision rules in finance by Genton, Ronchetti; Bugera, Uryasev, Zrazhevsky; García-Almanza, Tsang, Galván-López. The third area is banking, risk and macro-modelling. Contributions are three chapters on risk models by González-Hermosillo, Li; Iori, Deissenberg; Mitschele, Schlottmann, Seese, a chapter on network models in finance by Nagurney, Qiang and two chapters on stochastic models of finance and macroeconomics by Albanese, Trovato; Kendrick, Tucci, Amman.

The editors are most grateful to the anonymous referees for their contribution which helped to increase the quality of the papers substantially. Furthermore, the editors owe thanks to the following people for their support in proof reading and LATEX implementation: Vahidin Jeleskovic, Marianna Lyra, Mark Meyer, Christoph Preussner, Chris Sharpe, and Markus Spory.

Nicosia, London, and Gießen,   *Erricos J. Kontoghiorghes*
January 2008   *Berç Rustem*
                                                                Peter Winker

Contents

List of Contributors

Farid AitSahlia
University of Florida
Gainesville
Florida, 32611, USA
farid@ise.ufl.edu

Claudio Albanese
Level3finance
London, UK
claudio@level3finance.com

Amadeo Alentorn
Centre for Computational Finance
and Economic Agents (CCFEA) and
Old Mutual Asset Managers (UK)
Ltd
Amadeo.Alentorn@omam.co.uk

Hans. M. Amman
Utrecht School of Economics
Utrecht University
Heidelberlaan 8
3584 CS Utrecht, the Netherlands
amman@uu.nl

Vladimir Bugera
University of Florida
ISE, Risk Management and
Financial Engineering Lab
vladimir@bugera.com

Carl Chiarella
School of Finance and Economics
University of Technology, Sydney
P.O. Box 123,
Broadway NSW 2007 Australia
carl.chiarella@uts.edu.au

Georgios V. Dalakouras
University of Florida
Weil Hall
P.O. Box 116595
Gainesville, FL, 32611-6595, USA
gbdalako@ufl.edu

Christophe Deissenberg
Université de la
Méditerranée and GREQAM
Château Lafarge
Route des Milles,
13290 Les Milles, France.
christophe.deissenberg@
  univmed.fr

Nadima El–Hassan
School of Finance and Economics
University of Technology, Sydney
P.O. Box 123
Broadway NSW 2007 Australia
nadima.el-hassan@uts.edu.au

Edgar Galván-López
Department of Computer Science
University of Essex
Wivenhoe Park
Colchester CO4 3SQ, U.K.
egalva@essex.ac.uk

Alma Lilia García-Almanza
Department of Computer Science
University of Essex
Wivenhoe Park
Colchester CO4 3SQ, U.K.
algarc@essex.ac.uk

Marc G. Genton
University of Geneva
Bd du Pont-d'Arve 40
CH-1211 Geneva 4, Switzerland
Marc.Genton@metri.unige.ch

Brenda González-Hermosillo
International Monetary Fund
700 19th Street, NW
Washington, DC 20431
BGONZALEZ@imf.org

Giulia Iori
Department of Economics
City University
Northampton Square
London, EC1V 0HB
United Kingdom
g.iori@city.ac.uk

David A. Kendrick
Department of Economics
University of Texas
Austin
Texas 78712, USA
kendrick@eco.utexas.edu

Jaya Krishnakumar
University of Geneva
Bd du Pont-d'Arve 40
CH-1211 Geneva 4, Switzerland
jaya.krishnakumar@
  metri.unige.ch

Adam Kucera
Integral Energy, Australia
adam.kucera@integral.com.au

Daniel Kuhn
Department of Computing
Imperial College London
180 Queen's Gate
London SW7 2BZ, UK
dkuhn@doc.ic.ac.uk

Roy H. Kwon
University of Toronto
5 King's College Road
Toronto, Ontario, M5S3G8, Canada
rkwon@mie.utoronto.ca

Michele La Rocca
Department of Economics
and Statistics
University of Salerno
Via Ponte Don Melillo
84084 Fisciano (SA), Italy
larocca@unisa.it

Jenny X. Li
Department of Mathematics and
Department of Economics
The Pennsylvania State University
University Park, PA 16802
and Yunnan University of
Finance and Economics
li@math.psu.edu

Dietmar Maringer
Centre for Computational Finance
and Economic Agents (CCFEA)
University of Essex
Wivenhoe Park
Colchester CO4 3SQ
United Kingdom
dmaring@essex.ac.uk

Sheri Markose
Centre for Computational Finance
and Economic Agents (CCFEA)
University of Essex
Wivenhoe Park
Colchester CO4 3SQ
United Kingdom
scher@essex.ac.uk

Andreas Mitschele
Institute AIFB
Universität Karlsruhe (TH)
D-76128 Karlsruhe, Germany
mitschele@aifb.uni-karlsruhe.de

Anna Nagurney
Isenberg School of Management
University of Massachusetts
Amherst, Massachusetts 01003
nagurney@gbfin.umass.edu

David Neto
University of Geneva
Bd du Pont-d'Arve 40
CH-1211 Geneva 4, Switzerland
david.neto@metri.unige.ch

Panos M. Pardalos
University of Florida
Gainesville
Florida, 32611, USA
pardalos@ise.ufl.edu

Panos Parpas
Department of Computing
Imperial College London
180 Queen's Gate
London SW7 2BZ, UK
pp500@doc.ic.ac.uk

Cira Perna
Department of Economics
and Statistics
University of Salerno
Via Ponte Don Melillo
84084 Fisciano (SA), Italy
perna@unisa.it

Qiang Qiang
Isenberg School of Management
University of Massachusetts
Amherst, Massachusetts 01003
qqiang@som.umass.edu

Elvezio Ronchetti
University of Geneva
Bd du Pont-d'Arve 40
CH-1211 Geneva 4, Switzerland
Elvezio.Ronchetti@
  metri.unige.ch

Berç Rustem
Department of Computing
Imperial College London
180 Queen's Gate
London SW7 2BZ, UK
br@doc.ic.ac.uk

Frank Schlottmann
GILLARDON AG financial software
Research Department
Edisonstr. 2
D-75015 Bretten, Germany
frank.schlottmann@gillardon.de

Detlef Seese
Institute AIFB
Universität Karlsruhe (TH)
D-76128 Karlsruhe, Germany
seese@aifb.uni-karlsruhe.de

Yuan-Chyuan Sheu
University of Florida
Gainesville
Florida, 32611, USA
ycsheu@ufl.edu

Katja Specht
Pforzheim University of Applied
Sciences
Tiefenbronner Str. 65
75175 Pforzheim, Germany
katja.specht@hs-pforzheim.de

Manlio Trovato
Merrill Lynch
2 King Edward Street
London EC1A 1HQ and
Imperial College
Department of Mathematics
London SW7 2AZ, UK
manlio_trovato@ml.com

Edward P.K. Tsang
Department of Computer Science
University of Essex
Wivenhoe Park
Colchester CO4 3SQ, U.K.
edward@essex.ac.uk

Marco P. Tucci
Dipartimento di Economia Politica
Università di Siena
Piazza S. Francesco, 7

53100 Siena, Italy
tucci@unisi.it

Stan Uryasev
University of Florida
ISE, Risk Management and
Financial Engineering Lab
uryasev@ufl.edu

Peter Winker
Justus-Liebig-Universität Gießen
Licher Straße 64
35394 Gießen, Germany
Peter.Winker@
    wirtschaft.uni-giessen.de

Grigory Zrazhevsky
University of Florida
ISE, Risk Management and
Financial Engineering Lab

Part I

Portfolio Optimization and Option Pricing

Threshold Accepting Approach to Improve Bound-based Approximations for Portfolio Optimization

Daniel Kuhn[1], Panos Parpas[2], and Berç Rustem[3]

[1] Department of Computing, Imperial College London, 180 Queen's Gate, London SW7 2BZ, UK. dkuhn@doc.ic.ac.uk
[2] Department of Computing, Imperial College London, 180 Queen's Gate, London SW7 2BZ, UK. pp500@doc.ic.ac.uk
[3] Department of Computing, Imperial College London, 180 Queen's Gate, London SW7 2BZ, UK. br@doc.ic.ac.uk

Summary. A discretization scheme for a portfolio selection problem is discussed. The model is a benchmark relative, mean-variance optimization problem in continuous time. In order to make the model computationally tractable, it is discretized in time and space. This approximation scheme is designed in such a way that the optimal values of the approximate problems yield bounds on the optimal value of the original problem. The convergence of the bounds is discussed as the granularity of the discretization is increased. A threshold accepting algorithm that attempts to find the most accurate discretization among all discretizations of a given complexity is also proposed. Promising results of a numerical case study are provided.

Key words: Portfolio optimization, stochastic programming, time discretization, bounds, threshold accepting.

1 Introduction

Portfolio problems are prototypical examples of optimization problems under uncertainty. Finance researchers and investment practitioners remain fascinated by numerous questions around the inherent risk-return tradeoff, which implies that invested money can yield above average profits only if it is subject to the possibility of being lost. The modelling of financial risks and investor preferences as well as the solution of large-scale portfolio optimization problems are challenging tasks. A principal thesis of this article is the consensus that realistic portfolio problems must be cast in a multi-period setting and can only be tackled computationally. In fact, investment decisions are coupled in time through serial dependencies of returns, nonseparable utility functions, and, to a certain extent, through transaction costs, administrative fees, and taxes. Analytical solutions of such time-coupled models are usually not available. Numerical solutions, on the other hand, always rely on suitable approximations. In this paper we pursue the following strategy. Starting from a continuous-time continuous-state model, we discretize both the underlying planning horizon as well as the probability space of the random asset returns. This joint time and space discretization allows us to approximate the original investment problem by a sequence of multistage stochastic programs defined on finite scenario trees. The trees are designed in such a way that the solutions of the corresponding multistage stochastic programs provide bounds on the solution of the original problem. The effectiveness of discretization schemes depends on our ability to solve the resulting deterministic equivalent problems. Clearly, as the granularity of the discretization increases, an appropriate scheme should lead to a reduction in the error. From a practical point of view, however, we cannot solve approximate problems whose scenario trees exceed a certain size. For this reason, we attempt to optimize the tree generation by means of a heuristic algorithm. The heuristic we use is called *threshold accepting* and constitutes a variant of simulated annealing, see Dueck and Scheuer (1990). The basic idea is to perform a random walk on all possible scenario tree configurations of a given complexity. A new tree configuration is accepted if its quality exceeds the quality of the previous tree minus some threshold parameter. For our purposes, the complexity of a scenario tree is defined as the number of its nodes. Threshold accepting will allow us to find good scenario tree configurations and an optimal tradeoff between time and space discretization. Its advantage over other heuristics is that it requires less parameter tuning in order to compute acceptable solutions.

In the present paper, we generalize the bounding approximation scheme proposed in Kuhn (2007a,b) to stochastic programs with random technology matrices. Moreover, we develop a modelling framework in which the tradeoff between the granularity of the time and space approximations can be analyzed. Theoretical convergence results due to Kuhn (2007b) guarantee the asymptotic consistency of our discretized approximate stochastic programs as the cardinality of the underlying scenario trees increases. In this work, we go one step further and devise an efficient strategy to control the approximation error in situations when computational resources such as storage space and CPU power are limited. A novel contribution of this paper is the integrated use of bounding methods and stochastic search algorithms for finding optimal scenario trees. We show empirically that an intelligent scenario tree design can significantly reduce the approximation error, that is, the difference of upper and lower bounds, without increasing the tree size.

Since the pioneering work of Markowitz (1952), the literature on portfolio theory has grown dramatically. An overview of influential contributions up to 2001 is provided in Steinbach (2001). Much of the earlier literature focuses on investment situations in a single period setting. Threshold accepting approaches, for example, have been used by Gilli and Këllezi (2002) and Gilli et al. (2006) to solve nonconvex static portfolio optimization problems.

Note that in this article we propose a threshold accepting algorithm for optimally discretizing portfolio problems and not for solving them. Dynamic portfolio optimization in continuous time was popularized by Merton (1992). While most dynamic models maximize expected utility of terminal wealth and or intertemporal consumption, the traditional mean-variance approach due to Markowitz has received relatively little attention in the multi-period situation. A stochastic programming based approach to multi-period mean-variance analysis was proposed by Frauendorfer (1995) as well as Frauendorfer and Siede (2000). Gülpinar et al. (2002, 2004) refined this model and tested its performance for different scenario trees. Furthermore, they reformulated the model as a stochastic min-max problem, assuming that some parameters of the underlying return distributions are ambiguous, see Gülpinar and Rustem (2006). Recently, dynamic mean-variance portfolio problems without transaction costs have also been studied in a continuous-time framework. Analytical solutions were first obtained by Zhou and Li (2000) who employ an embedding technique and certain concepts of stochastic linear-quadratic control theory.

In the presence of market frictions, portfolio constraints, and or parameter instability etc., the analytical treatment of dynamic mean-variance portfolio problems is out of the question, and one has to resort to algorithmic solution procedures. Below, we will use stochastic programming techniques to solve the portfolio problems under consideration, see e.g. Birge and Louveaux (1997) or Kall and Wallace (1994). Suitable approximation schemes are necessary to keep the corresponding scenario trees of tractable size. In order to find solutions with provable performance guarantee, we use bounding methods to discretize the planning horizon and the probability space of the asset returns. Stage-aggregation and time-discretization methods with deterministic error bounds have previously been discussed in Wright (1994) and Kuhn (2007a). Bounding methods for scenario generation have first been proposed by Madansky (1960) and — in a game-theoretic setting — by Dupačová (1966). Extensions to more general problem classes are due to Edirisinghe and Ziemba (1994b,a), Frauendorfer (1994, 1996), and Kuhn (2007a,b).

The general outline of this article is as follows. In Section 2 we formulate a mean-variance portfolio selection problem in continuous time with transaction costs and portfolio constraints. In order to achieve computational tractability, we develop a scenario tree based approximation for this problem. We proceed in two steps: Section 3 discusses discretization of the underlying planning horizon, while Sections 4 and 5 address discretization of the underlying probability space. In both cases we carefully estimate the approximation error. The time discretization scheme of Section 3 is tailored to the portfolio problem under consideration. In contrast, the space discretization scheme of Section 5 applies to a broad class of stochastic programs detailed in Section 4. We propose a randomized threshold accepting algorithm to minimize the approximation error over all scenario trees of a given cardinality. Section 6 reports on numerical results.

2 Portfolio Problem

All random objects appearing in this article are defined on a complete probability space (Ω, Σ, P). By convention, random objects (i.e., random variables, random vectors, or stochastic processes) appear in boldface, while their realizations are denoted by the same symbols in normal script. Throughout our exposition, we consider a market of N investment opportunities or assets. We assume that the price $\boldsymbol{p}_n$ of asset n is driven by a geometric Brownian motion, that is,

$$\frac{\mathrm{d}\boldsymbol{p}_n(t)}{\boldsymbol{p}_n(t)} = \mu_n \, \mathrm{d}t + \mathrm{d}\boldsymbol{z}_n(t) \quad \text{for} \quad n = 1, \ldots, N \, ,$$

where the $\boldsymbol{z}_n$ are correlated Wiener processes whose variance rates may differ from 1. Concretely speaking, we assume that

$$\mathrm{cov}(\mathrm{d}\boldsymbol{z}_m, \mathrm{d}\boldsymbol{z}_n) = \mathrm{E}(\mathrm{d}\boldsymbol{z}_m \, \mathrm{d}\boldsymbol{z}_n) = \sigma_{mn} \, \mathrm{d}t \, .$$

The limitation to geometric Brownian motions with a time-invariant correlation structure is merely for expositional convenience. Note that our approach easily extends to more general stochastic models. We denote by $\mathbb{A} := \{\mathcal{A}^t\}_{t \geq 0}$ the filtration generated by the asset prices. The underlying σ-fields are thus given by

$$\mathcal{A}^t := \sigma(\boldsymbol{p}_n(s) \,|\, s \in [0, t], \, n = 1, \ldots, N) \, .$$

It is convenient to interpret $\mathcal{A}^t$ as the time-t information set of an investor who continuously observes the asset prices. Let us assume that this investor holds a dynamically rebalanced portfolio of the assets in our model economy. This means that for each n, the investor continuously buys and sells assets of type n, pursuing the general goal of accumulating wealth over time. We denote by $\boldsymbol{b}_n$ and $\boldsymbol{s}_n$ the cumulative purchase and sale processes, respectively. To be precise, we assume that $\boldsymbol{b}_n$ and $\boldsymbol{s}_n$ are elements of $C(\mathbb{A})$, the convex cone of right-continuous non-decreasing stochastic processes adapted to $\mathbb{A}$ that vanish at $t = 0$. Thus, $\boldsymbol{b}_n(t)$ stands for the amount of money used up to time t to buy assets of type n, while $\boldsymbol{s}_n(t)$ represents the amount of money obtained up to time t from sales of asset n. Moreover, we let $\boldsymbol{w}_n$ be the portfolio weight process of asset n, i.e., $\boldsymbol{w}_n(t)$ denotes the amount of money invested in asset n at time t. The processes $\boldsymbol{w}_n$, $\boldsymbol{b}_n$, and $\boldsymbol{s}_n$ are related through the asset balance equations

$$\mathrm{d}\boldsymbol{w}_n(t) = \boldsymbol{w}_n(t) \frac{\mathrm{d}\boldsymbol{p}_n(t)}{\boldsymbol{p}_n(t)} + \mathrm{d}\boldsymbol{b}_n(t) - \mathrm{d}\boldsymbol{s}_n(t) \, . \tag{1}$$

The total wealth process $\boldsymbol{\pi}$ is naturally defined via $\boldsymbol{\pi}(t) = \sum_{n=1}^{N} \boldsymbol{w}_n(t)$. Throughout this article we assume that proportional transaction costs are incurred whenever shares of the assets are traded. We denote by c_b and c_s

the transaction costs per dollar of shares bought and sold, respectively. The change of total wealth over a time interval of length dt is thus given by capital gains minus transaction costs, i.e.,

$$d\boldsymbol{\pi}(t) = \sum_{n=1}^{N} \boldsymbol{w}_n(t) \frac{d\boldsymbol{p}_n(t)}{\boldsymbol{p}_n(t)} - c_b\, d\boldsymbol{b}_n(t) - c_s\, d\boldsymbol{s}_n(t)\,. \tag{2}$$

Plugging the definition of $\boldsymbol{\pi}$ into (2) and using (1) we obtain

$$0 = \sum_{n=1}^{N} (1 + c_b)\, d\boldsymbol{b}_n(t) - (1 - c_s)\, d\boldsymbol{s}_n(t)\,. \tag{3}$$

The equations (1) and (3) are sufficient to describe the portfolio dynamics. Hence, the SDE (2) is not needed, and the process $\boldsymbol{\pi}$ can principally be eliminated. We will always impose the restrictions $\boldsymbol{w}_n(t) \geq 0$, which preclude the possibility to take short positions in the assets. These no short sales restrictions further guarantee that total wealth will always remain positive.

The performance of an investment portfolio of the above kind is usually not measured in absolute terms but rather relative to a benchmark portfolio. A natural choice for a benchmark is the market portfolio $\boldsymbol{\pi}^*(t) := \sum_{n=1}^{N} \boldsymbol{w}_n^*(t)$ with normalized initial value. By definition, the monetary portfolio weight of asset n in the market portfolio is given by $\boldsymbol{w}_n^*(t) := \boldsymbol{w}_n(0)\, \boldsymbol{p}_n(t)/\boldsymbol{p}_n(0)$. We will assume that the risk associated with a portfolio process $\boldsymbol{\pi}$, as perceived by our investor, amounts to

$$\int_0^T \mathrm{Var}\left(\boldsymbol{\pi}(t) - \boldsymbol{\pi}^*(t)\right) \alpha(dt)\,. \tag{4}$$

The involved probability measure $\alpha : \mathcal{B}([0,T]) \to [0,1]$ characterizes the temporal risk weighting. By using a deterministic benchmark and defining α as the Dirac measure concentrated at T, the risk functional (4) reduces to variance of terminal wealth. Portfolio performance, on the other hand, is measured by expected terminal wealth. As in the celebrated single-period Markowitz model (Markowitz, 1952) our investor faces a multi-objective decision problem. The conflicting objectives are to minimize risk while maximizing expected terminal wealth. Such problems are usually addressed by determining the set of Pareto optimal solutions, which in the current setting represent a family of *efficient portfolios*. We can find all efficient portfolios by minimizing risk under a parametric constraint of the form

$$\mathrm{E}\left(\boldsymbol{\pi}(T)\right) \geq e^{\varrho T}\, \mathrm{E}(\boldsymbol{\pi}^*(T))\,, \tag{5}$$

where the performance parameter ϱ is swept within a suitable interval. The investor will then pick the ϱ corresponding to that efficient portfolio which is best suited to his or her individual risk preferences. Moreover, the investment strategy associated with a particular efficient portfolio can be found by solving the following stochastic optimization problem.

$$\underset{b_n, s_n \in C(\mathbb{A})}{\text{minimize}} \int_0^T \text{Var}\left(\sum_{n=1}^N w_n(t) - w_n^*(t)\right) \alpha(dt)$$

$$\left.\begin{aligned}
\text{s.t. } &\mathrm{d}w_n(t) = w_n(t)\, \mathrm{d}p_n(t)/p_n(t) + \mathrm{d}b_n(t) - \mathrm{d}s_n(t) \;\forall t, n \\
&0 = \sum_{n=1}^N (1 + c_b)\, \mathrm{d}b_n(t) - (1 - c_s)\, \mathrm{d}s_n(t) \qquad \forall t \\
&0 \le \mathrm{E}(\sum_{n=1}^N w_n(T) - e^{\varrho T}\, w_n^*(T)) \\
&0 \le w_n(t), \qquad\qquad\qquad\qquad\qquad\qquad \forall t, n
\end{aligned}\right\} \begin{aligned}(\mathcal{P}_\text{c})\\ P\text{-a.s.}\end{aligned}$$

The presence of portfolio constraints and transaction costs severely limits analytical tractability of problem $\mathcal{P}_\text{c}$, which we consider as a prototype model that will eventually be generalized in several directions. For example, we plan to incorporate additional portfolio constraints and alternative *one-sided* risk measures. Moreover, we intend to replace our crude asset price model by a more sophisticated one that accounts for heteroscedasticity, fat tails, parameter instability, etc. Against this background, we should seek a computational approach for solving problem $\mathcal{P}_\text{c}$. This requires discretization of the problem with respect to time and (probability) space.

3 Time Discretization

To work towards computational tractability of problem $\mathcal{P}_\text{c}$, we first simplify its temporal complexity. The goal of this section is to elaborate a stochastic optimization problem in discrete time that approximates the continuous-time problem $\mathcal{P}_\text{c}$. To this end, we select a set of ordered time points $0 = t_1 < \cdots < t_H < t_{H+1} = T$ and assume that portfolio rebalancing is restricted to these discrete dates. For notational convenience, we will sometimes use an additional time point $t_0 := 0$. The optimal value of the approximate problem $\mathcal{P}$ with this extra restriction will provide an upper bound on the optimal value of the original problem $\mathcal{P}_\text{c}$.

In order to formulate $\mathcal{P}$ as a discrete-time problem, we have to introduce additional notation. Let $w_{n,h}^-$ be the capital invested in asset n at time $t_h - 0$ before reallocation of funds, and let $w_{n,h}^+$ be the capital invested in asset n at time $t_h + 0$ after portfolio rebalancing. Furthermore, denote by $b_{n,h}$ and $s_{n,h}$ the amount of money used to buy and sell assets of type n at time t_h, respectively. In agreement with these definitions, $w_{n,h}^*$ stands for the capitalization of asset n in the market portfolio at time t_h — note that the market portfolio is never rebalanced. By convention, initial wealth is normalized, i.e., $\sum_{n=1}^N w_{n,1}^- = 1$. Recall also that the deterministic initial endowments satisfy $w_{n,1}^- = w_{n,1}^*$. Finally, it is convenient to introduce separate variables for the expected values of the individual asset positions. Thus, we define

$\bar{w}_{i,h}^+ = \mathrm{E}(w_{i,h}^+)$ and $\bar{w}_{i,h}^* = \mathrm{E}(w_{i,h}^*)$. By using the notation introduced so far, the objective function of the time-discretized version of problem $\mathcal{P}_c$ can be represented as

$$\sum_{h=1}^{H} \int_{I_h} \mathrm{Var}\left(\sum_{n=1}^{N}(w_{n,h}^+ - w_{n,h}^*)\frac{\boldsymbol{p}_n(t)}{\boldsymbol{p}_n(t_h)}\right)\alpha(\mathrm{d}t)\,, \tag{1}$$

where $I_h := [t_h, t_{h+1})$ for $h < H$ and $I_H := [t_H, t_{H+1}]$ denote the time intervals between consecutive rebalancing dates. This expression can be further simplified by using the constants

$$\Delta_{mn,h} := \int_{I_h} \mathrm{E}\left(\frac{\boldsymbol{p}_m(t)}{\boldsymbol{p}_m(t_h)}\right)\mathrm{E}\left(\frac{\boldsymbol{p}_n(t)}{\boldsymbol{p}_n(t_h)}\right)\alpha(\mathrm{d}t) = \int_{I_h} e^{(\mu_m+\mu_n)(t-t_h)}\alpha(\mathrm{d}t)$$

and

$$\Lambda_{mn,h} := \int_{I_h} \mathrm{Cov}\left(\frac{\boldsymbol{p}_m(t)}{\boldsymbol{p}_m(t_h)}, \frac{\boldsymbol{p}_n(t)}{\boldsymbol{p}_n(t_h)}\right)\alpha(\mathrm{d}t)$$

$$= \int_{I_h} e^{(\mu_m+\mu_n)(t-t_h)}\left(e^{\sigma_{mn}(t-t_h)} - 1\right)\alpha(\mathrm{d}t)\,.$$

After some elementary manipulations, the objective function (1) takes the form

$$\mathrm{E}\left(\sum_{h=1}^{H}\sum_{m,n=1}^{N}(w_{m,h}^+ - w_{m,h}^*)(\Lambda_{mn,h} + \Delta_{mn,h})(w_{n,h}^+ - w_{n,h}^*)\right. \tag{2}$$

$$\left. -(\bar{w}_{m,h}^+ - \bar{w}_{m,h}^*)\Delta_{mn,h}(\bar{w}_{n,h}^+ - \bar{w}_{n,h}^*)\right).$$

Next, we reformulate the constraints of the time-discretized version of problem $\mathcal{P}_c$. Our reformulation invokes vectors of price relatives

$$\boldsymbol{\xi}_h := (\boldsymbol{\xi}_{1,h}, \dots, \boldsymbol{\xi}_{N,h})\,, \quad \text{where} \quad \boldsymbol{\xi}_{n,h} := \boldsymbol{p}_n(t_h)/\boldsymbol{p}_n(t_{h-1})\,.$$

Price relatives relate to the time periods between successive rebalancing dates. In our setting, they are serially independent. We define $\mathcal{F}^h := \sigma(\boldsymbol{\xi}_1, \dots, \boldsymbol{\xi}_h)$ to be the σ-algebra induced by the first h price relatives. All constraints discussed below are assumed to hold almost surely with respect to the probability measure P. A set of static constraints applies for each $h = 1, \dots, H$.

$$w_{n,h}^+ = w_{n,h}^- + b_{n,h} - s_{n,h}$$
$$(1 + c_b)\sum_{n=1}^{N} b_{n,h} = (1 - c_s)\sum_{n=1}^{N} s_{n,h}$$
$$\bar{w}_{n,h}^+ = \mathrm{E}(w_{n,h}^+)\,, \ \bar{w}_{n,h}^* = \mathrm{E}(w_{n,h}^*) \tag{3a}$$
$$w_{n,h}^+ \ge 0\,, \ b_{n,h} \ge 0\,, \ s_{n,h} \ge 0$$
$$w_{n,h}^-\,, \ w_{n,h}^+\,, \ \bar{w}_{n,h}^+\,, \ \bar{w}_{n,h}^*\,, \ b_{n,h}\,, \ s_{n,h} \text{ are } \mathcal{F}^h\text{-measurable}$$

The following dynamic constraints couple neighboring decision stages and therefore only hold for $h = 1, \ldots, H - 1$.

$$w^-_{n,h+1} = w^+_{n,h}\,\boldsymbol{\xi}_{n,h+1}, \quad w^*_{n,h+1} = w^*_{n,h}\,\boldsymbol{\xi}_{n,h+1} \tag{3b}$$

The terminal wealth constraint represents a single static constraint associated with decision stage H.

$$\sum_{n=1}^{N} \bar{w}^+_{n,H} \geq e^{\varrho T} \sum_{n=1}^{N} \bar{w}^*_{n,H} \tag{3c}$$

In summary, we denote by $\mathcal{P}$ the problem of minimizing the objective function (2) over all investment strategies subject to (3). Any strategy feasible in $\mathcal{P}$ corresponds to a strategy feasible in $\mathcal{P}_c$ with the same objective value. Therefore, the approximate problem $\mathcal{P}$ provides an upper bound on the optimal value of the original problem $\mathcal{P}_c$. Even though $\mathcal{P}$ has a finite number of decision stages, it involves a continuum of scenarios and thus remains computationally intractable.

4 Multistage Stochastic Programs

Discrete-time stochastic optimization problems involving a large or infinite number of scenarios are usually addressed by approximating the underlying random data by a discrete process that involves only a modest number of scenarios. This amounts to discretizing the state space of the data generating process. The stochastic program associated with the discrete approximate process has a finite number of variables and constraints, thus principally allowing for numerical solution.

Stochastic programming research puts much effort in determining whether the optimal values of the approximate and original problems are close to each other and whether the optimal solution of the approximate problem can be related in a quantitative way to some near-optimal solution of the original problem. In this work we employ a discretization method which provides deterministic error bounds on the optimal value of the original problem. Moreover, we outline how the solution of the approximate problem can be transformed to a policy that is implementable in reality and achieves an objective value between those bounds. As our method applies to a broad problem class, we explain it by referring to an abstract cost minimization problem under uncertainty. This generalized stochastic program should incorporate expected value constraints, which are also present in the portfolio optimization problem of Section 3. As usual, we assume that decisions are selected at different time points (or stages) indexed $h = 1, \ldots, H$ when new information about the underlying random parameters becomes available. To keep our exposition reasonably simple, we now introduce some notational conventions applying to all discrete-time stochastic processes considered below.

Definition 4.1 *We say that ζ is a discrete-time stochastic process with state space Z if $\zeta = (\zeta_1, \ldots, \zeta_H)$ and $Z = \times_{h=1}^{H} Z_h$ such that each random vector ζ_h maps (Ω, Σ) to the Borel space $(Z_h, \mathcal{B}(Z_h))$ and each Z_h is a convex closed subset of some finite-dimensional Euclidean space. Moreover, we define combined random vectors $\zeta^h := (\zeta_1, \ldots, \zeta_h)$ valued in $Z^h := \times_{i=1}^{h} Z_i$ for all $h = 1, \ldots, H$.*

As an example we mention the process $\xi := (\xi_1, \ldots, \xi_H)$, where ξ_h represents the vector of price relatives introduced in Section 3. As stock prices never drop below zero, ξ constitutes a discrete-time stochastic process in the sense of Definition 4.1, whose state space is given by the nonnegative orthant of $\mathbb{R}^{NH}$. Furthermore, we can introduce a decision process x associated with problem $\mathcal{P}$ in Section 3. It is defined through

$$x_h := \mathrm{vec}((w_{n,h}^-, w_{n,h}^+, \bar{w}_{n,h}^+, w_{n,h}^*, \bar{w}_{n,h}^*, b_{n,h}, s_{n,h}) \,|\, n = 1, \ldots, N),$$

where the operator 'vec' returns the concatenation of its arguments. Thus, x_h has dimension $n_h := 7N$. As the constraints (3) preclude negative x_h, the state space of the process x can be chosen to be the nonnegative orthant of $\mathbb{R}^{7NH}$.

In order to present the most general form of our approximation method, we assume that the multistage stochastic program under consideration is driven by two exogenous stochastic processes η and ξ with state spaces Θ and Ξ, respectively. We assume that η impacts the objective function of $\mathcal{P}$, whereas ξ appears in the constraints. For notational convenience, we also introduce the combined data process $\zeta := (\eta, \xi)$ with state space $Z := \Theta \times \Xi$. Furthermore, we let $\mathcal{F}^h := \sigma(\zeta^h)$ be the information that is available at stage h by observing the data process. We will frequently use the shorthand notation $\mathcal{F} := \mathcal{F}^H$ and denote by $\mathbb{F} := \{\mathcal{F}^h\}_{h=1}^{H}$ the filtration generated by ζ. The process of price relatives shows up only in the constraints of the portfolio problem of Section 3. In agreement with the current convention, it is therefore denoted by ξ. There is no (or only a fictitious deterministic) process η that affects the objective function of that problem.

With the above conventions, we can formulate an abstract multistage stochastic program with expected value constraints as follows.

$$\begin{aligned} &\underset{x \in X(\mathbb{F})}{\text{minimize}} \ \mathrm{E}\left(c(x, \eta)\right) \\ &\text{s.t.} \quad \mathrm{E}(f_h(x, \xi) \,|\, \mathcal{F}^h) \leq 0 \quad P\text{-a.s. } \forall h = 1, \ldots, H \end{aligned} \qquad (\mathcal{P})$$

Minimization is over a convex set of stochastic processes x, all of which share the same state space $X \subset \times_{h=1}^{H} \mathbb{R}^{n_h}$. These processes are commonly called strategies, policies, or decision processes. The set of admissible strategies is defined as

$$X(\mathbb{F}) := \{x \in \times_{h=1}^{H} \mathcal{L}^\infty(\Omega, \mathcal{F}^h, P; \mathbb{R}^{n_h}) \,|\, x(\omega) \in X \text{ for } P\text{-a.e. } \omega \in \Omega\}.$$

Note that all strategies in $X(\mathbb{F})$ are adapted to $\mathbb{F}$, that is, they are *non-anticipative* with respect to the underlying data process, see e.g. Rockafellar and Wets (1978). The objective criterion and the (explicit) constraints in $\mathcal{P}$ are determined through a real-valued cost function $c : X \times \Theta \to \mathbb{R}$ and a vector-valued constraint function $f_h : X \times \Xi \to \mathbb{R}^{m_h}$ for each stage index $h = 1, \ldots, H$. We emphasize that problem $\mathcal{P}$ accommodates expected value constraints. The corresponding constraint functions are required to be nonpositive in expectation (instead of almost everywhere), where expectation is conditional on the stagewise information sets. In order to elaborate our space discretization scheme, we impose the following regularity conditions:

(C1) c is convex in x, concave in η, and continuous;
(C2) f_h is representable as $f_h(x, \xi) = \tilde{f}_h((1, x) \otimes (1, \xi))$, where $\tilde{f}_h$ is convex, continuous, and constant in $x_i \otimes \xi_j$ for all $1 \leq j \leq i \leq H$, $h = 1, \ldots, H$;
(C3) X is convex and compact;
(C4) ζ is a serially independent process with compact state space Z.

Here, the operator '$\otimes$' stands for the usual dyadic product of vectors. The assumptions (C1) and (C2) allow for linear multistage stochastic programs which exhibit randomness in the objective function coefficients, the right hand side vectors, and the technology matrices (but not in the recourse matrices). As reasonable numerical solutions are always bounded, assumption (C3) is nonrestrictive. The compactness requirement in (C4) can always be enforced by truncating certain extreme scenarios of the data process ζ that have a negligible effect on the solution of the stochastic program $\mathcal{P}$. Moreover, the serial independence requirement can often be circumvented by rewriting the original data process as a Rosenblatt transformation of a serially independent noise process, see Rosenblatt (1952). In this case, any serial dependencies can be absorbed in the definition of the cost and constraint functions.

For the further discussion, we introduce the set

$$Y(\mathbb{F}) := \{ \boldsymbol{y} \in \times_{h=1}^{H} \mathcal{L}^1(\Omega, \mathcal{F}^h, P; \mathbb{R}^{m_h}) \,|\, \boldsymbol{y} \geq 0 \ P\text{-a.s.} \},$$

which comprises all nonnegative integrable *dual* decision processes adapted to the filtration $\mathbb{F}$. By using $Y(\mathbb{F})$, the stochastic optimization problem $\mathcal{P}$ *with* explicit constraints can be recast as a min-max problem *without* explicit constraints.

Lemma 1. (Wright, 1994, § 4) *Under the conditions (C1)–(C3) we have*

$$\inf \mathcal{P} = \inf_{\boldsymbol{x} \in X(\mathbb{F})} \sup_{\boldsymbol{y} \in Y(\mathbb{F})} \mathrm{E} \left(c(\boldsymbol{x}, \boldsymbol{\eta}) + \sum_{h=1}^{H} \boldsymbol{y}_h \cdot f_h(\boldsymbol{x}, \boldsymbol{\xi}) \right). \tag{1}$$

The min-max formulation (1) of problem $\mathcal{P}$ will be very helpful for the development of our space discretization scheme in Section 5.

5 Space Discretization

Recall that we assume ζ to be a serially independent process with compact state space Z. Besides that we make no assumption concerning its distribution.

We now construct another process $\zeta^u = (\eta^u, \xi^u)$ such that the component processes η^u and ξ^u have state spaces Θ and Ξ, respectively. Thus, ζ^u has the same state space $Z = \Theta \times \Xi$ as the original process ζ. Furthermore, ζ^u has a discrete distribution and approximates ζ in a sense that will be explained below.

The construction of ζ^u goes along the lines of (Kuhn, 2007a, § 4). To keep this article self-contained, we briefly sketch the basic procedure. Essentially, ζ^u is constructed by specifying its conditional distribution given ζ. Then, the joint distribution of ζ and ζ^u is uniquely determined by the product measure theorem (Ash, 1972, Theorem 2.6.2). There always exists a rich enough sample space (Ω, Σ, P) on which both ζ and ζ^u are defined. In fact, we are free to set $\Omega = Z \times Z$ and $\Sigma = \mathcal{B}(Z \times Z)$, while P may be identified with the joint distribution of ζ and ζ^u.

The distribution of ζ^u conditional on $\zeta = \zeta$ is obtained from the product measure theorem by combining the conditional distributions $\{P_h^u\}_{h=1}^H$, where P_h^u stands for the distribution of ζ_h^u conditional on $\zeta = \zeta$ and $\zeta^{u,h-1} = \zeta^{u,h-1}$. Thus, construction of the approximate process ζ^u reduces to specifying P_h^u for each stage index h. In order to do so, we select a disjoint set partition $\mathcal{Z}_h$ of Z_h which depends measurably on $\zeta^{u,h-1}$ and satisfies for all $\zeta^{u,h-1} \in Z^{h-1}$

$$Z_h = \bigcup_{W \in \mathcal{Z}_h(\zeta^{u,h-1})} W, \quad W \cap W' = \emptyset \quad \text{for} \quad W \neq W' \in \mathcal{Z}_h(\zeta^{u,h-1}).$$

We say that the set partition $\mathcal{Z}_h$ depends measurably on $\zeta^{u,t-1}$ if $\mathcal{Z}_h(\zeta^{u,t-1})$ consists of disjoint sets $W_h^i(\zeta^{u,t-1})$, $i = 1, \ldots, I_h$, for all $\zeta^{u,h-1} \in Z^{h-1}$, and each $W_h^i : Z^{t-1} \rightrightarrows Z_h$ constitutes a measurable multifunction. Moreover, we require each $W \in \mathcal{Z}_h(\zeta^{u,h-1})$ to be representable as $W = U \times V$ where $U \subset \Theta_h$ and $V \subset \Xi_h$ are bounded nondegenerate (not necessarily closed) simplices. For a bounded nondegenerate simplex V, an extreme point $e \in \text{ext cl} V$, and a vector $v \in \text{cl} V$ we denote by $P_V(e|v)$ the convex weight of v with respect to e. By definition, we thus have

$$\sum_{e \in \text{ext cl} V} P(e|v) = 1 \quad \text{and} \quad \sum_{e \in \text{ext cl} V} e P(e|v) = v.$$

Moreover, we introduce a combined vector $\zeta_h^V(e) = (\eta_h^V(e), \xi_h^V(e))$ defined through

$$\xi_h^V(e) := e \quad \text{and} \quad \eta_h^V(e) := \frac{\mathrm{E}(\eta_h\, P_V(e|\xi_h)\, 1_W(\zeta_h))}{\mathrm{E}(1_W(\zeta_h))}.$$

Without much loss of generality, we may assume that the denominator in the rightmost expression is always nonzero. Using the notation introduced so far, we can specify the conditional distribution P_h^u. For $\zeta \in Z$ and $\zeta^{u,h-1} \in Z^{h-1}$ we set

$$P_h^u(\cdot|\zeta, \zeta^{u,h-1}) := \sum_{\substack{W \in \mathcal{Z}_h(\zeta^{u,h-1}) \\ W = U \times V}} \sum_{e \in \text{ext cl} V} P_V(e|\xi_h)\, 1_W(\zeta_h)\, \delta_{\zeta_h^V(e)}(\cdot), \quad (1)$$

where $\delta_{\zeta_h^V(e)}$ denotes the Dirac measure concentrated at $\zeta_h^V(e)$. The approximate process ζ^u obtained by combining the P_h^u in the appropriate way has several intriguing properties. First, by averaging (1) over all ζ_h, one verifies that the marginal distribution of ζ^u constitutes a *barycentric scenario tree* in the sense of Frauendorfer (1994, 1996). As opposed to the traditional construction of barycentric scenario trees, the present approach allows us to view ζ^u and ζ as correlated processes on a joint probability space. In the reminder of this section we will elaborate some of the advantages of this new perspective.

From now on we denote by $\mathbb{F}^u$ the filtration generated by ζ^u, that is, $\mathbb{F}^u := \{\mathcal{F}^{u,h}\}_{h=1}^H$ where $\mathcal{F}^{u,h} := \sigma(\zeta^{u,h})$, and we use the convention $\mathcal{F}^u := \mathcal{F}^{u,H}$.

Lemma 2. (Kuhn, 2007a, § 4) *The following relations hold for suitable versions of the conditional expectations, respectively.*

$$\mathrm{E}(\boldsymbol{x}|\mathcal{F}) \in X(\mathbb{F}) \text{ for all } \boldsymbol{x} \in X(\mathbb{F}^u) \tag{2a}$$

$$\mathrm{E}(\boldsymbol{y}|\mathcal{F}^u) \in Y(\mathbb{F}^u) \text{ for all } \boldsymbol{y} \in Y(\mathbb{F}) \tag{2b}$$

$$\mathrm{E}(\boldsymbol{\xi}^u|\mathcal{F}) = \boldsymbol{\xi} \tag{2c}$$

$$\mathrm{E}(\boldsymbol{\eta}|\mathcal{F}^u) = \boldsymbol{\eta}^u \tag{2d}$$

The relations (2) are crucial for our main result on the approximation of discrete-time stochastic programs; see Theorem 1 below. A rigorous proof of Lemma 2 is provided in (Kuhn, 2007a, § 4). Here, we only give some intuition. The inclusions (2a) and (2b) are related to the serial independence of $\{\zeta_h\}_{h=1}^H$ and the recursive construction of the $\{\zeta_h^u\}_{h=1}^H$. Relation (2c) follows immediately from the definition of P_h^u. In fact, we have

$$\mathrm{E}(\boldsymbol{\xi}_h^u|\zeta, \zeta^{u,h-1}) = \int_{Z_h} \boldsymbol{\xi}_h^u\, P_h^u(\mathrm{d}\zeta_h^u|\zeta, \zeta^{u,h-1}) \tag{3}$$

$$= \sum_{\substack{W \in \mathcal{Z}_h(\zeta^{u,h-1}) \\ W = U \times V}} 1_W(\zeta_h) \sum_{e \in \text{ext cl } V} e\, P_V(e|\boldsymbol{\xi}_h) = \boldsymbol{\xi}_h \quad P\text{-a.s.}$$

The law of iterated conditional expectations then allows us to conclude that $\mathrm{E}(\boldsymbol{\xi}_h^u|\zeta) = \boldsymbol{\xi}_h$ for all h, which is equivalent to (2c). The last relation (2d) is proved in a similar manner. We first use Bayes' theorem to evaluate the conditional expectation of $\boldsymbol{\eta}_h$ given $\{\zeta_i\}_{i \neq h}$ and $\zeta^{u,h}$ and then employ the law of iterated conditional expectations.

It should be emphasized that there is considerable flexibility in the construction of ζ^u since there are many different ways to specify the set partitions $\mathcal{Z}_h$. If all of these partitions contain only sets of diameter less than ϵ, then the supremum distance of ζ and ζ^u is at most ϵ, that is, $\|\zeta - \zeta^u\|_\infty \leq \epsilon$.

It proves useful to introduce another discrete process $\zeta^l = (\boldsymbol{\eta}^l, \boldsymbol{\xi}^l)$ such that $\boldsymbol{\eta}^l$ and $\boldsymbol{\xi}^l$ are valued in Θ and Ξ, respectively. Again, ζ^l is supposed to approximate the original data process ζ. We construct ζ^l in exactly the same way as ζ^u, but the roles of $\boldsymbol{\eta}$ and $\boldsymbol{\xi}$ are interchanged. Again, the construction is very flexible, relying on a family of disjoint set partitions $\mathcal{Z}_h$ that depend on the history of the approximate process. When constructing ζ^l, we reuse the same family of set partitions that was used for the construction of ζ^u. This is not necessary for our argumentation but leads to simplifications. By choosing these partitions appropriately, for every given tolerance $\epsilon > 0$ we can construct a discrete process ζ^l with $\|\zeta - \zeta^l\|_\infty \le \epsilon$.

The induced filtration $\mathbb{F}^l$ is constructed as usual, that is, $\mathbb{F}^l := \{\mathcal{F}^{l,h}\}_{h=1}^H$ where $\mathcal{F}^{l,h} := \sigma(\zeta^{l,h})$, and we use the convention $\mathcal{F}^l := \mathcal{F}^{l,H}$. By permutation symmetry, the following result is an immediate consequence of Lemma 2.

Corollary 5.1 *The following relations hold for suitable versions of the conditional expectations, respectively.*

$$\mathrm{E}(\boldsymbol{x}|\mathcal{F}^l) \in X(\mathbb{F}^l) \text{ for all } \boldsymbol{x} \in X(\mathbb{F}) \tag{4a}$$

$$\mathrm{E}(\boldsymbol{y}|\mathcal{F}) \in Y(\mathbb{F}) \text{ for all } \boldsymbol{y} \in Y(\mathbb{F}^l) \tag{4b}$$

$$\mathrm{E}(\boldsymbol{\xi}|\mathcal{F}^l) = \boldsymbol{\xi}^l \tag{4c}$$

$$\mathrm{E}(\boldsymbol{\eta}^l|\mathcal{F}) = \boldsymbol{\eta} \tag{4d}$$

If we replace the true data process ζ by ζ^u and the true filtration $\mathbb{F}$ by $\mathbb{F}^u$ in $\mathcal{P}$, then we obtain an approximate optimization problem denoted $\mathcal{P}^u$. Another approximate problem $\mathcal{P}^l$ is obtained by substituting ζ^l for ζ and $\mathbb{F}^l$ for $\mathbb{F}$. Note that replacing the filtrations has a primal and a dual effect, that is, after substitution, the primal decisions as well as the constraints are adapted to the approximate filtration. Notice that Lemma 1 remains valid for problems $\mathcal{P}^u$ and $\mathcal{P}^l$ with their corresponding data processes and filtrations. In the remainder of this section we will prove that the optimal values of $\mathcal{P}^l$ and $\mathcal{P}^u$ bracket the optimal value of the original problem $\mathcal{P}$. To this end, we first establish two technical lemmas.

Lemma 3. *The following relations hold for suitable versions of the conditional expectations and for all $1 \le i < j \le H$.*

(i) $\mathrm{E}(\boldsymbol{x}_i \otimes \boldsymbol{\xi}_j^u|\mathcal{F}) = \mathrm{E}(\boldsymbol{x}_i|\mathcal{F}) \otimes \boldsymbol{\xi}_j$ *for all* $\boldsymbol{x} \in X(\mathbb{F}^u)$
(ii) $\mathrm{E}(\boldsymbol{x}_i \otimes \boldsymbol{\xi}_j|\mathcal{F}^l) = \mathrm{E}(\boldsymbol{x}_i|\mathcal{F}^l) \otimes \boldsymbol{\xi}_j^l$ *for all* $\boldsymbol{x} \in X(\mathbb{F})$

Proof. Select arbitrary time indices i and j such that $1 \le i < j \le H$, and let $\boldsymbol{x}$ be an element of $X(\mathbb{F}^u)$. Then, we find

$$\begin{aligned}
\mathrm{E}(\boldsymbol{x}_i \otimes \boldsymbol{\xi}_j^u|\mathcal{F}) &= \mathrm{E}(\mathrm{E}(\boldsymbol{x}_i \otimes \boldsymbol{\xi}_j^u|\mathcal{F} \wedge \mathcal{F}^{u,j-1})|\mathcal{F}) \\
&= \mathrm{E}(\boldsymbol{x}_i \otimes \mathrm{E}(\boldsymbol{\xi}_j^u|\mathcal{F} \wedge \mathcal{F}^{u,j-1})|\mathcal{F}) \\
&= \mathrm{E}(\boldsymbol{x}_i \otimes \boldsymbol{\xi}_j|\mathcal{F}) \\
&= \mathrm{E}(\boldsymbol{x}_i|\mathcal{F}) \otimes \boldsymbol{\xi}_j,
\end{aligned}$$

where the third equality follows from (3). The above argument proves assertion (i). Next, let $\boldsymbol{x}$ be an element of $X(\mathbb{F})$. By assumption, $\boldsymbol{x}_i$ is a function of $\boldsymbol{\xi}^i$, which is independent of $\boldsymbol{\xi}_j$. Assertion (ii) immediately follows from (4c) and independence of $\boldsymbol{x}_i$ and $\boldsymbol{\xi}_j$. $\qquad\square$

Lemma 4. *The following relations hold for suitable versions of the conditional expectations and for all $1 \leq h \leq H$.*

(i) $\mathrm{E}(f_h(\boldsymbol{x}, \boldsymbol{\xi}^u)|\mathcal{F}) \geq f_h(\mathrm{E}(\boldsymbol{x}|\mathcal{F}), \boldsymbol{\xi})$ for all $\boldsymbol{x} \in X(\mathbb{F}^u)$
(ii) $\mathrm{E}(f_h(\boldsymbol{x}, \boldsymbol{\xi})|\mathcal{F}^l) \geq f_h(\mathrm{E}(\boldsymbol{x}|\mathcal{F}^l), \boldsymbol{\xi}^l)$ for all $\boldsymbol{x} \in X(\mathbb{F})$

Proof. Select $\boldsymbol{x} \in X(\mathbb{F}^u)$. By using condition (C2) and the conditional Jensen inequality, we find

$$
\begin{aligned}
\mathrm{E}(f_h(\boldsymbol{x}, \boldsymbol{\xi}^u)|\mathcal{F}) &= \mathrm{E}(\tilde{f}_h((1, \boldsymbol{x}) \otimes (1, \boldsymbol{\xi}^u))|\mathcal{F}) \\
&\geq \tilde{f}_h(\mathrm{E}((1, \boldsymbol{x}) \otimes (1, \boldsymbol{\xi}^u)|\mathcal{F})) \\
&= \tilde{f}_h(\mathrm{E}((1, \boldsymbol{x})|\mathcal{F}) \otimes (1, \boldsymbol{\xi})) \\
&= f_h(\mathrm{E}(\boldsymbol{x}|\mathcal{F}), \boldsymbol{\xi}) \, .
\end{aligned}
$$

Note that the equality in the third line follows from Lemma 3(i) and the independence of $\tilde{f}_h$ in $x_i \otimes \xi_j$ for all $i \geq j$. This establishes (i). Assertion (ii) is proved in a similar manner. □

Armed with the above preliminary results, we are now ready to state the main theorem of this section.

Theorem 1. *Assume that the problems $\mathcal{P}^l$ and $\mathcal{P}^u$ are solvable with finite optimal values. If $\boldsymbol{x}^u$ solves $\mathcal{P}^u$, then $\hat{\boldsymbol{x}} := \mathrm{E}(\boldsymbol{x}^u|\mathcal{F})$ is feasible in $\mathcal{P}$ and*

$$
\inf \mathcal{P}^l \leq \inf \mathcal{P} \leq \mathrm{E}(c(\hat{\boldsymbol{x}}, \boldsymbol{\eta})) \leq \inf \mathcal{P}^u \, .
$$

Proof. Our argumentation relies on ideas from the proofs of Theorems 1 and 2 in Kuhn (2007a) as well as Theorem 5.1 in Kuhn (2007b). A preliminary calculation yields

$$
\begin{aligned}
\inf \mathcal{P} &\geq \inf_{\boldsymbol{x} \in X(\mathbb{F})} \sup_{\boldsymbol{y} \in Y(\mathbb{F}^l)} \mathrm{E}\left(c(\boldsymbol{x}, \mathrm{E}(\boldsymbol{\eta}^l|\mathcal{F})) + \sum_{h=1}^{H} \mathrm{E}(\boldsymbol{y}_h|\mathcal{F}) \cdot f_h(\boldsymbol{x}, \boldsymbol{\xi}) \right) \\
&\geq \inf_{\boldsymbol{x} \in X(\mathbb{F})} \sup_{\boldsymbol{y} \in Y(\mathbb{F}^l)} \mathrm{E}\left(c(\boldsymbol{x}, \boldsymbol{\eta}^l) + \sum_{h=1}^{H} \boldsymbol{y}_h \cdot f_h(\boldsymbol{x}, \boldsymbol{\xi}) \right) \\
&= \inf_{\boldsymbol{x} \in X(\mathbb{F})} \sup_{\boldsymbol{y} \in Y(\mathbb{F}^l)} \mathrm{E}\left(\mathrm{E}(c(\boldsymbol{x}, \boldsymbol{\eta}^l)|\mathcal{F}^l) + \sum_{h=1}^{H} \boldsymbol{y}_h \cdot \mathrm{E}(f_h(\boldsymbol{x}, \boldsymbol{\xi})|\mathcal{F}^l) \right) \, .
\end{aligned}
$$

The first inequality follows from Lemma 1 as well as the relations (4b) and (4d). Moreover, the second inequality uses the conditional Jensen inequality, which applies since the cost function is concave in its second argument, while $\boldsymbol{x}$ and $\boldsymbol{\xi}$ are $\mathcal{F}$-measurable. From the above chain of inequalities follows

$$
\begin{aligned}
\inf \mathcal{P} &\geq \inf_{\boldsymbol{x} \in X(\mathbb{F})} \sup_{\boldsymbol{y} \in Y(\mathbb{F}^l)} \mathrm{E}\left(c(\mathrm{E}(\boldsymbol{x}|\mathcal{F}^l), \boldsymbol{\eta}^l) + \sum_{h=1}^{H} \boldsymbol{y}_h \cdot f_h(\mathrm{E}(\boldsymbol{x}|\mathcal{F}^l), \boldsymbol{\xi}^l) \right) \\
&\geq \inf_{\boldsymbol{x} \in X(\mathbb{F}^l)} \sup_{\boldsymbol{y} \in Y(\mathbb{F}^l)} \mathrm{E}\left(c(\boldsymbol{x}, \boldsymbol{\eta}^l) + \sum_{h=1}^{H} \boldsymbol{y}_h \cdot f_h(\boldsymbol{x}, \boldsymbol{\xi}^l) \right) \, .
\end{aligned}
$$

Here, the first inequality holds by Lemma 4 (ii) and the conditional Jensen inequality, which applies since the cost function is convex in its first argument, while y and η^l are $\mathcal{F}^l$-measurable. Note that the second inequality is due to assumptions (4a). In conclusion, we thus have shown $\inf \mathcal{P}^l \leq \inf \mathcal{P}$.

Next, we use the fact that x^u is an element of $X(\mathbb{F}^u)$, implying via (2a) that the conditional expectation $\mathrm{E}(x^u|\mathcal{F})$ is an element of $X(\mathbb{F})$.

$$
\begin{aligned}
\inf \mathcal{P} &\leq \sup_{y \in Y(\mathbb{F})} \mathrm{E}\left(c(\mathrm{E}(x^u|\mathcal{F}), \eta) + \sum_{h=1}^{H} y_h \cdot f_h(\mathrm{E}(x^u|\mathcal{F}), \xi) \right) \\
&\leq \sup_{y \in Y(\mathbb{F})} \mathrm{E}\left(\mathrm{E}(c(x^u, \eta)|\mathcal{F}) + \sum_{h=1}^{H} y_h \cdot \mathrm{E}(f_h(x^u, \xi^u)|\mathcal{F}) \right) \qquad (5) \\
&= \sup_{y \in Y(\mathbb{F})} \mathrm{E}\left(c(x^u, \eta) + \sum_{h=1}^{H} y_h \cdot f_h(x^u, \xi^u) \right).
\end{aligned}
$$

The second inequality in (5) uses Lemma 4 (i) and the conditional Jensen inequality, while the equality relies on the law of iterated conditional expectations. Another application of the conditional Jensen inequality yields

$$
\begin{aligned}
\inf \mathcal{P} &\leq \sup_{y \in Y(\mathbb{F})} \mathrm{E}\left(c(x^u, \mathrm{E}(\eta|\mathcal{F}^u)) + \sum_{h=1}^{H} \mathrm{E}(y_h|\mathcal{F}^u) \cdot f_h(x^u, \xi^u) \right) \\
&\leq \sup_{y \in Y(\mathbb{F}^u)} \mathrm{E}\left(c(x^u, \eta^u) + \sum_{h=1}^{H} y_h \cdot f_h(x^u, \xi^u) \right) \\
&= \inf_{x \in X(\mathbb{F}^u)} \sup_{y \in Y(\mathbb{F}^u)} \mathrm{E}\left(c(x, \eta^u) + \sum_{h=1}^{H} y_h \cdot f_h(x, \xi^u) \right).
\end{aligned}
$$

Here, the second inequality holds by the assumptions (2b) and (2d), entailing a relaxation of the dual feasible set. The last line of the above expression corresponds to $\inf \mathcal{P}^u$, which is finite by assumption. This implies that the supremum over $Y(\mathbb{F})$ in the first line of (5) is also finite, and $y = 0$ is an optimal solution. Thus, $\hat{x} = \mathrm{E}(x^u|\mathcal{F})$ is feasible in $\mathcal{P}$, and the corresponding objective value $\mathrm{E}(c(\hat{x}, \eta))$ satisfies the postulated inequalities. □

As the marginal distributions of ζ^l and ζ^u are discrete, the approximate problems $\mathcal{P}^l$ and $\mathcal{P}^u$ constitute finite-dimensional convex programs with a finite number of constraints. Thus, they principally allow for numerical solution. By Theorem 1, the optimal values of the approximate problems provide an a priori estimate of the minimal cost that is principally achievable. However, the optimal strategies corresponding to $\mathcal{P}^l$ and $\mathcal{P}^u$ are only given for discrete scenarios and are therefore not implementable almost surely. It is a priori unclear how the optimal solutions of the approximate problems can be used to determine a near-optimal strategy for the original problem. Theorem 1 provides a particularly satisfactory answer to this question by proposing a policy $\hat{x}$ which is implementable in every possible scenario and whose expected cost is bracketed by $\inf \mathcal{P}^l$ and $\inf \mathcal{P}^u$. Note that $\mathrm{E}(c(\hat{x}, \eta))$ represents an a posteriori estimate of the minimal cost which is achievable in reality. It can conveniently be calculated by Monte Carlo simulation. In fact, since x^u is finitely supported, evaluation of $\hat{x}$ for an arbitrary realization of ζ reduces to the evaluation of a finite sum and poses no computational challenges. Generically, $\mathrm{E}(c(\hat{x}, \eta))$ constitutes a better upper bound on the true objective value than $\inf \mathcal{P}^u$.

Note that there is considerable flexibility in constructing the discrete approximate processes. In fact, we are entirely free in choosing the set partitions $\mathcal{Z}_h$ for $h = 1, \ldots, H$, and each choice corresponds to a pair of discrete processes ζ^u and ζ^l, which provide upper and lower bounds on the true optimal objective value. By making the diameters of the polytopes in all partitions uniformly small, we can construct discrete stochastic processes that approximate the original process arbitrarily well with respect to the $\mathcal{L}^\infty$-norm. Thus, for each $J \in \mathbb{N}$ we can introduce two discrete processes $\zeta_J^u = (\eta_J^u, \xi_J^u)$ and $\zeta_J^l = (\eta_J^l, \xi_J^l)$ of the above kind such that

$$\lim_{J \to \infty} \|\zeta_J^u - \zeta\|_\infty = \lim_{J \to \infty} \|\zeta_J^l - \zeta\|_\infty = 0 .$$

In the remainder, let $\mathcal{P}_J^u$ and $\mathcal{P}_J^l$ be the approximate stochastic programs associated with the discrete processes ζ_J^u and ζ_J^l, respectively. If we assume that the original stochastic program $\mathcal{P}$ satisfies an unrestrictive strict feasibility condition, then for $J \to \infty$ we have

$$\inf \mathcal{P}_J^u \to \inf \mathcal{P} \quad \text{and} \quad \inf \mathcal{P}_J^l = \inf \mathcal{P} . \tag{6}$$

The proof of this convergence result follows the lines of (Kuhn, 2007b, § 6) and is omitted for brevity of exposition. For each $J \in \mathbb{N}$ for which $\mathcal{P}_J^u$ and $\mathcal{P}_J^l$ are solvable, we can further introduce a policy $\hat{x}_J$ as in Theorem 1. Combining (6) and Theorem 1, we conclude that $\hat{x}_J$ is feasible in $\mathcal{P}$ and that for $J \to \infty$ we have

$$\mathrm{E}(c(\hat{x}_J, \eta)) \to \inf \mathcal{P} .$$

In spite of these satisfying theoretical results, the approximation error $E_J := \inf \mathcal{P}_J^u - \inf \mathcal{P}_J^l$ cannot be made arbitrarily small in practice due to limited computer power. It is a major challenge to select the set partitions $\mathcal{Z}_h$ recursively for $h = 1, \ldots, H$ in such a way that the approximation error becomes small for manageable problem sizes. Ideally, we would like to proceed as follows: find the least number of discretization points given an acceptable level of error. Of course, such a scheme is difficult to implement since computing the error for a given configuration of discretization points involves the solution of two large-scale multistage stochastic programs, which is extremely time consuming. To overcome this difficulty, we use a heuristic method based on a stochastic search algorithm that can find a good scenario tree configuration without requiring a large number of function evaluations. The heuristic we use is known under the name 'threshold accepting', see Dueck and Scheuer (1990). The basic idea is to perform a random search over all possible scenario tree configurations of a given complexity. For our purposes, the complexity of a tree is defined as the number of its nodes. The algorithm can be described as follows:

Step 1 Set $J = 1$, and let $\mathcal{Z}_J := \{\mathcal{Z}_{J,h}\}_{h=1}^H$ be stagewise set partitions defining the scenario trees of the upper and lower approximations, respectively. Let E_J be the error (i.e., the gap between upper and lower bounds) associated with $\mathcal{Z}_J$ and choose a threshold $\tau > 0$.

Step 2 If the error E_J is smaller than some predefined tolerance, then stop. Otherwise, perform small random perturbations (described in § 6) in the current configuration $\mathcal{Z}_J$ to obtain $\widehat{\mathcal{Z}}$.

Step 3 Solve the deterministic equivalent problems associated with $\widehat{\mathcal{Z}}$ and define the error $\widehat{E}$ in the obvious way.

Step 4 Define the error difference $\Delta E := E_J - \widehat{E}$ and set

$$(\mathcal{Z}_{J+1}, E_{J+1}) := \begin{cases} (\widehat{\mathcal{Z}}, \widehat{E}) & \text{if } \Delta E > -\tau, \\ (\mathcal{Z}_J, E_J) & \text{otherwise.} \end{cases}$$

Step 5 If more than u iterations are performed without a change in configuration, then lower the threshold τ. Increase J by one unit and go to Step 2.

To keep our numerical experiments simple, we assume here that the disjoint set partitions are not path-dependent. This simplification will be relaxed in future work to fully exploit the flexibility of the bounding approximation scheme presented in this section. The advantage of threshold accepting is that it is easy to implement and requires little tuning. Put differently, its performance is relatively insensitive to the choice of the algorithm control parameters. Popular alternative heuristics tend to be more difficult to control. The choice of a suitable annealing schedule, for instance, is nontrivial but crucial for the practical convergence of simulated annealing algorithms.

6 Case Study

In Section 5 we described how sequential refinements of the probability space will eventually lead to a solution of the time-discretized problem. In this section we discuss the practical implementation of such a scheme. To this end, we assume that the original portfolio problem is approximated by a given time-discretized problem $\mathcal{P}$ with H decision stages. In the remainder, we keep H fixed and focus on space discretization. In order to generate computationally tractable approximations for $\mathcal{P}$, we cover the support of the random vector $\boldsymbol{\xi}_h$

by a compact simplex Z_h for each $h = 1, \ldots, H$. Clearly, this leaves us some
flexibility. A simple approach is to truncate the lognormal distribution of the
price relatives outside some bounded confidence region which covers most of
the probability mass. The truncated distribution must be re-normalized for
consistency reasons. Note that the error introduced by truncation and renor-
malization is generally small and may be disregarded. Next, we enclose the
confidence region of the price relatives in a suitable simplex. A graphical illustration for a problem with $N = 2$ assets is shown in Figure 1.

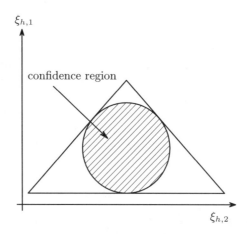

Fig. 1. Simplicial covering of a confidence region with high probability mass.

The coarsest space discretizations use the expected values and the extreme
points of the simplices as the realizations of $\boldsymbol{\xi}^l$ and $\boldsymbol{\xi}^u$, respectively. Solving the
expected value problem $\mathcal{P}^l$ yields a lower bound. The problem $\mathcal{P}^u$ involving
the extreme scenarios, that is, the vertices of the simplices with probabilities
given by (1), results in an upper bound.

In order to improve the bounds, we refine the trivial set coverings $\mathcal{Z}_{1,h} =$
$\{Z_h\}$ for each $h = 1, \ldots, H$ by subdividing some or all of the involved simplices. One possibility to partition an N-dimensional simplex is by choosing
an interior point and introducing $N + 1$ subsimplices with all meet at this
point. A natural choice is the barycenter of the simplex (we always work with
the *classical* barycenter corresponding to a uniform mass distribution). For
example, when the simplex in Figure 1 is partitioned, the three subsimplices
A,B and C in Figure 2(a) are created. When simplex C is further refined,
then three subsubsimplices C1, C2 and C3 are created, see Figure 2(b). As
an illustration, we consider a small problem with three assets, and a planning
horizon comprising three time periods. The problem is discretized by using
the techniques described in Section 5, and the resulting deterministic equivalent problems are solved by using the OOQP solver by Gertz and Wright
(2003). All numerical experiments are performed on a Linux machine with

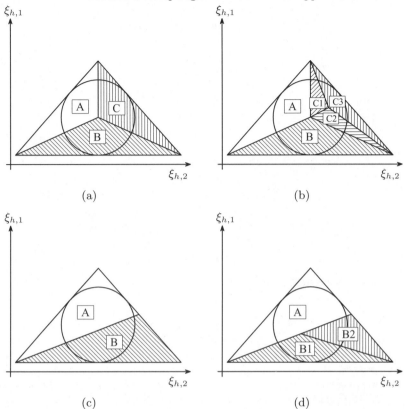

Fig. 2. Refinement of simplicial partition.

a 2.0Ghz processor and 2GB of RAM. The convergence of the upper and lower bounds due to a naive brute-force refinement strategy is shown in Figure 3. In this example, the discretization factor refers to the depth of recursive simplicial subdivisions. For example, Figure 1 corresponds to the discretization factor $J = 1$, while Figure 2(a) corresponds to $J = 2$. If in Figure 2(b) we further subdivided all the subsimplices (not just C), then the resulting covering would be assigned the discretization factor $J = 3$. Problems with higher discretization factors cannot be solved due to the fact that problem size increases rapidly with the branching factor of the underlying scenario tree. This *curse of dimensionality* is even more acute in problems with a large number of assets or a large number of time periods. In fact, the number of nodes in the approximate stochastic programs amounts to $(f - 1)^{-1}(f^H - 1)$, where f denotes the branching factor of the underlying scenario tree. Observe that f can be expressed as a function of the discretization factor J and the number of assets N. The scenario tree of the upper bounding problem has $f = N + 1 + N^{-1}((N + 1)^{J-1} - 1)$, while the branching factor of the lower

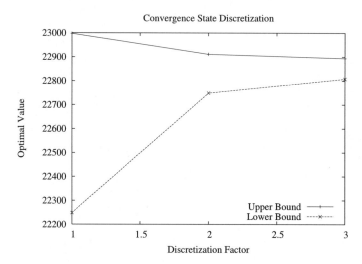

Fig. 3. Convergence of bounds due to brute-force refinement strategy.

bounding problem is given by $f = N^{-1}((N+1)^J - 1)$. In order to determine the number of variables and constraints in the approximate stochastic programs, we multiply the number of nodes by some terms proportional to N.

As argued at the end of Section 5, we may expect to improve the approximation error if the brute force refinement strategy is replaced by a more sophisticated one based on a threshold accepting algorithm. Before we report our initial experience with this advanced approach, we discuss some practical issues relating to its implementation. We first explain how to generate small perturbations of a scenario tree configuration. For a given configuration we use a Poisson process to randomly select the subsimplices that will be merged with other subsimplices. By sampling from another Poisson process, we also determine the simplices that will be further partitioned. The use of Poisson processes ensures that certain discretization points will randomly 'depart' and 'arrive' in the scenario tree. However, the number of nodes of the tree may never exceed a certain budget. Our sampling is designed to make 'arrivals' of new discretization points more likely in early decision stages, while 'departures' of existing discretization points are more likely in late stages. This accounts for the general consensus in stochastic programming literature that scenario trees should have a higher branching factor in the initial periods, see Hochreiter and Pflug (2007). In our implementation, the threshold parameter τ is initially set to 100 and reduced by 10% whenever a reduction is required. The threshold is reduced if the scenario tree remains unchanged for three iterations. After five iterations with no change, the algorithm terminates. The threshold accepting heuristic is described in detail in Gilli and Winker (2007).

The main goal of our numerical studies is to find scenario trees that are numerically tractable yet have a small error. Our first numerical experiments

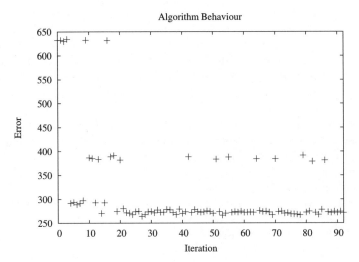

Fig. 4. Error reduction by threshold accepting algorithm (barycenter splitting).

show that a threshold accepting scheme can be beneficial. In Figure 4 we show how the algorithm reduces the error as it attempts to find an optimal discretization. The error is reduced dramatically after only a few iterations. From Figure 4 one concludes that for the problem under consideration there exist at least two local minima. The algorithm appears to be able to jump from one cluster of local minima to the next. However, the approximation error never drops below 250 in the first 100 iterations.

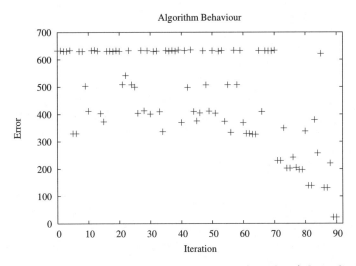

Fig. 5. Error reduction by threshold accepting algorithm (edge splitting).

The approximation error can be further reduced by using an alternative partitioning strategy. Instead of introducing a new point at the barycenter of an existing simplex, thereby subdividing it into $N + 1$ subsimplices, we may introduce a new point in the middle of one of its edges. Here, we use the standard convention that an edge of a simplex is the line segment between two of its vertices. In doing so, the initial simplex is split into two subsimplices only, see Figures 2(c) and 2(d). 'Edge splitting' has the advantage that the diameters of the subsimplices can be made uniformly small, which is a prerequisite for the theoretical convergence result in Section 5. The 'barycenter splitting' technique considered before fails to have this property since edges are never split. However, it produces $N+1$ sub-simplices per additional vertex, which is superior to the mere two subsimplices for edge splitting. We tested our threshold accepting algorithm also in conjunction with the edge splitting strategy, while randomly selecting the edges to be cut in half. Figure 5 indicates that the approximation error decreases slower if edge splitting is used instead of barycenter splitting. However, the approximation error is virtually eliminated after 90 iterations although tree size does not increase after the tenth iteration.

Our preliminary results suggest that randomized algorithms (e.g. threshold accepting, TABU search, genetic algorithms, simulated annealing) can be used to find the best bounding scenario tree from a set of trees that do not exceed a certain size. In the future we will allow the simplicial partitions of the state spaces to be path dependent. This increased flexibility will be exploited to achieve a specified error tolerance with even smaller scenario trees and, in addition, to address more complex portfolio problems with many decision stages. Future research will also focus on keeping the time discretization flexible when searching for optimal scenario trees. However, much more work is required in order to find efficient and theoretically appealing methods for solving this problem.

Acknowledgments

Daniel Kuhn thanks the Swiss National Science Foundation for financial support. The authors also wish to acknowledge support from EPSRC grants GR/T02560/01 and EP/C513584/1.

References

Ash, R.B.: 1972, *Real Analysis and Probability*, Probability and Mathematical Statistics, Academic Press, Berlin.

Birge, J.R. and Louveaux, F.: 1997, *Introduction to Stochastic Programming*, Springer-Verlag, New York.

Dueck, G. and Scheuer, T.: 1990, Threshold accepting: a general purpose optimization algorithm appearing superior to simulated annealing, *J. Comput. Phys.* **90**(1), 161–175.

Dupačová, J.: 1966, On minimax solutions of stochastic linear programming problems, *Časopis pro Pěstování Matematiky* **91**, 423–429.

Edirisinghe, N.C.P. and Ziemba, W.T.: 1994a, Bounding the expectation of a saddle function with application to stochastic programming, *Math. Oper. Res.* **19**, 314–340.

Edirisinghe, N.C.P. and Ziemba, W.T.: 1994b, Bounds for two-stage stochastic programs with fixed recourse, *Math. Oper. Res.* **19**, 292–313.

Frauendorfer, K.: 1994, Multistage stochastic programming: Error analysis for the convex case, *Z. Oper. Res.* **39**(1), 93–122.

Frauendorfer, K.: 1995, The stochastic programming extension of the Markowitz model, *Int. J. Neural Mass-Parallel Comput. Inform. Syst.* **5**, 449–460.

Frauendorfer, K.: 1996, Barycentric scenario trees in convex multistage stochastic programming, *Math. Program.* **75**(2), 277–294.

Frauendorfer, K. and Siede, H.: 2000, Portfolio selection using multi-stage stochastic programming, *Central European J. Oper. Res.* **7**, 277–290.

Gertz, E.M. and Wright, S.J.: 2003, Object-oriented software for quadratic programming, *ACM Trans. Math. Software* **29**(1), 58–81.

Gilli, M. and Këllezi, E.: 2002, The Threshold Accepting Heuristic for Index Tracking, *in* P. Pardalos and V.K. Tsitsiringos (eds), *Financial Engineering, E-Commerce and Supply Chain*, Applied Optimization Series, Kluwer Academic Publishers, Boston, pp. 1–18.

Gilli, M., Këllezi, E. and Hysi, H.: 2006, A data-driven optimization heuristic for downside risk minimization, *Journal of Risk* **8**(4), 1–18.

Gilli, M. and Winker, P.: 2007, Heuristic optimization methods in econometrics, *in* E.J. Kontoghiorghes and D. Belsley (eds), *Handbook of Computational Econometrics*, Handbooks series of Computing and Statistics with Applications, Elsevier. In preparation.

Gülpinar, N. and Rustem, B.: 2006, Worst-case robust decisions for multiperiod mean-variance portfolio optimization, *European J. Oper. Res.* . In press.

Gülpinar, N., Rustem, B. and Settergren, R.: 2002, Multistage stochastic programming in computational finance, *in* E. J. Kontoghiorghes, B. Rustem and S. Siokos (eds), *Computational Methods in Decision-Making, Economics and Finance*, Vol. 74 of *Applied Optimization*, Kluwer Academic Publishers, pp. 33–45.

Gülpinar, N., Rustem, B. and Settergren, R.: 2004, Optimization and sinmulation approaches to scenario tree generation, *J. Econ. Dyn. Control* **28**(7), 1291–1315.

Hochreiter, R. and Pflug, G.C.: 2007, Financial scenario generation for stochastic multi-stage decision processes as facility location problems, *Ann. Oper. Res.* **156**(1), 257–272.

Kall, P. and Wallace, S.W.: 1994, *Stochastic Programming*, John Wiley & Sons, Chichester.

Kuhn, D.: 2007a, Aggregation and discretization in multistage stochastic programming, *Math. Program. A* . Online First.

Kuhn, D.: 2007b, Convergent bounds for stochastic programs with expected value constraints, *The Stochastic Programming E-Print Series (SPEPS)* .

Madansky, A.: 1960, Inequalities for stochastic linear programming problems, *Manage. Sci.* **6**, 197–204.

Markowitz, H.M.: 1952, Portfolio selection, *Journal of Finance* **7**(1), 77–91.

Merton, R.C.: 1992, *Continuous Time Finance*, Basil Blackwell, Cambridge, MA.

Rockafellar, R.T. and Wets, R.J.-B.: 1978, The optimal recourse problem in discrete time: L^1-multipliers for inequality constraints, *SIAM J. Control Optimization* **16**, 16–36.

Rosenblatt, M.: 1952, Remarks on a multivariate transformation, *Ann. Math. Stat.* **23**(3), 470–472.

Steinbach, M.C.: 2001, Markowitz revisited: mean-variance models in financial portfolio analysis, *SIAM Rev.* **43**(1), 31–85.

Wright, S.E.: 1994, Primal-dual aggregation and disaggregation for stochastic linear programs, *Math. Oper. Res.* **19**(4), 893–908.

Zhou, X.Y. and Li, D.: 2000, Continuous time mean-variance portfolio selection: A stochastic LQ framework, *Appl. Math. Optim.* **42**, 19–33.

Risk Preferences and Loss Aversion in Portfolio Optimization

Dietmar Maringer

Centre for Computational Finance and Economic Agents (CCFEA), University of Essex, Wivenhoe Park, Colchester CO4 3SQ, United Kingdom.
dmaring@essex.ac.uk

Summary. Traditionally, portfolio optimization is associated with finding the ideal trade-off between return and risk by maximizing the expected utility. Investors' preferences are commonly assumed to follow a quadratic or power utility function, and asset returns are often assumed to follow a Gaussian distribution. Investment analysis has therefore long been focusing on the first two moments of the distribution, mean and variance. However, empirical asset returns are not normally distributed, and neither utility function captures investors' true attitudes towards losses. The impact of these specification errors under realistic assumptions is investigated. As traditional optimization techniques cannot deal reliably with the extended problem, the use of a heuristic optimization approach is suggested. It is found that loss aversion has a substantial impact on what investors consider to be an efficient portfolio and that mean-variance analysis alone can be utterly misguiding.

Key words: Loss aversion, risk aversion, portfolio optimization, heuristics, differential evolution.

1 Assets and Asset Selection

1.1 Properties of Asset Returns

Portfolio optimization and understanding of investment decisions have come a long way over the last decades. Following the seminal work of Markowitz (1952), the standard paradigm in financial modeling has long been that the first two moments, mean and variance, are sufficient to describe asset returns and capture the preferences of a rational risk averse investor. In this framework the investment decision is based on the trade-off between higher mean and higher variance of the returns. The underlying assumption for this approach is that either the returns are normally distributed, or that investors have a quadratic utility function. Unfortunately, neither seems to hold for the real world.

One of the salient properties of normal distribution is that any observation from minus to plus infinity is assigned a positive probability. Negative log returns up to minus infinity require that losses up to the amount of initial investment can occur, but not more so. Hence it does apply, e.g., to bonds, stocks, or options when held in a long position, but not to futures or stocks and options when held in a short position. Arbitrarily large positive returns can be achieved for long positions when prices can increase without any limit. This is true (at least in principle) for assets such as stocks, futures or call options. Put options already have a theoretical upper limit on their value (though this bound is unlikely to be reached anyway unless the underlying defaults). For assets such as bonds, however, there is a much clearer upper bound on potential profits: a long position in a simple bond will earn the yield to maturity if it does not default or less if it does; additional deviations due to changes in the general interest rate structure might cause slight, but no dramatic deviations from this. Hence, the assumption of symmetric deviations from the expected value which make positive and negative deviations of the same magnitude equally likely, will by definition apply only to a small set of assets such as stocks. And even for those assets, skewed empirical returns are not uncommon.

The even bigger problem with the normality assumption is that is does not capture extreme events appropriately. Looking at the Dow Jones Industrial Average's daily log returns, there were two days with losses bigger than 5% within the 1684 days considered for the computational study in this chapter; for the observed mean and variance, the normal distribution predicts that it ought to take about 1.5 million days for these two extreme events to happen. For individual stocks, excess kurtosis and the gaps between predicted and actually observed frequencies are typically even larger. Hence, looking only at the first two moments of return distributions leaves out crucial information. As a consequence, other approaches have become more popular, such as including student's t-distribution (which, in particular with low degrees of freedom, does exhibit some excess kurtosis) or, more recently, extreme value theory (Gilli and Këllezi, 2006); alternatively, empirical distributions and historical sampling methods are widely used and sanctioned by regulatory authorities as are alternative risk measures (Basel Committee on Banking Supervision, 2003; Gilli et al., 2006).

1.2 Investor Preferences

At the same time, the usual assumptions about risk preferences imply that investors actually might prefer deviations from a normal distribution. A common assumption in modeling an investor's preferences is diminishing marginal utility of wealth; this implies that larger deviations on the positive side are required to offset her for losses. In other words, investors are assumed to prefer positively skewed returns. For the same reason, variance and excess kurtosis are not desirable: both measure deviations from the mean but ignore the sign;

with marginally diminishing utility, losses lower the utility more than profits of the same magnitude would increase it. As a consequence, higher variance and kurtosis, respectively, are accepted only if they are rewarded with an increase in the mean payoff. All things considered, the representative risk averse rational investor should find an investment that optimizes the trade-off between its expected return and expected deviations from it.

More recent work in behavioral finance suggests that utility functions should also capture a more human property: investors do not only dislike losses, sometimes they also put more emphasis on it than pure statistics or a traditional framework of rational investors would suggest. In their seminal studies, Kahneman and Tversky (1979) found that losses play a more prominent part in making decisions than traditional utility analysis would predict. Hence, decisions are merely driven by loss aversion and the prospect of ending up with a lower than the current wealth, and not just by risk aversion which captures any deviation – positive and negative alike – from the expected wealth. Prospect theory, which arose from these findings, assumes that investors put additional weight on losses, meaning that they either overestimate the likelihood or the magnitude of losses. Empirical work found these effects not only for ex ante decision problems (Thaler et al., 1997), but also for dynamic portfolio revisions where investors tend to sell winners and to keep losers (Odean, 1998). These findings have also coincided with (and, arguably, supported) the advent of new risk measures that measure only losses, such as Value at Risk and conditional Value at Risk, or take the asymmetries between profits and losses into account, such as the Omega risk measure (Gilli et al., 2006).

1.3 Consequences for Asset Selection

If neither the asset returns nor the decision makers' preferences agree with the assumptions of traditional optimization problems, disagreement on what is an optimal solution can be expected. Behavioral finance has made progress in identifying more realistic models of preference and choice, there exists little evidence of how this would (or should) affect the actual investment selection process in a realistic market. This contribution aims to answer to this question by analyzing the consequences from deviating preferences and properties. Based on an empirical computational study for Dow Jones stock data, it is investigated how the choice of utility function, level of risk aversion and, in addition, level of loss aversion affects the investment process. The results show that these differences in preferences not only lead to different ideas of "efficient" portfolios, but also that investors' portfolios will differ substantially in their stylized facts and properties.

The remainder of this chapter is organized as follows. Section 2 presents the financial background in more and formalizes the optimization problems, Section 3 offers a method that can solve them. Section 4 presents an empirical study, and Section 5 concludes.

2 Portfolio Optimization under Loss Aversion

2.1 Loss Aversion, Risk Aversion and Utility Analysis

In myopic portfolio optimization, a popular choice for modeling investors' preferences is the quadratic utility function,

$$\mathcal{U}_q(w) = w - \frac{b}{2}w^2, \qquad 0 < b < 1/w, \tag{1}$$

where w is the wealth and b is the level of risk aversion. With this function, higher moments are either ignored or assumed not to exist – as is the case under a normal distribution. One peculiarity of this function is that it is not defined for wealth levels $w < 1/b$: the function's first derivative, i.e., the marginal utility of wealth, would become negative which clearly contradicts the usual assumption of non-satiation and investors preferring more wealth to less. A simple remedy is to assume zero marginal utility for wealth beyond this bliss point and set the utility to $\mathcal{U}_q(w) = 1/(2b) \ \forall w \geq 1/b$.

Alternatively, utility functions such as the power utility

$$\mathcal{U}_p(w) = \left(w^{1-\gamma}\right) \cdot (1 - \gamma)^{-1}, \qquad \gamma > 0 \tag{2}$$

implicitly do take higher moments into account. $(1 - \gamma)$ measures the risk aversion, with γ also being the constant coefficient of relative risk aversion, $-w \cdot \mathcal{U}_p''(w)/\mathcal{U}'(w)$. For $\gamma = 1$, the function turns into the log utility function (or Kelly criterion) where the log of the wealth, $\mathcal{U}_l(w) = \ln(w)$, is considered.

However, individuals tend to be more sensitive to falling below a certain level of wealth than exceeding it (Samuelson, 1963). Typically, this reference point is the current level of wealth, w_0, which makes the decision maker loss averse. A simple way to account for this phenomenon is to use modified future levels of wealth, m, that enhance the losses:

$$m = w_0 + \underbrace{(w - w_0)\Im_{w \geq w_0}}_{\text{profit}} + \lambda \underbrace{(w - w_0)\Im_{w < w_0}}_{\text{loss}}$$

$$= w + (\lambda - 1)(w - w_0)\Im_{w < w_0},$$

where $\lambda \geq 1$ is the degree of loss aversion; the indicator function $\Im_{w < w_0}$ ($\Im_{w \geq w_0}$) is 1 (0) if a loss is occurred and $w < w_0$, and 0 (1) otherwise. When using the quadratic or power utility function, the utility curves remain concave, but have a kink at w_0 and are steeper left of it than their "traditional" counterpart. Alternatively, the levels of wealth can be left unchanged but the probabilities for incurring losses and profits are altered. In either way, the left-hand part of the curve is over-emphasized when the expected utility is computed.

In their seminal experiments, Kahneman and Tversky (1979) (KT) find that individuals are not only loss averse, but that the preferences actually form an S-shaped value curve of changes in wealth, $x = w - w_0$,

$$v(x) = \begin{cases} x^\alpha & x \geq 0 \\ -\lambda(-x)^\beta & x < 0 \, . \end{cases} \qquad (3)$$

Furthermore, they assume decision makers use decision weights, π_i, which are nonlinear transforms of the probabilities of outcome i, p_i. According to their *prospect theory*, the prospective utility of a gamble G can then be modeled as

$$V(G) = \sum_i \pi_i v(x_i) \, . \qquad (4)$$

They find for their experiments that the parameters for the value function are $\alpha = \beta = 0.88$ and $\lambda = 2.25$.

2.2 Operational Aspects of Portfolio Optimization

Investors are assumed to maximize their expected utility. In perfect markets with well behaved return distributions, investors prefer a combination of a (universal) risky portfolio with a risk free asset. Depending on their level of risk aversion, they choose different fractions of wealth to invest in either of these (Tobin, 1958). In the absence of a risk free asset, however, and with distributions that do not follow the usual parametric distributions, solutions become less straightforward. The problem becomes even more demanding when practical constraints are considered. While aspects such as transaction costs, taxes, liquidity risk, etc. can be approximated by modifying the (distribution of) returns, constraints on the asset weights (i.e., the assets' fractions within the risky portfolio) can cause severe challenges to numerical approaches. Not uncommonly, weights for individual assets can have upper limits, $x_i \leq x^u$ in order to encourage diversification and reduce excessive exposure to a single asset's risk. At the same time, there can also be minimum weights for included assets, $x_i = 0 \vee x_i \geq x^\ell$, to avoid over-fragmentation and keep costs to a reasonable level.

In the presence of such weight constraints, the portfolio optimization problem for a maximizer of expected utility can be stated as follows:

$$\max_{\mathbf{x}} E(\mathcal{U}(w, \lambda)) \qquad (5)$$

subject to

$$w_i = \sum_{a=1}^{N} w_0 \exp(r_{a,i})x_a$$

$$m = w + (\lambda - 1)(w - w_0)\Im_{w<w_0} \tag{6}$$

$$E(\mathcal{U}(w, \lambda)) = \sum_{i=1}^{I} p_i \mathcal{U}(w_i, \lambda)$$

$$\mathcal{U}(w, \lambda) = \begin{cases} \text{quadratic utility: } \mathcal{U}_q(m) = \begin{cases} m - \frac{b}{2}m^2 & b < m/2 \\ m/2 & \text{otherwise} \end{cases} \\ \text{power utility: } \quad \mathcal{U}_p(m) = \begin{cases} \frac{m^{1-\gamma}}{1-\gamma} & \gamma \neq 1 \\ \ln(m) & \text{otherwise} \end{cases} \\ \text{prospect theory: } \mathcal{U}_{PT}(w) = (w - w_0)^\alpha \Im_{w \geq w_0} - \lambda(w_0 - w)^\beta \Im_{w<w_0} \end{cases}$$

$$\sum_i x_i = 1$$

$$x_i : \begin{cases} x_i = 0 \\ x^\ell \leq x_i \leq x^u \end{cases} \quad \forall i$$

$$\Im_{w<w_0} = \begin{cases} 1 & w < w_0 \\ 0 & \text{otherwise} \end{cases}, \quad \Im_{w \geq w_0} = 1 - \Im_{w<w_0}.$$

As with many financial optimization problems, this problem, too, is non-convex, and standard non-linear optimization methods cannot be guaranteed to find the global optimum. Hence, alternative methods such as heuristics or non-deterministic approaches have gained popularity since they can deal with non-convexity and discontinuities.

3 Heuristic Methods for Portfolio Optimization

3.1 A Brief History of Heuristics in Finance

Heuristic methods have become an increasingly popular alternative to traditional optimization methods. One salient feature of these methods is the inclusion of non-deterministic elements, another the occasional acceptance of impairments (despite the obvious preference for improvements). Because of these two principles, the search process is no longer deterministic – which can be very beneficial: local optima can be overcome more easily. Also, restarts do not necessarily produce the same reported result; if the search converges to an inferior solution once, another run can identify another optimum – ideally the global one. Restarts will therefore increase the likelihood of identifying the true optimum, which is not the case for traditional deterministic search methods. Furthermore, heuristics are often designed as general principles (meta-heuristics), that can easily be adapted to all sorts of problems and constraints. An introduction to the use of heuristics in finance can be found in Gilli et al. (2008). For applications of deterministic methods, see, e.g., Konno (2005).

A common traditional approach is to find a suitable trajectory through the search space that leads to the optimum. Gradient search, e.g., takes its steps from first order conditions and, in the case of a maximization problem, moves the current solution towards the steepest ascend. This simple deterministic rule, however, has difficulties coping with rough search spaces that have several optima; often, the search ends in the local optimum closest to the starting point which is not necessarily the global optimum. A simple yet effective remedy is the inclusion of noise: adding noise with expected value zero to the gradient will lead to diversions in the trajectory, and local optima can be circumvent and left. Parpas and Rustem (2006) and Maringer and Parpas (2008) apply this approach to a portfolio selection problem, where updates of the current solutions contain deterministic (i.e., the gradient) and non-deterministic (i.e., noise) elements and where the magnitude of noise is gradually reduced and replacement decisions are based on improvements only.

Introduced by Kirkpatrick et al. (1983), simulated annealing (SA) is a heuristic that abandons the deterministic ingredients all together. Suggested changes are randomly picked based on a neighborhood definition (i.e., adding noise to the current solution), and acceptance is based on a stochastic rule. This rule takes into account the magnitude of the change in the objective function and how progressed the search is: Other things being equal, larger impairments are less likely to be accepted than smaller ones or improvements. When several iterations have passed, the algorithm should have moved to a favorable region of the search space (due to its preference for improvements over impairments), and it becomes less tolerant in accepting impairments. Hence, in the course of multiple iterations the search process becomes more and more like a greedy uphill search.

Dueck and Scheuer (1990) suggest threshold accepting (TA) where SA's random generation of new solutions is kept, but the stochastic acceptance principle is replaced with a deterministic rule and any improvement is accepted as well as any impairment that does not exceed a given threshold. This threshold is lowered in the course of iterations which, again, makes it more intolerant to impairments. More on TA and applications to economics and econometrics can be found in Winker (2001) and Winker and Maringer (2007); applications of SA and TA in finance are presented in Maringer (2005).

TA has been successfully applied to different portfolio optimization problems. For example, Gilli and Këllezi (2002) show how to use it for index tracking problems with demanding constraints, and Gilli et al. (2006) offer a significant and helpful extension to the usability and practical application of TA by suggesting a data driven approach for tuning one of the ingredients, the threshold sequence, and showing how to tackle investment decisions under different downside risk measures. In these contributions, Gilli and colleagues managed not only to answer successfully challenging and demanding problems that would be unsolvable with traditional methods; they also helped to establish the use of heuristics in finance as well as in econometrics.

An alternative class of (meta-) heuristics is based on evolutionary principles and natural evolvement. Evolutionary strategies (Rechenberg, 1965, 1973) and the more popular genetic algorithms (Holland, 1975) are typical and widely used examples. The main principles are mimicking natural processes: New solutions are produced by slightly modifying existing ones ("mutation") and, in the case of "populations" of candidate solutions, combining them into new ones ("cross-over"). For both approaches, randomness is the key element: which part of the solution is altered to which extent is decided non-deterministically, and so is which "parent" solutions contribute what to their offsprings. Following the "survival of the fittest principles," new solutions will replace current ones if they outperform them – yet this decision, too, can be subject to a tournament which adds another stochastic element to the process. Fogel (2001) offers a detailed account of evolutionary methods and general applications, while financial applications of evolutionary methods can be found, e.g., in Brabazon and O'Neill (2006) or Maringer (2005).

3.2 Differential Evolution

A recent addition to the class of evolutionary heuristics is a method called differential evolution (DE). Suggested by Storn and Price (1995, 1997), DE uses a population of P vectors, $\mathbf{v}_p$, $p = 1 \ldots P$, where the N real valued elements of the vectors represent the objective variables. The basic idea is to produce a new solution for each current vector $\mathbf{v}_{p_0}$, where the new solution is a combination of four distinct current solutions. This involves the following steps: First, three different vectors are randomly chosen from the current population. One vector, $\mathbf{v}_{p_1}$, is used as the base vector to which the weighted difference of two other vectors $\mathbf{v}_{p_2}$ and $\mathbf{v}_{p_3}$ is added. The combined solution is then $\mathbf{v}_k = \mathbf{v}_{p_1} + F \cdot (\mathbf{v}_{p_2} - \mathbf{v}_{p_3})$. Finally, this combined solution is crossed-over with a fourth solution, $\mathbf{v}_{p_0}$. This procedure of producing a new solution is repeated for each member p_0 of the current population. Once all the crossed-over solutions have been generated, they replace their respective parent, $\mathbf{v}_{p_0}$, if the objective value is better.

As long as different vectors do not "agree" on the values of the elements, the difference vector $(\mathbf{v}_{p_2} - \mathbf{v}_{p_3})$ will have non-zero elements, and genuinely new solutions will be produced. If, however, one or more of the elements of the difference vector are small or even zero, the new solution will inherit the (more or less) unchanged corresponding values of the base vector. The new solutions will then move towards what they consider the best values for the elements; these reference points can change in the course of multiple iterations as the solutions are steadily improved. Eventually, they will agree, however, and (graphically speaking) flock at or around a point in the solution space which they consider the global optimum.

Diversity within the population is salient to avoid premature convergence. For example, additional perturbation can be achieved by adding some noise to the difference vector, by adding a second difference vector, etc. Alternatively, to reinforce good solutions, the best solution found so far, the so-called elitist, can be chosen as the base vector or as the vector $\mathbf{v}_k$ is crossed over with. A detailed presentation of variants and how to apply these and further extension can be found in Price et al. (2005). In Maringer and Parpas (2008), DE is applied to a portfolio optimization problem and compared against an alternative method, Stochastic Differential Equations, while Maringer and Meyer (2008) compare it against threshold accepting in an application to model selection. Both studies find that DE is a competitive method capable of solving demanding optimization problems.

One of the advantages of DE is that the number of technical parameters is rather low. In its typical version with the difference vector consisting of two current elements and a cross-over with a fourth current solution, only the population size, the scaling factor F and the cross-over probability need to be pre-specified. If noise is used to add diversity, it is common to add normally distributed random values with expected value zero and small standard deviation. Also it is claimed to be easy to use as it needs little or no parameter tuning. These favorable property was also experienced in this application; a more detailed discussion of calibrating DE can be found in Maringer (2008).

DE's way of generating new solutions has been designed for search in a continuous space. In order to deal with the discontinuities of the search space due to the constraints on the weights, a repair function is introduced that maps the candidate solutions $\mathbf{v}$ into a feasible solution of weights, $\mathbf{x}$. Following the approach suggested in Maringer and Oyewumi (2007), this repair function first assigns the minimum positive weight, x^ℓ, to all assets where the values of the corresponding elements of $\mathbf{v}$ exceed this value. If this would leave fewer than $n^{\min} = 1/x^u$ assets, then the $n^{\min}$ assets with the highest values in $\mathbf{v}$ are picked. Next, the weights of the included assets are increased in proportion to the values in $\mathbf{v}$ until the weights add up to one and no asset weights violates the upper limit. In preliminary experiments, this repair function was tested against several alternatives. It was found that the function itself is computationally cheap as in most of the cases the mechanism simply picks the largest positive elements and scales them such that the consumption constraint is met. Furthermore, it does not encourage premature convergence: while the values for the included elements tend to be similar to their repaired counterparts, those of the other elements tend to converge either to zero or to some arbitrary negative value. Given the production algorithm for new candidate solutions, the latter case can cause that these elements become positive again in the candidate solution and the corresponding assets are included in the portfolio, and local optima can be escaped. In addition to upper and lower limits, Maringer and Oyewumi (2007) also consider a cardinality constraint which limits the number of different assets in a portfolio with, say, k. Other

---

Algorithm 1: Pseudocode for utility maximization with Differential Evolution.

---

1 randomly initialize population of vectors $\mathbf{v}_p, p = 1 \ldots P$;
2 **repeat**
3 | %% generate new solutions $\widetilde{\mathbf{v}}_p$;
4 | **for** *current solutions* $\mathbf{v}_p, p = 1 \ldots P$ **do**
5 | | randomly pick $c_1 \neq c_2 \neq c_3 \neq p$;
6 | | **for** *all elements i in the solution vector* **do**
7 | | | with probability $\pi_1 : z_1[i] \leftarrow N(0, \sigma_1)$ else $z_1[i] \leftarrow 0$;
8 | | | with probability $\pi_2 : z_2[i] \leftarrow N(0, \sigma_2)$ else $z_2[i] \leftarrow 0$;
9 | | | randomly pick $u[i] \sim U(0, 1)$;
10 | | | **if** $u[i] < \pi$ **then**
11 | | | | $\widetilde{v}_p[i] \leftarrow v_p[i]$;
12 | | | **else**
13 | | | | $\widetilde{v}_p[i] \leftarrow v_{c_1}[i] + (F + z_1[i]) \cdot (v_{c_2}[i] - v_{c_3}[i] + z_2[i])$;

14 | %% select new population: replace if improvement;
15 | **for** *current solutions* $\mathbf{v}_p, p = 1 \ldots P$ **do**
16 | | map solutions into asset weights, $\mathbf{v}_p \rightarrow \mathbf{x}_p, \widetilde{\mathbf{v}}_p \rightarrow \widetilde{\mathbf{x}}_p$;
17 | | **if** $E(\mathcal{U}(w(\widetilde{\mathbf{x}}_p), \lambda)) > E(\mathcal{U}(w(\mathbf{x}_p), \lambda))$ **then**
18 | | | $\mathbf{v}_p \leftarrow \widetilde{\mathbf{v}}_p$;

19 **until** *halting criterion met* ;

---

things equal, the repair mechanism then picks only the k elements with the biggest values if there are more than k positive ones.

For the generation of new solutions, the inclusion of noise was found helpful. More specifically, the i-th element of the new vector was computed according to $v_{n,i} = v_{p_1,i} + (F + z_1) \cdot (v_{p_2,i} - v_{p_3,i} + z_2)$ where z_1 and z_2 are either zero (with probabilities of 1 and 2 percent, respectively) or normally distributed random variables with expected values of zero and standard deviations of 0.02. Typically, the population size was set to 50, and the number of iterations to 2500. Preliminary experiments suggest that the results did not differ substantially for other values of the technical parameters. For all problems, up to 50 restarts were performed and the best of the reported results were used for the analysis in the following section. Algorithm 1 summarizes the pseudocode.

4 Empirical Study

4.1 Data

The empirical study is based on the stocks included in the Dow Jones Industrial Average (DJIA). Using adjusted daily prices downloaded from `finance.yahoo.com` for 2 March 2000 to 17 November 2006, 1684 log returns where computed. Figure 1 contains scatter plots of the moments of the asset returns; points on the same vertical level refer to the same asset. Some of the assets were not included in any of the optimized portfolios and are represented by dots; circles depict assets that are included in at least one of the optimized portfolios. It is noteworthy that it is not the assets with the highest volatility or kurtosis or most negative skewness that are avoided; also, in some cases assets that are dominated in the mean-volatility space are included. The main reason for this is the structure of the (higher) co-moments.

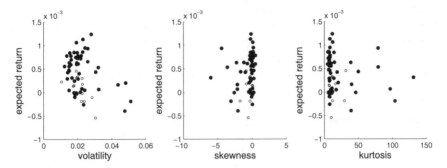

Fig. 1. Assets in the mean-volatility, mean-skewness and mean-kurtosis space. Dots: assets included in at least one portfolio; circles: assets never included.

4.2 Portfolios under Risk Aversion and Loss Aversion

In traditional utility analysis, an investor's utility of a (future) level of wealth, w is irrespective of his initial wealth w_0. In this case, the investor is concerned with the expected value of future wealth and its expected distribution, but there is no explicit distinction whether or not a certain level of wealth is a loss or a gain. Hence, the investor is risk averse, but there is no additional loss aversion. In terms of optimization problem (5), this implies a parameter of loss aversion of $\lambda = 1$, and no correction is required to estimate the expected utility.

Figure 2 depicts the moments of the arithmetic returns of optimized portfolios for a quadratic utility (Figure 2(a), line) and a power utility (Figure 2(b), dots) maximizer with different parameters of risk aversion. As expected, decreasing risk aversion (i.e., decreasing parameters b and γ, respectively) leads to portfolios with higher expected returns, yet also higher volatilities. In this region, fewer, but more profitable and riskier assets are included, as can be seen from the cumulated weights. For both utility functions, the efficient sets are the usual curves in mean-volatility space with slight kinks due to the weight constraints.

Quadratic utility is oblivious to skewness and kurtosis; power utility, on the other hand, is affected by higher moments. Not surprisingly, the power utility maximizer therefore prefers portfolios with higher skewness and/or lower kurtosis. The graphs in Figure 2(c) show that for cases with a rather low level of risk aversion (high return portfolios), the two investors hardly differ in their investment choice. The higher the risk aversion (low return portfolios), however, the more pronounced the deviations become. Though hardly noticeable, when the level of risk aversion is very high, the power utility maximizer (PUM) is even prepared to accept a slightly higher volatility than the quadratic utility maximizer (QUM), if this comes with more favorable values for the higher moments. When returns are perfectly normally distributed, these deviations will not manifest themselves. However, in the presence of higher order moments PUM and QUM might consider each other's optimized portfolios as inefficient; a look beyond the usual mean-volatility framework, however, shows that neither of them behaves irrationally – the differences are due to different, yet perfectly rational, preferences and, consequently, they have different notions of what an efficiency means.

These differences become even more apparent when investors are not only risk averse, but also loss averse. When the initial wealth, w_0 influences the utility of a future level of wealth, then the same future level of wealth w will give the investor a higher utility if it comes with a gain than if it actually comes with a loss: $\mathcal{U}(w|w_0) > \mathcal{U}(w|w_0') \ \forall w_0 < w < w_0'$. With concave utility functions, decision makers will then be even more keen on avoiding losses. Primarily, this means a reduction of risk in terms of volatility. If, however, the utility function can look beyond the second moment, than investors will try to avoid not just any, but specifically negative deviations from the expected value; i.e., they will try to increase the (positive) skewness. Also, they want to avoid extreme events and will therefore aim to reduce the kurtosis of their portfolios.

Figure 3 depicts the moments of optimized portfolios for different utility functions, levels of risk aversion and levels of loss aversion. As can be seen immediately from the graphs for the mean-volatility space (left), the high return / high volatility regions of the previous graph are missing, and both QUM and PUM are now in regions which, in the absence of loss aversion, would have been chosen only with a substantially higher level of risk aversion. This effect is the stronger, the higher the loss aversion. Since the modified wealth levels (equation (6)) include the level of loss aversion and emphasize the losses accordingly, the efficient lines for quadratic utility maximizers now deviate from the one without loss aversion. Again, the constraints on minimum and maximum weights for included assets cause kinks in the efficient lines (here: in a mean-volatility space); the actual effects of these constraints, however, are not too big. Looking at the higher moments, however, shows QUM's lack of concern with deviations from the expected value that are not captured by volatility: By definition, higher risk aversion demands lower volatility, but not necessarily higher skewness or lower kurtosis.

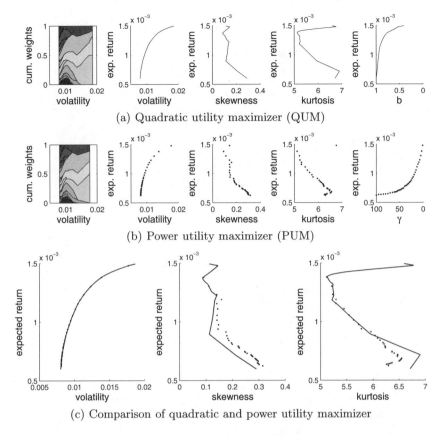

(a) Quadratic utility maximizer (QUM)

(b) Power utility maximizer (PUM)

(c) Comparison of quadratic and power utility maximizer

Fig. 2. Portfolios optimized under quadratic utility (line) and power utility (dots) without loss aversion ($\lambda = 1$), projected into the mean-volatility, mean-skewness and mean-kurtosis space.

For power utility maximizers, the picture looks different. PUM with higher risk aversion and higher loss aversion, too, prefer lower volatility. However, they care more about higher moments than QUM. Notably, for any given mean return, their portfolios never have lower skewness and higher kurtosis than the respective portfolios of QUM. However, they are prepared to accept higher volatility than QUM because the benefits of the higher skewness and the reduced kurtosis offsets.

Eventually, this concern with higher moments leads to the effect that PUM with increasing risk aversion start increasing their portfolios' volatilities. Looking at the mean-volatility diagram alone might suggest irrationality or inefficiency. Looking at the higher moments, however, shows that this is perfectly rational. With higher risk aversion, PUM are inclined to avoid (extreme) losses. Loss aversion emphasizes the losses, and the way the modified (or perceived) wealth levels, m, are computed (equation (6)) adds negative skewness.

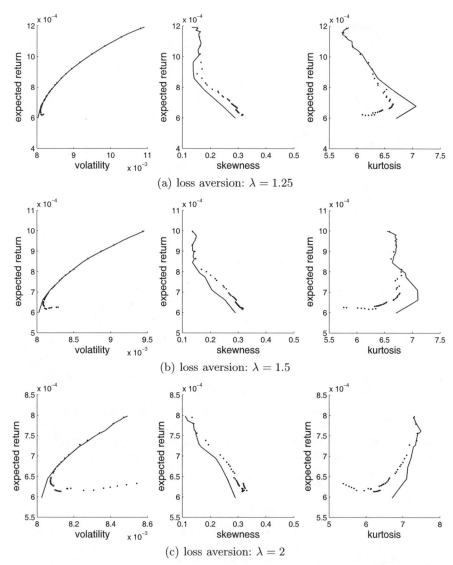

(a) loss aversion: $\lambda = 1.25$

(b) loss aversion: $\lambda = 1.5$

(c) loss aversion: $\lambda = 2$

Fig. 3. Moments of portfolios optimized under quadratic utility (line) and power utility (dots) and different levels of loss aversion, λ, projected into the mean-volatility, mean-skewness and mean-kurtosis space.

In the optimization process, these distortions are to be counterbalanced, and a potential increase in skewness and reduction of kurtosis outweighs a mere volatility reduction. To a small degree, this can be observed already in the results for the case without loss aversion but a very high level of risk aversion; when there is loss aversion in addition, this effect can become prevailing.

As can be seen from the results, the effects of loss aversion are different for different risk preferences. Obviously, when some of the investors look at different (higher) moments of return distributions, they will have different views about efficient and optimal solutions. When only looking at the mean and volatility, power utility maximizers with loss aversion might look as if they become risk lovers, while in actual facts they become even more risk averse but are increasingly concerned with the higher moments of their portfolios. Volatility is no longer sufficient to describe all aspects of risk, and basing the definition of "effectiveness" on this risk measure alone might lead to flawed conclusions.

It must be emphasized that all the empirical cases considered so far are for investors with strictly convex utility functions who are always risk averse and rational. What the consequences for optimal portfolios and their higher moments are when, as suggested by prospect theory, decision makers become risk seeking when occurring losses, is discussed in the next subsection.

4.3 Portfolios under Prospect Theory

When investors no longer have a decreasing marginal utility for any level of wealth, then their utility curve becomes convex in certain regions – which characterizes risk lovers. In Kahneman and Tversky (1979), this is found for losses: investors dislike bigger losses, yet the additional disutility is decreasing. In addition, however, investors are more sensitive to losses than they are to profits of the same magnitude; their utility of losses is therefore weighted with a loss aversion coefficient λ. The resulting utility curve is S-shaped with a kink at the origin, and, assuming the convexity for losses is equal to the concavity is profits ($\alpha = \beta$), it can be modeled by

$$\mathcal{U}_{PT}(w) = (w - w_0)^\alpha \Im_{w \geq w_0} - \lambda (w_0 - w)^\beta \Im_{w < w_0}$$
$$= (|w - w_0|)^\beta \left(1 - (1 + \lambda)\Im_{w < w_0}\right) .$$

Furthermore, it is assumed that the decision weights are equal to the probabilities of the different states, i.e., $\pi_i = p_i$.

If $\lambda = 1$ then the investor is not loss averse. When the parameter β is equal to one, the utility function is a straight line, and the investor will simply maximize the expected wealth. However, when β is lowered the curvature on either side increases. This has two effects: on the right hand side, bigger profits earn less and less additional utility and the investor will accept lower expected returns if the distribution is more favorable. On the left hand side, however, bigger losses cause less and less additional concern, and, other things equal, investors with low β are less concerned with lower or even negative skewness. Also, kurtosis is now less important: extreme events provide virtually the same (dis-)utility as events that are not quite so extreme, hence increasing the kurtosis has no substantial effect if none of the other moments is affected. Figure 4(a) illustrates these effects. Note that the lower and upper limits on the weights of included assets cause discontinuities.

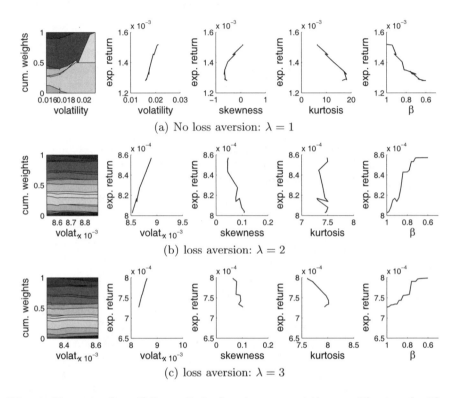

Fig. 4. Moments of portfolios optimized under prospect theory without and with λ, projected into the mean-volatility, mean-skewness and mean-kurtosis space.

When loss aversion is introduced, however, the situation changes; this can be seen from Figures 4(b) and 4(c). When they are loss averse, investors concentrate on low risk portfolios, and the actual value of the parameter β, governing the curvature of the utility function, looses importance: with the same level of loss aversion, an investor with high risk aversion (low β) will choose a portfolio quite similar to that of an investor who is risk neutral. While in the absence of loss aversion, the risk neutral aims for the highest returns (and accepts higher volatility), it is now investors with stronger curvature who are prepared to accept higher risk. Hence, even though parts of their utility curve is concave indicating they are risk seeking, their loss aversion prevents them from choosing more risky portfolios, and they select better diversified portfolios than they would in the absence of loss aversion.

5 Conclusion

Over the last decades, the importance of behavioral aspects in finance have widely been recognized. One of the insights of this stream of research is that simple utility functions are not sufficient to capture the actual preferences of investors. In particular, they do not capture that investors take their current level of wealth into account when evaluating the utility of future levels of wealth; that moving from profits to losses triggers a higher level of sensitivity; and that eventually they cease to care about even bigger losses as much as they cease to care about even bigger profits. The latter property is addressed in prospect theory where the profits are evaluated with the concave utility function of a risk averse, while losses are subject to a convex utility function of a risk loving investor. The former property can be considered by not only looking at the risk aversion, but also at the loss aversion of a decision maker.

In this chapter, the consequences of different attitudes towards risk and losses are investigated for portfolio optimization where additional practical constraints on the asset weights where introduced. Based on an empirical study, different utility functions and levels of loss aversions are tested and the properties of the resulting portfolios are investigated. It is found that higher order moments are needed to explain the choices: Even when investors are strictly risk averse, a higher level of risk aversion might lead to an investment with more, not less volatility. What seems irrational at first sight, is actually perfectly rational as under loss aversion, investors are more sensitive towards skewness and kurtosis, and increasing the desirable positive skewness and decreasing kurtosis might prevail a mere volatility reduction strategy once the level of risk aversion is big enough. Loss aversion even outbalances some of the preferences of investors who become risk seeking when they face losses.

The results found in this chapter help to understand seemingly irrational behavior and underline the inadequacy of the mean-volatility framework for many real life investment problems. Furthermore, it is shown that demanding optimization problems can be solved with heuristic optimization techniques. However, the results also open some new questions. A broader computational study for different markets and data sets shall help to generalize the findings, while additional constraints on asset selection can analyze the practical implications. Furthermore, the inclusion of other types of assets such as bonds and derivatives can provide additional insights on the consequences for financial engineering. Also, the workings and convergence behavior of the heuristic search method should be investigated in more detail. All this, however, has to be left to future research.

Acknowledgements

The author is grateful for numerous invaluable discussions with Manfred Gilli on heuristics, optimization, econometrics and financial modeling, for encouragement and support. Financial support from the EU Commission through MRTN-CT-2006-034270 COMISEF is gratefully acknowledged.

References

Basel Committee on Banking Supervision: 2003, Consultative document: The new Basel capital accord, Bank for International Settlement, available from: www.bis.org.

Brabazon, A. and O'Neill, M.: 2006, *Biologically inspired algorithms for financial modeling*, Springer, New York.

Dueck, G. and Scheuer, T.: 1990, Threshold accepting: A general purpose algorithm appearing superior to simulated annealing, *Journal of Computational Physics* **90**, 161–175.

Fogel, D.B.: 2001, *Evolutionary Computation: Toward a New Philosophy of Machine Intelligence*, 2nd edn, IEEE Press, New York, NY.

Gilli, M. and Këllezi, E.: 2002, The threshold accepting heuristic for index tracking, *in* Panos Pardalos and Vassilis Tsitsiringos (eds), *Financial engineering, E-commerce, and supply chain*, Kluwer Academic Press, Bosten, Mass.

Gilli, M. and Këllezi, E.: 2006, An application of extreme value theory for measuring financial risk, *Computational Economics* **27**(2-3), 207–228.

Gilli, M., Këllezi, E. and Hysi, H.: 2006, A data-driven optimization heuristic for downside risk minimization, *The Journal of Risk* **8**(3), 1–18.

Gilli, M., Maringer, D. and Winker, P.: 2008, Applications of heuristics in finance, *in* D. Seese, C. Weinhardt and F. Schlottmann (eds), *IT and Finance*, Springer. forthcoming.

Holland, J.H.: 1975, *Adaption in Natural and Artificial Systems*, University of Michigan Press, Ann Arbor, MI.

Kahneman, D. and Tversky, A.: 1979, Prospect theory: An analysis of decision under risk, *Econometrica* **47**, 263–291.

Kirkpatrick, S., Gelatt, C.D. and Vecchi, M.P.: 1983, Optimization by simulated annealing, *Science* **220**(4598), 671–680.

Konno, H.: 2005, Applications of global optimization to portfolio analysis, *in* C. Audet, P. Hansen and G. Savard (eds), *Essays and Surveys in Global Optimization*, Springer, pp. 195–210.

Maringer, D.: 2005, *Portfolio Management with Heuristic Optimization*, Springer-Verlag.

Maringer, D.: 2008, Constrained index tracking under loss aversion using differential evolution, *in* A. Brabazon and M. O'Neill (eds), *Natural Computing in Computational Economics and Finance*, Springer, chapter Constrained

Index Tracking under Loss Aversion using Differential Evolution. forthcoming.

Maringer, D. and Meyer, M.: 2008, Smooth transition autoregressive models – new approaches to the model selection process, *Studies in Nonlinear Dynamics and Econometrics* . forthcoming.

Maringer, D. and Oyewumi, O.: 2007, Index tracking with constrained portfolios, *Intelligent Systems in Accounting, Finance and Management* **15**, 57–71.

Maringer, D. and Parpas, P.: 2008, Global optimization of higher moments in portfolio selection, *Journal of Global Optimization* . forthcoming.

Markowitz, H.M.: 1952, Portfolio selection, *The Journal of Finance* **7**(1), 77–91.

Odean, T.: 1998, Are investors reluctant to realize their losses?, *Journal of Finance* **53**, 1775–1798.

Parpas, P. and Rustem, B.: 2006, Global optimization of the scenario generation and portfolio selection problems, *in* M. Gavrilova, O. Gervasi, V. Kumar, C.J.K. Tan, D. Taniar, A. Laganà, Y. Mun and H. Choo (eds), *Computational Science and Its Applications – ICCSA 2006, Proceedings Part III*, Vol. 3982 of *Lecture Notes in Computer Science*.

Price, K., Storn, R. and Lampinen, J.: 2005, *Differential Evolution: A Practical Approach to Global Optimization*, Springer.

Rechenberg, I.: 1965, Cybernetic solution path of an experimental problem, *Library Translation 1122*, Royal Aircraft Establishment.

Rechenberg, I.: 1973, *Evolutionsstrategie: Optimierung technischer Systeme nach Prinzipien der biologischen Evolution*, Fromman-Holzboog Verlag, Stuttgart.

Samuelson, P.A.: 1963, Risk and uncertainty: A fallacy of large numbers, *Scientia* **XCVIII**, 108–113.

Storn, R. and Price, K.: 1995, Differential evolution – a simple and efficient adaptive scheme for global optimization over continuous spaces, *Technical report*, International Computer Science Institute, Berkeley.

Storn, R. and Price, K.: 1997, Differential evolution – a simple and efficient heuristic for global optimization over continuous spaces, *Journal of Global Optimization* **11**(4), 341–359.

Thaler, R.H., Tversky, A., Kahneman, D. and Schwartz, A.: 1997, The effect of myopia and loss aversion on risk taking: An experimental test, *Quaterly Journal of Economics* **112**, 647–661.

Tobin, J.: 1958, Liquidity preference as behavior towards risk, *Review of Economic Studies* **26**(1), 65–86.

Winker, P.: 2001, *Optimization Heuristics in Econometrics. Applications of Threshold Accepting*, John Wiley & Sons, ltd., Chichester et al.

Winker, P. and Maringer, D.: 2007, The threshold accepting optimization algorithm in economics and statistics, *in* E.J. Kontoghiorges and C. Gatu (eds), *Advances in Computational Economics, Finance and Management Science*, Kluwer, pp. 107–125.

Generalized Extreme Value Distribution and Extreme Economic Value at Risk (EE-VaR)

Amadeo Alentorn[1] and Sheri Markose[2]

[1] Centre for Computational Finance and Economic Agents (CCFEA), Old Mutual Asset Managers (UK) Ltd,

[2] Centre for Computational Finance and Economic Agents (CCFEA), Department of Economics, University of Essex (corresponding author). scher@essex.ac.uk

Summary. In 2000, Ait-Sahalia and Lo have argued that Economic VaR (E-VaR) calculated under option market implied risk neutral density (RND) is a more relevant measure of risk than historically based VaR. As industry practice requires VaR at high confidence level of 99%, Extreme Economic Value at Risk (EE-VaR) based on the Generalized Extreme Value (GEV) distribution has been proposed as a new risk measure. This follows from a GEV option pricing model developed by Markose and Alentorn in 2005 which shows that the GEV implied RND can accurately capture negative skewness and fat tails, with the latter explicitly determined by the market implied tail index. Here, the term structure of the GEV based RNDs is estimated which permits the calibration of an empirical scaling law for EE-VaR, and thus, obtain daily EE-VaR for any time horizon. Backtesting results for the FTSE 100 index from 1997 to 2003, show that EE-VaR has fewer violations than historical VaR. Further, there are substantial savings in risk capital with EE-VaR at 99% as compared to historical VaR corrected by a factor of 3 to satisfy the violation bound. The efficiency of EE-VaR arises because an implied VaR estimate responds quickly to market events and in some cases even anticipates them. In contrast, historical VaR reflects extreme losses in the past for longer.

Key words: Economic Value-at-Risk, EE-VaR, empirical scaling law, term structure of implied RNDs.

1 Introduction

Value-at-Risk (VaR) has become the most popular measure for risk management. Value-at-Risk, denoted by $\mathrm{VaR}(q, k)$, is an estimate, for a given confidence level q, of the maximum that can be lost from a portfolio over a given time horizon k. An alternative measure of risk is the Economic VaR (E-VaR) proposed by Ait-Sahalia and Lo (2000) and calculated under the option-implied risk neutral density. It has been argued that E-VaR is a more general measure of risk, since it incorporates investor risk preferences, demand-supply effects, and market implied probabilities of losses or gains (Panigirtzoglou and Skiadopoulos, 2004). E-VaR can be seen as a forward looking measure to quantify market sentiment about the future course of financial asset prices, whereas historical or statistically based VaR (S-VaR) is backward looking, based on the historical data. With the development in 1993 of the traded option implied VIX index for the S&P-500 returns volatility over a 30 calendar day horizon, the so called "investor fear gauge", a significant move toward the use of a market implied rather than a historical measure of risk in practical aspects of risk management has occurred. Policy makers such as the Bank of England use traded option implied risk neutral density, volatility and quantile measures to gauge market sentiment regarding future asset prices.[3]

Given the industry standard for 10 day VaR at high confidence levels of 99%, it is important to correctly model the distribution of the extreme values of asset returns, as it is well known that the probability distributions of asset returns are not Gaussian especially at short time horizons (Cont, 2001). In the management of risk, the modelling of asymmetries and the asymptotic behaviour of the tails of the distribution of losses is important. Extreme value theory is a robust framework to analyse the tail behaviour of distributions. Extreme value theory has been applied extensively in hydrology, climatology and also in the insurance industry (Embrechts et al., 1997)). Despite early work by Mandelbrot (1963) on the possibility of fat tails in financial data and evidence on the inapplicability of the assumption of log normality in option pricing, a systematic study of extreme value theory for financial modelling and risk management has only begun recently. Embrechts et al. (1997) is a comprehensive source on extreme value theory and applications.[4] Dacorogna et al. (2001) develop a VaR estimate based on the extreme value Pareto distribution for the tails of the distribution which is then empirically estimated from high frequency data using a bootstrap method for the Hill estimator.

In this paper, we propose Extreme Economic Value at Risk (EE-VaR) as a new risk measure, which is calculated from an implied risk neutral density that is based on the Generalized Extreme Value (GEV) distribution. It has been shown in Markose and Alentorn (2005) that the GEV option pricing model not only accurately captures the negative skewness and higher kurtosis

---

[3] For the VIX index see www.cboe.com/micro/vix/ and for the Bank of England option traded implied probability density functions, volatility and quantile measures see, www.bankofengland.co.uk/statistics/impliedpdfs/. In particular, market risk premia for a given holding period is estimated as payoffs from volatility swaps which effectively take the difference between realized volatility and the option implied volatility.

[4] Embrechts et al. (1999) and Embrechts (2000) consider the potential and limitations of extreme value theory for risk management. Without being exhaustive here, De Haan et al. (1994) and Danielsson and de Vries (1997) study quantile estimation. Bali (1991) uses the GEV distribution to model the empirical distribution of returns. McNeil and Frey (2000) gives an extensive overview of extreme value theory for risk management. See also Dowd (2002, pp. 272–284).

of the implied risk neutral density (RND), but it also delivers the market implied tail index that governs the tail shape. It is important to note that the GEV does not pose a priori restrictions on the tail shape as the GEV distribution encompasses the thin and short tailed class of the Gumbel and Weibull , respectively, along with the fat tailed Fréchet. The Gumbel class includes the normal, exponential, gamma and log normal while the Weibull include distributions such as the uniform and beta. Examples of fat tailed distributions that belong to the Fréchet class are Pareto, Cauchy, Student-t and mixture distributions. Indeed, one of the main findings from Alentorn and Markose (2006) and Alentorn (2007) is that the daily implied tail shape parameter estimated without maturity effects from the GEV RND model indicates that market perception of fat tailed behaviour of extreme events is interspersed with thin and short tailed Gumbel and Weibull values. Even during extreme events, though the implied tail index results in fat tails for the GEV RND based returns, at all times the first four moments were bounded. Hence, the assumption of the GEV parametric model for the RND overcomes problems, associated with the estimation of the risk neutral density function to flexibly include extreme values and fat tails, which are often encountered with many non-parametric methods and with the use of parametric models such as the Gaussian.[5] In this paper, we will focus on estimating the term structure of the GEV based implied RNDs, which allows us to calibrate an empirical scaling law for EE-VaR at different confidence levels, and thus, to obtain the daily EE-VaR for any time horizon, without having to employ the widely used but incorrect square root of time scaling rule.

There is a vast literature on the analysis of information implied from option markets. One of the areas that has received the most attention is the study of the implied volatility surfaces, such as in Day and Craig (1988), Ncube (1996), Dumas et al. (1998) and others. The great majority of studies of implied distributions have focused on the analysis of the distributions at a single point in time for event studies, such as Bates (1991) for the study of the 1987 crash, Gemmill and Saflekos (2000) for the study of British elections, Melick and Thomas (1996) for the analysis of oil prices during the Gulf war crisis. Starting with the study of the day to day dynamics of implied volatility surfaces (see Cont (2001)), recently, Clews et al. (2000) and Panigirtzoglou and Skiadopoulos (2004) have developed a framework for the analysis of dynamics of implied RND functions.

---

[5] In order to estimate risks at high confidence levels, such as 99% - many non-parametric methods for RND estimation fail to capture tail behaviour of distributions because of sparse data for options traded at very high or very low strike prices. Hence, parametric models have become unavoidable. This, however, replaces sampling error by model error. Markose and Alentorn (2005) have argued that as the GEV distribution encompasses the 3 main classes of tail behaviour, it mitigates model error and further there is parsimony in the number of parameters necessary to define the distribution.

A problem encountered when looking at the daily dynamics of RNDs, or RND implied measures such as volatility or their associated quantile values,[6] the E-VaR, is the time to maturity and the contract switch effects (Melick and Thomas, 1998). RNDs are usually constructed using the options with shortest time to maturity. Since options have a fixed expiry date, this means that both the time horizon of the RND and the holding period of the underlying asset change with time to maturity. The degree of uncertainty decreases as the expiry date approaches. Uncertainty jumps up again when the option with the shortest time to maturity expires, and we switch to options with the next expiration date. For instance, given that options on the FTSE 100 index expire on the third Friday of the expiry month, the jump would occur on the third Monday of the expiry month. Note also that option prices with less than 5 working days to maturity are usually excluded. Thus, the problem associated with obtaining constant horizon RNDs and option implied values for VaR or volatility for the underlying assets from traded options is non-trivial. Clearly, the use of E-VaR for risk management is feasible only if it can be calculated and reported daily for a constant time horizon or holding period that is required.

With regard to the traded option implied E-VaR, to our knowledge, there are only three previous studies that have carried out an empirical analysis of E-VaR and two of these study the daily constant horizon E-VaR. Ait-Sahalia and Lo (2000) estimated the E-VaR for a 126 day horizon. Clews et al. (2000) have suggested a semi-parametric methodology that can remove maturity effects in the construction of constant horizon RNDs. The methodology consists of interpolating the Black-Scholes implied volatility surface in delta space at a given time horizon, and then deriving the implied RND by calculating the second derivative of the call pricing function, using the Breeden and Litzenberger (1978) result. This methodology is used by the Monetary Instruments and Markets Division at the Bank of England to report daily E-VaR values for the FTSE 100 index at confidence levels ranging from 5% to 95% for the FTSE 100, for a 3 month constant horizon RND. However, with this methodology, it is not possible to construct a constant time horizon implied RND for a time horizon shorter than the shortest maturity available, given that the implied volatility surface in delta space is non-linear. Panigirtzoglou and Skiadopoulos (2004) looked at the E-VaR calculated at 95% confidence level for constant horizons of 1, 3 and 6 months for every 14 days during the year 2001. However, the problem of reporting daily E-VaR at short constant horizons such as 10 days remains and typically semi-parametric methods for RND extraction fail to report E-VaR at 99% confidence level.

In this paper, we focus on obtaining a daily estimate of a constant time horizon GEV based E-VaR using a discrete term structure of RNDs. In Sec-

---

[6] Alentorn and Markose (2006) give an extensive survey of the studies done on removing maturity effects on implied volatility and higher moments of the RND. Here, we focus on the quantile values, the E-VaR.

tion 2, the new methodology we propose proceeds by first constructing a daily discrete term structure of implied RNDs, using option prices of all maturities available and a cross section of strikes for each maturity. Hence, there is a RND for each maturity available for traded options in a given day. Assuming the parametric GEV model for the RND, we calculate the EE-VaR at different confidence levels as the quantile values for the RND for each available maturity. We exploit the linear behaviour of quantile values vis-a-vis the holding period, k, in the log-log scale to derive an empirical scaling law for different confidence levels, q.[7] One of the advantages of this linear relationship is that it allows us to both interpolate and extrapolate from the available maturities and obtain daily E-VaR values for any constant horizon from 1 day to m days and can be used regardless of the method for extracting the discrete RNDs. To test the robustness of our methodology, we use the daily 90 day E-VaR reported by the Bank of England for the 95% confidence level to compare the performance of the GEV implied EE-VaR and also E-VaRs obtained from parametric RNDs for the Black Scholes and the mixture of two lognormals. We then proceed to report a 10 day EE-VaR which is easily done with our method regardless of the time horizon of the closest maturity option contracts. We analyse the performance of the EE-VaR for different confidence levels, different time horizons, and for a large dataset, and compare it with the performance of historical VaR and the Black-Scholes E-VaR. In this paper, we perform an in depth analysis of the daily EE-VaR performance for over 7 years, using daily closing index option prices on the FTSE 100 from 1997 to 2003. This is the first paper to do this and the empirical implementation and results are reported in Sections 3 and 4. Backtesting results, based on the FTSE 100 index from 1997 to 2003, show that the EE-VaR has fewer violations than historical VaR. Note that statistical VaR is done for a 1 day return and then scaled by the square root of time rule. The 10 day S-VaR when corrected by a multiplication factor of 3, to satisfy the violation bound, requires substantially more risk capital than EE-VaR. This saving in risk capital with EE-VaR at high confidence levels of 99% arises because an implied VaR estimate responds quickly to market events and in some cases even anticipate them. In contrast, VaR estimates based on historical data reflect extreme losses in the past for longer.

---

[7] The empirical evidence for the scaling parameter b in the relationship, $VaR(q,k) = VaR(q,1)k^b$, which is linear in logs has been studied by Hauksson et al. (2001), Menkens (2004) and Provizionatou et al. (2005) in the context of historical VaR. Also, Dacorogna et al. (2001) derived an extreme value based VaR scaling law for high frequency Forex data. Here, we investigate the scaling relationship for implied VaR, rather than for historical VaR.

2 Model and Methodology

2.1 Extraction of GEV Based RND from Option Prices

A large number of methods have been proposed for extracting implied distributions from option prices since the seminal work of Breeden and Litzenberger (1978), (see Jackwerth (1999) for an extensive survey). In this paper we use the methodology proposed by Markose and Alentorn (2005) based on the Generalized Extreme Value (GEV) distribution.

Let S_t denote the underlying asset price at time t. The European call option C_t is written on this asset with strike K and maturity T. We assume the interest rate r is constant. Following the Harrison and Pliska (1981) result on the arbitrage free European call option price, there exists a risk neutral density (RND) function, $g(S_T)$, such that the equilibrium call option price can be written as:

$$C_t\left(K\right) = E_t^Q[e^{-r(T-t)}\max\left(S_T - K, 0\right)]$$
$$= e^{-r(T-t)} \int_K^\infty \left(S_T - K\right) g\left(S_T\right) dS_T. \tag{1}$$

Also, the following martingale condition holds for the stock price:

$$S_t = e^{-r(T-t)} E_t^Q[S_T]. \tag{2}$$

Here, $E_t^Q[.]$ is the risk-neutral expectation operator, conditional on all information available at time t, and $g(S_T)$ is the risk-neutral density function of the underlying at maturity. Note that the GEV option pricing model in Markose and Alentorn (2005) is based on the assumption that negative returns, L_T, as defined in equation (3) below, follow a GEV distribution:

$$L_T = -R_T = -\frac{S_T - S_t}{S_t} = 1 - \frac{S_T}{S_t}. \tag{3}$$

The GEV distribution, in the form in von Mises (1936) (Reiss and Thomas, 2001, pp. 16–17) which incorporates a location parameter μ, a scale parameter σ, and a tail shape parameter ξ, is defined by:

$$F_{\xi,\mu,\sigma}(x) = \exp\left(-\left(1 + \frac{\xi}{\sigma}(x - \mu)\right)^{-1/\xi}\right), \xi \neq 0, \tag{4}$$

with

$$1 + \frac{\xi}{\sigma}(x - \mu) > 0,$$

and

$$F_{0,\mu,\sigma}\left(x\right) = \exp\left(-\exp\left(\frac{x - \mu}{\sigma}\right)\right), \xi = 0. \tag{5}$$

The tail shape parameter $\xi = 0$ yields thin tailed distributions with the so called tail index $1/\xi = \alpha$ being equal to infinity, implying that all moments of this class of distributions exist. When $\xi < 0$, the GEV distribution class is Weibull. The fat tailed Fréchet distributions arise when $\xi > 0$ and note $\xi > 0.25$ is sufficient to imply infinite kurtosis. The RND function $g(S_T)$ in (1) for the underlying asset price given that L_T is assumed to satisfy the GEV density function (Reiss and Thomas, 2001, pp. 16–17) is given by:[8]

$$
g(S_T) = \frac{1}{S_t\sigma}\left(1 + \frac{\xi(L_T - \mu)}{\sigma}\right)^{-1-1/\xi} \exp\left(-\left(1 + \frac{\xi(L_T - \mu)}{\sigma}\right)^{-1/\xi}\right),
$$
(6)

with

$$
1 + \frac{\xi}{\sigma}(L_T - \mu) = 1 + \frac{\xi}{\sigma}\left(1 - \frac{S_T}{S_t} - \mu\right) > 0.
$$
(7)

Note if the above condition in (7) is not satisfied, the GEV density function is not defined on the real line. When $\xi > 0$ and the distribution for L_t is fat tailed, condition (7) implies that the GEV density function for the price is truncated on the right, that is, the probability that the price will rise above this truncation value is zero. On the other hand, when the $\xi < 0$ and L_t is Weibull class, the GEV density function for S_T is truncated on the left implying that the price will not fall below the truncation value. Markose and Alentorn (2005) find that while this did affect the limits of integration for the option price equation in (1), the closed form solution for the call (and put) option for all cases of $\xi \neq 0$ is identical. Omitting the proof , which can be found in Markose and Alentorn (2005), the closed form GEV RND based call option price is given by:

$$
C_t(K) = e^{-r(T-t)}\{\frac{-S_t\sigma}{\xi}\Gamma(1 - \xi, H^{-1/\xi})
$$
$$
- (S_t(1 - \mu + \frac{\sigma}{\xi}) - K)(-e^{-H^{-1/\xi}})\},
$$
(8)

where $H = 1 + \frac{\xi}{\sigma}\left(1 - \frac{K}{S_t} - \mu\right)$ and $\Gamma(1 - \xi, H^{-1/\xi}) = \int_{H^{-1/\xi}}^{\infty} z^{-\xi}e^{-z}dz$ is the incomplete Gamma function.

The structural GEV parameters ξ, μ and σ can be estimated by minimizing the sum of squared errors (SSE) between the analytical solution of the GEV option pricing equations in (8) and the observed traded option prices with strikes K_i, as given in (9) below:

$$
SSE(t) = \min_{\xi,\mu,\sigma}\left\{\sum_{i=1}^{N}\left(C_t(K_i) - \widetilde{C_t}((K_i))\right)^2\right\}.
$$
(9)

---

[8] Note the relationship between the density function for L_t, $f(L_t)$, and that for the underlying, $g(S_T)$, is given by the general formula $g(S_T) = f(L_T)|\frac{\partial L_T}{\partial S_T}| = f(L_T)\frac{1}{S_t}$.

For purposes of comparison, we use the above method to back out the respective implied parameters for the Black-Scholes model and also the RND from the mixture of two lognormals (MLN) first constructed by Ritchey (1990).

At the estimation stage, we use the data on the index futures contract with the same maturity as the options and as the futures price at maturity yields, $F_T = S_T$, the no arbitrage martingale condition in (2) enables us to substitute out $E^Q(S_T)$ by using $F_{t,T} = E^Q(S_T)$. This also vitiates the need for data on the dividend yield rate. The optimization problem in (10), was performed using the non-linear least squares algorithm from the optimization toolbox in MatLab. A more detailed analysis of the estimation results, including time series of implied parameters, pricing performance and comparison of results of the GEV model with other parametric models, can be found in Alentorn and Markose (2006) and Alentorn (2007). As already noted in the Introduction, the daily implied tail shape parameters ξ, for the sample period ranged between -0.2 and $+0.22$.

2.2 EE-VaR Calculation from GEV RND

The quantile for the GEV distribution, i.e., the VaR value associated with a given confidence level q, is given as a function of the three GEV parameters (Dowd, 2002, p. 274):

$$VaR = \mu - \frac{\sigma}{\xi}\left[1 - (-\log(q))^{-\xi}\right], \xi \neq 0, \tag{10}$$

and

$$VaR = \mu - \sigma \log\left[\log(1/q)\right], \xi = 0. \tag{11}$$

On substituting the implied GEV parameters from daily traded option prices for a given maturity horizon, the extreme economic value at risk (EE-VaR) is calculated from (10) and (11).

The results obtained using EE-VaR will be compared with E-VaR values under the Gaussian assumptions of the Black-Scholes model and that of the mixture of two lognormals. The quantile of the normal distribution is used to calculate the E-VaR values for the Gaussian case using the Black-Scholes implied volatility. The MLN method models the RND as a weighted sum of two lognormals, and is given by:

$$f(S_T) = ph(S_T \mid \mu_1 T, \sigma_1\sqrt{T}) + (1-p)h(S_T \mid \mu_2, \sigma_2\sqrt{T}). \tag{12}$$

The MLN RND has been extensively used in the literature, given that it is very flexible, and allows the modelling of different levels of skewness, as well as bimodal densities. However, compared to the GEV RND it has five unknown parameters $\theta = \{\mu_1, \mu_2, \sigma_1, \sigma_2, p\}$, the means of each lognormal function μ_1 and μ_2, the standard deviations σ_1 and σ_2, and the weighting coefficient p. We obtain the set of implied parameters $\hat{\theta}$ by the method in (9). Then, E-VaR is calculated as the quantile of the MLN density, which consists of a weighted sum of the two inverse cumulative distribution functions, H, and is given by:

$$\text{E-VaR}(q, k) = \hat{p}H^{-1}\left(q \mid \hat{\mu}_1, \hat{\sigma}_1, T\right) + (1 - \hat{p})H^{-1}\left(q \mid \hat{\mu}_2, \hat{\sigma}_2, T\right). \qquad (13)$$

Some authors, such as Shiratsuka (2001) and Melick (1999), argue that the values for the higher quantiles of implied RNDs are very sensitive to the choice of RND estimation technique, since the range of strike prices that are actually traded is very limited and the tails of the estimated implied RND vary depending on the procedure employed. Table 1 shows the percentage number of days between 1997 and 2003 with traded put options with strike below each of the confidence levels.

Table 1. Percentage number of days with put option prices with strikes below each of the confidence levels FTSE-100 traded options (1997-2003).

Confidence level	Percentage number of days
70%	94%
80%	86%
90%	68%
95%	51%
99%	22%

Hence, we will also compare the quantile values obtained from the parametric RND models with those at the highest confidence level of 95% reported by the Bank of England which uses the semi-parametric RND method discussed earlier.

3 Data Description

The data used in this study are the daily settlement prices of the FTSE 100 index call and put options published by the London International Financial Futures and Options Exchange (LIFFE). These settlement prices are based on quotes and transactions during the day and are used to mark options and futures positions to market. Options are listed at expiry dates for the nearest three months and for the nearest March, June, September and December. FTSE 100 options expire on the third Friday of the expiry month. The FTSE 100 option strikes are in intervals of 50 or 100 points depending on time-to-expiry, and the minimum tick size is 0.5.

The period of study was from 1997 to 2003, so there were 28 expiration dates (7 years with 4 contracts per year). This period includes some events, such as the Asian crisis, the LTCM crisis and the 9/11 attacks, which resulted in a sudden fall of the underlying FTSE 100 index, and will be useful to analyze the performance of the methods under extreme events. The average number of maturities available with more than 3 options traded in our sample (1997-2003) is displayed in Table 2. On average across all years, we have 5.33 different maturities each day.

Table 2. Average number of maturities available FTSE-100 traded options (1997-2003).

Year	Average number of maturities available
1997	3.96
1998	4.57
1999	5.19
2000	5.49
2001	5.84
2002	6.19
2003	6.09
Average	5.33

The LIFFE exchange quotes settlement prices for a wide range of options, even though some of them may have not been traded on a given day. In this study, we only consider prices of traded options, that is, options that have a non-zero volume. The data was also filtered to exclude days when the cross-sections of options had less than three option strikes, since a minimum of three strikes is required to estimate the three parameters of the GEV model. Also, options whose prices were quoted as zero or that had less than 5 days to expiry were eliminated. Finally, option prices were checked for violations of the monotonicity condition. Monotonicity requires that the call (put) prices are strictly decreasing (increasing) with respect to the exercise price.

The risk-free rates used are the British Bankers Association's 11 a.m. fixings of the 3-month Short Sterling London InterBank Offer Rate (LIBOR) rates from the website www.bba.org.uk. Even though the 3-month LIBOR market does not provide a maturity-matched interest rate, it has the advantages of liquidity and of approximating the actual market borrowing and lending rates faced by option market participants (Panigirtzoglou and Ski-adopoulos, 2004).

4 Empirical Modelling and Results on Implied RNDs

4.1 Term Structure of RNDs

To calculate the EE-VaR, ideally, one would use a RND implied by options with time to maturity exactly equal to the time horizon we are interested. That is, to calculate the 10 day EE-VaR we would use prices from options that mature in 10 days to obtain an implied RND, and calculate the quantile of that density at the confidence level required. However, in practice, we only have options that expire every month during the next three months, and also, options that expire in March, June, September and December. In the original study of Markose and Alentorn (2005), at each trading day, only the RND implied by the closest to maturity contracts for which futures contracts were available (March, June, September and December) was extracted. Here, we propose, on a daily basis, the extraction of an RND for each of the maturities with a sufficient number of traded option prices. Then, using this discrete set of RNDs, each with a different maturity, we can construct what we call a term structure of implied RNDs. This term structure can be visualized as a 3 dimensional chart that displays, for a given day, how the implied RNDs vary across different maturities. For purposes of illustration, Figure 1 displays the implied RND term structure for a typical day, 21 August 2001, using the GEV model. Note from Figure 1 that the main feature of the term structure, which is independent of the RND extraction method used, is that the peakedness of the RNDs decreases as the time horizon increases. This term structure of implied RNDs will be used in the following section to obtain constant time horizon E-VaRs.

Table 3 displays the actual EE-VaR values. As one would expect, the EE-VaR values increase both with confidence level and with time horizon. Also, note how the number of options prices available decreases as time to maturity increases, that is, the options with the closest to maturity dates are the ones that have the widest range of traded strikes.

Table 3. EE-VaR values for each available maturity and at different confidence levels on 21 August 01.

Expiry month	Days to maturity	Number options	EE-VaR				
			70%	80%	90%	95%	99%
Sep-01	31	44	2.4%	4.4%	7.4%	10.1%	15.6%
Oct-01	59	31	3.1%	5.9%	10.2%	14.0%	21.7%
Nov-01	87	13	3.7%	7.2%	12.7%	17.5%	27.4%
Dec-01	122	16	4.2%	8.5%	15.0%	20.8%	32.6%
Mar-01	213	13	5.7%	11.4%	20.1%	27.7%	42.8%
Jun-01	304	10	6.9%	13.7%	23.8%	32.4%	49.0%

4.2 Empirical Scaling of EE-VaR

One of the requirements of the Basel accord is that banks should report the daily 10 day VaR at 99% confidence level of their portfolios. However, there are some difficulties with estimating the 10 day VaR, due to the need for a long time series in order to compute the 10 day returns, and then, calculate the quantiles of their distribution. In practice, the square root of time scaling rule is widely used to scale up the 1 day VaR to the 10 day VaR. This scaling rule is only appropriate for time series that have Gaussian properties, but it

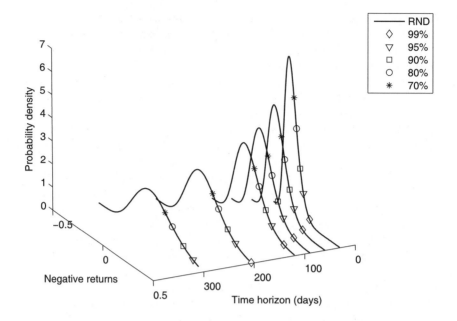

Fig. 1. Term Structure of GEV based implied RNDs and EE-VaRs on August, 21th, 2001.

has been well established in the literature for a long time (see Fama (1965) and Mandelbrot (1967)), that financial data is non-Gaussian. Following the wide spread use of VaR as a risk measure and reporting requirement, there have been several recent studies that looked at the problem of scaling VaR, such as McNeil and Frey (2000), Hauksson et al. (2001), Kaufmann and Patie (2003), Zigrand and Danielsson (2003), Menkens (2004) and Provizionatou et al. (2005).

In this study, we are faced with a similar problem, but instead of having to scale up the 1 day E-VaR, we need to scale down from the maturities available, to 10 day and 1 day E-VaR. Without resorting to scaling, we would only be able to calculate the 10 day E-VaR for only one day each month, the day when there are exactly 10 days to maturity for the closest to maturity contract (in the case of FTSE 100 data, it would be around the first Friday of each month, since contracts mature in the third Friday of the month). Following a similar approach as in Hauksson et al. (2001) and Menkens (2004), we have identified an empirical scaling law for EE-VaR against time horizon that is linear in a log-log scale:

$$\log\left(\text{EE-VaR}\left(k,q\right)\right) = b\left(q\right)\log\left(k\right) + c\left(q\right), \qquad (14)$$

where k is the number of days, $c(q)$ is the 1-day EE-VaR value (given that $\log(1) = 0$), and the slope $b(q)$ is the EE-VaR scaling parameter for a given confidence level q. Once we estimate the parameters $b(q)$ and $c(q)$ for a given day and for a given confidence level q, we can obtain the k-day EE-VaR value as follows:

$$\text{EE-VaR}(k, q) = 10^{\hat{b}(q)\log(k) + \hat{c}(q)}. \tag{15}$$

Figure 2 displays the EE-VaR values obtained from the RNDs in Figure 1 above, using the linear regression line from equation (14).

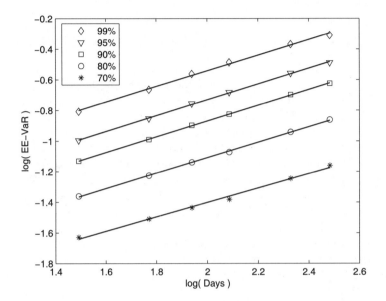

Fig. 2. Log-log plot for 21 August 01, with the estimated linear scaling rule for each confidence level.

Table 4. Regression coefficients b and c for 21 August 2001.

Confidence level	b	c
70%	0.47	-2.18
80%	0.50	-2.09
90%	0.51	-1.91
95%	0.52	-1.82
99%	0.51	-1.73
Average	0.50	-1.95

While we report the average of the full set of daily scaling coefficients, b, implied from the term structure of EE-VaR for the sample period, what Table 4 indicates is how the E-VaR based scaling coefficients differ from scaling in historical VaR. Hauksson et al. (2001) report these to be around 0.43

(though it is not clear what confidence level this is for) while Provizionatou et al. (2005) report scaling coefficients which range from 0.47 to 0.45 for the 70% and 99% confidence levels, respectively. As will be seen, in general the market implied VaR scales more vigorously with time at higher quantiles. However, the size of the scaling coefficients in Table 4 should not be confused with implying unbounded second and higher moments of the RND functions as the implied tail parameters ξ at all times for the sample period showed that up to 4 moments exist.

4.3 Improving the Estimation of the Linear Scaling Law by Using WLS

The linear regression estimated to obtain the time scaling for EE-VaR can be affected by EE-VaR values calculated from a RND constructed from very few option prices. The EE-VaR estimates in such cases will have very wide confidence intervals. As an example, take the data and regression for 12 Nov 97, shown in Figure 3. The R^2 of the OLS regression was 64.8%, a very poor fit. The EE-VaR value furthest away from maturity was obtained from an RND estimated using only 4 option prices, and thus the confidence intervals of the EE-VaR estimate are much wider than the EE-VaR values obtained for closer maturities, which are based on RNDs extracted using around 25 contracts.

One method to solve this issue is to use a Weighted Least Squares (WLS) regression, with the number of option prices available at each maturity relative to total prices as weights for the EE-VaR values.

$$Weighted\,R^2 = 1 - \frac{\sum_{i=T_1}^{T_N} w_i \left(y_i - \hat{y}_i\right)^2}{\sum_{i=T_1}^{T_N} w_i \left(y_i - \overline{y}_i\right)^2}, with \sum_{i=T_1}^{T_N} w_i = 1, \qquad (16)$$

$$w_i = \frac{NumberOfPriceAtMaturity\_i}{TotalNumberOfPrices}. \qquad (17)$$

Table 5. Average R^2 for different quantiles, and number of days with different ranges of R^2.

Confidence level	70%	80%	90%	95%	99%
Average R^2	87.9%	97.9%	98.8%	98.7%	97.9%
Number of days with $R^2 > 99\%$	429	1152	1410	1320	901
$99\% \geq R^2 > 90\%$	860	510	282	375	742
$R^2 \leq 90\%$	444	71	41	38	90

Table 6 shows the average weighted R^2 at each confidence level. Note how the fitting performance increases with confidence level, while it is lowest at 96.8% for the lowest quantile of 70%.

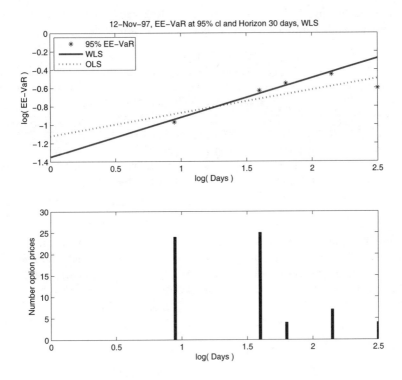

Fig. 3. Example of linear regression using OLS vs. WLS for a day when there are some maturities with very few option prices available.

Table 6. Average weighted R^2 at each quantile.

Confidence level	70%	80%	90%	95%	99%
Weighted R^2	96.8%	99.4%	99.6%	99.5%	99.3%

4.4 Full Regression Results for Scaling in EE-VaR

The average regression coefficients b and c across the 1733 days in our sample period are displayed in Table 7. We can see that the slope b increases with the confidence interval, and that intercept, i.e., the 1 day EE-VaR also increases with confidence level, as one would expect. The standard deviation of the estimates at different confidence levels is fairly constant. We also report the percentage number of days where b was found to be statistically significantly different from one half.[9] On average, irrespective of the confidence level, we

---

[9] Newey and West (1987) heteroskedasticity and autocorrelation consistent standard error was used to test the null hypothesis H0: $b = 0.5$. We employed this

found that in around 50% of the days the scaling was significantly different from 0.5, the scaling implied by the square root of time rule.

Table 7. Average regression coefficients b, c and $\exp(c)$ across the 1733 days in the sample, for the GEV case. *: The percentage number of days where b was statistically significantly different from 0.5. •: Standard deviation of b.

Confidence level q	$b(q)$	c	E-VaR$(1,q) = \exp(c)$
70%	0.41(51.3%) * (0.12)•	−2.24 (0.26)•	0.6%
80%	0.48(51.1%) * (0.09)•	−2.04 (0.23)•	0.9%
90%	0.51(52.4%) * (0.09)•	−1.84 (0.23)•	1.4%
95%	0.53(52.1%) * (0.09)•	−1.73 (0.23)•	1.9%
99%	0.56(52.6%) * (0.11)•	−1.57 (0.25)•	2.7%

Figure 4 displays the time series of the b estimates (the slope of the scaling law) for the GEV case at the 95% confidence level. Even though the average value of b at this confidence level is 0.53 (see Table 7), it appears to be time varying and takes values that range from 0.2 to 1.

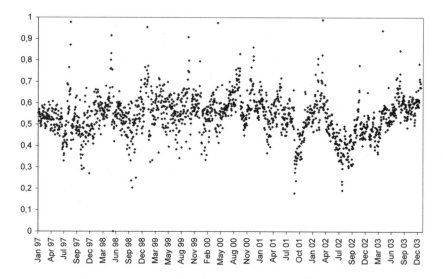

Fig. 4. Time series of b estimates for the GEV model at 95% confidence level.

methodology when testing the statistical significance of the estimated slope b, because E-VaR estimates are for overlapping horizons, and therefore are autocorrelated. The Newey-West lag adjustment used was $n - 1$.

4.5 Comparison of EE-VaR with other Models for E-VaR

Figure 5 displays the time series of 90 day E-VaR estimates at the 95% confidence level for each of the three parametric methods for RND extraction (combined with their respective empirical scaling regression results) together with the estimates from the Bank of England (BoE) non-parametric method. The 90 day FTSE returns are also displayed in Figure 5. Table 8 shows the sample mean and standard deviation of each of the four E-VaR time series. If we use the BoE values as the benchmark, we can see that on average, the mixture of lognormals method overestimates E-VaR, while the Black-Scholes method underestimates it. Among the three parametric methods, the GEV method yields the time series of E-VaRs closest to the BoE one. This can be seen both from Table 8 and also in the Figure 5, where the BoE time series practically overlaps the GEV E-VaR time series. This confirms that the GEV based VaR calculation is equivalent to the semi-parametric one used by the Bank of England but also has the added advantage, as we will see in the next section, of being capable of providing 10 day E-VaRs at the industry standard 99% confidence level.

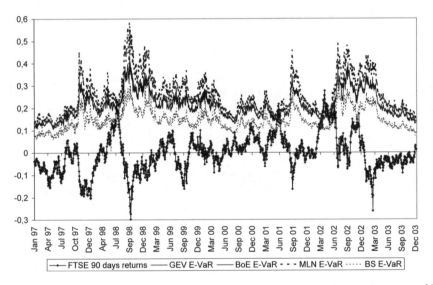

Fig. 5. BoE E-VaR vs. GEV E-VaR vs. mixture lognormals for 90 days at 95% confidence level.

4.6 EE-VaR in Risk Management: Backtesting Results

The performance of a VaR methodology is usually assessed in terms of its backtesting performance. Here, we compare the backtesting performance of EE-VaR against statistical VaR (S-VaR) estimated using the historical method

Table 8. BoE E-VaR vs. GEV E-VaR vs. mixture lognormals for 90 days at 95% confidence level.

Method	Sample mean	Sample standard deviation
BoE method	0.217	5.9%
GEV model	0.212	6.0%
Mixture lognormals	0.259	8.5%
Black Scholes	0.147	4.7%

and scaled using the square root of time, and also against the E-VaR estimated under the Black-Scholes assumptions.

The historical 1 day VaR at, say 99% confidence level, is obtained by taking the third highest loss in the time window of the previous 250 trading days, that is, in the previous year. Then, to obtain the 10 day VaR, the most commonly used technique is to scale up the 1 day VaR using the square root of time rule. The percentage number of violations at each confidence level q needs to be below the benchmark $(1 - q)$. For example, at the 99% confidence level, the percentage of days when the predicted VaR is exceeded by the market should be 1%. If there are fewer violations than the benchmark 1%, it means that the VaR estimate is too conservative, and it could impose an excessive capital requirement for the banks. On the other hand, if there are more violations than the benchmark 1%, the VaR estimates are too low, and the capital set aside by the bank would be insufficient.

We use our method to calculate the GEV and Black-Scholes based E-VaR values at constant time horizons of 1, 10, 30, 60, and 90 days. Tables 9, 10 and 11 show the results in terms of percentage number of violations for EE-VaR, S-VaR, and Black-Scholes E-VaR, respectively. The values highlighted in bold indicate that the benchmark percentage number of violations has been exceeded. On the other hand, the non highlighted values indicate that the benchmark percentage number of violations at the given confidence level has not been exceeded. The last row of each table displays the average percentage number of violations across maturities for each confidence level. In the row labelled "Benchmark", we can see the target percentage number of observations at each confidence level.

We can see that of all the three methods, EE-VaR yields the least number of cases where the benchmark percentage violations is exceeded. However, it exceeds the benchmark at all confidence levels for the 1 day horizon. The S-VaR also exceeds the benchmark at all confidence levels for the 1 day horizon, but additionally, it exceeds it for the 10 day horizon at 97% and 99% confidence levels, and for the 30 day horizon at almost all confidence levels. The Black-Scholes based E-VaR is the worst of all methods, exceeding the benchmark in 14 out of the 25 cases.

Looking at the averages across horizons for the three methods, we see that EE-VaR and S-VaR yield similar results at the higher quantiles (98% and 99%), but EE-VaR appears further away from the benchmark than S-VaR at the other confidence levels, indicating that it may be too conservative. On average, the Black-Scholes based E-VaR exceeds the benchmark percentage number of violations at all but the lowest confidence level, which indicates that it substantially underestimates the probability of downward movements at high confidence levels.

Table 9. Percentage violation of EE-VaR.

Horizon	Confidence level				
(days)	95%	96%	97%	98%	99%
Benchmark	5%	4%	3%	2%	1%
1	**6.2%**	**5.5%**	**4.4%**	**3.8%**	**3.1%**
10	2.4%	1.8%	1.5%	1.3%	0.6%
30	2.8%	2.0%	1.3%	0.5%	0.1%
60	2.5%	2.2%	1.3%	0.7%	0.0%
90	2.9%	2.3%	1.5%	0.8%	0.0%
Average	3.4%	2.8%	2.0%	1.4%	0.8%

Table 10. Percentage violation of statistical VaR(S-VaR).

Horizon	Confidence level				
(days)	95%	96%	97%	98%	99%
Benchmark	5%	4%	3%	2%	1%
1	**5.9%**	**4.7%**	**3.8%**	**2.5%**	**1.5%**
10	4.6%	4.0%	**3.1%**	1.7%	**1.4%**
30	**5.3%**	**4.2%**	**3.3%**	**2.1%**	0.8%
60	4.3%	3.6%	2.8%	1.1%	0.3%
90	4.5%	2.8%	1.9%	0.7%	0.0%
Average	4.9%	3.9%	3.0%	1.6%	0.8%

Table 11. Percentage violation of Black-Scholes based E-VaR.

Horizon	Confidence level				
(days)	95%	96%	97%	98%	99%
Benchmark	5%	4%	3%	2%	1%
1	**6.8%**	**5.9%**	**5.1%**	**4.4%**	**3.5%**
10	3.6%	3.2%	2.3%	2.0%	1.0%
30	4.0%	3.8%	**3.3%**	**2.4%**	**1.3%**
60	3.9%	3.3%	2.9%	**2.4%**	**1.7%**
90	4.6%	**4.1%**	**3.7%**	**3.0%**	**2.0%**
Average	4.6%	**4.1%**	**3.5%**	**2.8%**	**1.9%**

Figure 6 shows the time series of 10 day FTSE returns, 10 day EE-VaR and 10 day S-VaR, both at 99% confidence level. We have chosen to plot the VaR of these particular set of (q, k) values as the 10 day VaR at 99% confidence level is one of most relevant VaR measures for practitioners, given the regulatory reporting requirements. We can see how the S-VaR is violated more times (25 times, or 1.4% of the time) than the EE-VaR (10 times, or 0.6%) by the 10 day FTSE return.

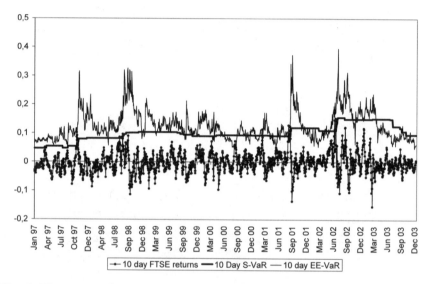

Fig. 6. Time series of the 10 day FTSE returns, 10 day EE-VaR and 10 day S-VaR, both at a 99% confidence level.

4.7 E-VaR vs. Historic VaR×3

The Basle Committee on Banking Supervision explains in the "Overview of the Amendment to the Capital Accord to Incorporate Market Risks" (Basel Committee on Banking Supervision, 1996) that the multiplication factor, ranging from 3 to 4 depending on the backtesting results of a bank's internal model, is needed to translate the daily value-at-risk estimate into a capital charge that provides a sufficient cushion for cumulative losses arising from adverse market conditions over an extended period of time. Moreover, it is also designed to account for potential weaknesses in the modelling process. Such weaknesses exist because:

- Market price movements often display patterns (such as "fat tails") that differ from the statistical simplifications used in modelling (such as the assumption of a "normal distribution").

- The past is not always a good approximation of the future (for example volatilities and correlations can change abruptly).
- VaR estimates are typically based on end-of-day positions and generally do not take account of intra-day trading risk.
- Models cannot adequately capture event risk arising from exceptional market circumstances.

It is interesting to note from Figure 7 that the E-VaR values in periods of market turbulence (Asian Crisis, LTCM crisis and 9/11) are similar to the historical S-VaR values when multiplied by a factor of 3. Thus, the multiplication factor 3 for a 10 day S-VaR appears to justify the reasoning that it will cover extreme events. In contrast, as we are modelling extreme events explicitly under EE-VaR such ad hoc multiplication factors are unnecessary.

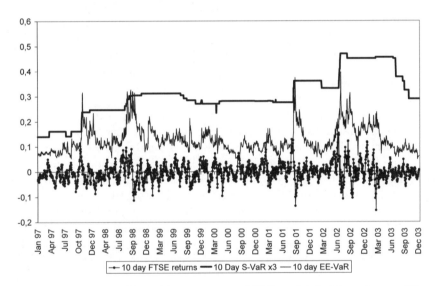

Fig. 7. 10 day at 99% E-VaR vs. statistical VaR multiplied by a factor of 3.

4.8 Capital Requirements

The average capital requirement based on the EE-VaR, S-VaR and S-VaR×3 for the 10 day VaR at different confidence level is displayed in Table 12. Here, the capital requirement is calculated as a percentage of the value of a portfolio that replicates the FTSE 100 index. What is very clear is that EE-VaR when compared to S-VaR×3 shows substantial savings in risk capital, needing only on average 1.7% more than S-VaR to give cover for extreme events. S-VaR×3 needs over 2.36 times as much capital for risk cover at 99% level. Figure 7 clearly gives the main drawback of historically derived VaR estimates where the impact of large losses in the past result in high VaR for some 250 days at a time. The EE-VaR estimates are more adept at incorporating market data information contemporaneously.

Table 12. Average daily capital requirement based on a 10 day horizon at different confidence levels, with standard deviations in brackets.

| | Confidence level | | | | | Aver- |
	95%	96%	97%	98%	99%	age
S-VaR	6.5%	7.3%	7.8%	9.1%	10.1%	8.2%
	(1.9%)	(2.3%)	(2.4%)	(2.9%)	(2.8%)	(2.5%)
S-VaR×3	19.5%	21.9%	23.4%	27.3%	30.3%	24.48%
	(5.7%)	(6.9%)	(7.2%)	(8.7%)	(8.4%)	(7.38%)
EE-VaR	8.1%	8.7%	9.6%	10.8%	12.7%	9.9%
	(3.1%)	(3.3%)	(3.7%)	(4.2%)	(5.0%)	(3.9%)

5 Conclusions

We propose a new risk measure, Extreme Economic Value at Risk (EE-VaR), which is calculated from an implied risk neutral density that is based on the Generalized Extreme Value (GEV) distribution. In order to overcome the problem of maturity effect, arising from the fixed expiration of options, we have developed a new methodology to estimate a constant time horizon EE-VaR by deriving an empirical scaling law in the quantile space based on a term structure of RNDs. Remarkably, the Bank of England semi-parametric method for RND extraction and the constant horizon implied quantile values estimated daily for a 90 day horizon coincides closely with the EE-VaR values at 95% confidence level showing that GEV model is flexible enough to avoid model error displayed by the Black-Scholes and mixture of lognormal models. The main difference between the Bank's method and the one that relies on an empirical linear scaling law of the E-VaR based on a daily term structure of the GEV RND is that shorter than 1 month E-VaRs and in particular daily 10 day E-VaR can be reported in our framework. This generally remains problematic in the RND extraction method based on the implied volatility surface in delta space as it is non-linear in time to maturity and also it cannot reliably report E-VaR for high confidence levels of 99%.

Based on the backtesting and capital requirement results, it is clear that there is a trade-off between the frequency of benchmark violations of the VaR value and the amount of capital required. The 10 day EE-VaR gives fewer cases of benchmark violations, but yields higher capital requirements compared to the 10 day S-VaR. However, when the latter is corrected by a multiplication factor of 3, to satisfy the violation bound, the risk capital needed is more than 2.3 times as much as EE-VaR for the same cover at extreme events. This saving in risk capital with EE-VaR at high confidence levels of 99% arises because an implied VaR estimate responds quickly to market events and in some cases even anticipate them.

While the power of such a market implied risk measure is clear, both as an additional tool for risk management to estimate the likelihood of extreme outcomes and for maintaining adequate risk cover, the EE-VaR needs further testing against other market implied parametric models as well as S-VaR methods. This will be undertaken in future work.

Acknowledgements

We are grateful for comments from two anonymous referees which have improved the paper. We acknowledge useful discussions with Thomas Lux, Olaf Menkens, Christian Schlag, Christoph Schleicher, Radu Tunaru and participants at the FMA 2005 at Siena and CEF 2005 in Washington DC. We have benefited from discussions with Manfred Gilli, and also financial support from the EU Commission through MRTN-CT-2006-034270 COMISEF is gratefully acknowledged.

References

Ait-Sahalia, Y. and Lo, A.W.: 2000, Nonparametric risk management and implied risk aversion, *Journal of Econometrics* **94**, 9–51.

Alentorn, A.: 2007, Option pricing with the generalized extreme value distribution and applications, *PhD thesis*, University of Essex.

Alentorn, A. and Markose, S.: 2006, Removing maturity effects of implied risk neutral densities and related statistics, *Discussion Paper 609*, Department of Economics, University of Essex.

Bali, T.G.: 1991, The generalized extreme value distribution, *Economics Letters* **79**, 423–427.

Basel Committee on Banking Supervision: 1996, *Amendment to the Capital Accord to Incorporate Market Risks*.

Bates, D.S.: 1991, The Crash of '87: Was it expected? The evidence from options market, *Journal of Finance* **46**, 1009–1044.

Breeden, D. and Litzenberger, R.: 1978, Prices of state-contingent claims implicit in option prices, *Journal of Business* **51**, 621–651.

Clews, R., Panigirtzoglou, N. and Proudman, J.: 2000, Recent developments in extracting information from option markets, *Bank of England Quarterly Bulletin* **40**, 50–60.

Cont, R.: 2001, Empirical properties of asset returns: stylized facts and statistical issues, *Quantitative Finance* **1**, 223–236.

Dacorogna, M.M., Müller, U.A., Pictet, O.V. and de Vries, C.G.: 2001, Extremal Forex returns in extremely large data sets, *Extremes* **4**, 105–127.

Danielsson, J. and de Vries, C.G.: 1997, Tail index estimation with very high frequency data, *Journal of Empirical Finance* **4**, 241–257.

Day, T.E. and Craig, L.M.: 1988, The behaviour of the volatility implied in prices of stock index options, *Journal of Financial Economics* **22**, 103–122.

De Haan, L., Cansen, D.W., Koedijk, K. and de Vries, C.G.: 1994, Safety first portfolio selection, extreme value theory and long run asset risks, *in* J. Galambos et al. (ed.), *Extreme value theory and applications*, Kluwer, Dordrecht, pp. 471–487.

Dowd, K.: 2002, *Measuring Market Risk*, Wiley, Chichester.

Dumas, B., Fleming, J. and Whalley, R.E.: 1998, Implied volatility functions: Empirical tests, *Journal of Finance* **53**(6), 2059–2106.

Embrechts, P.: 2000, Extreme value theory: Potential and limitations as an integrated risk management tool, *Derivatives Use, Trading & Regulation* **6**, 449–456.

Embrechts, P., Klüppelberg, C. and Mikosch, T.: 1997, *Modelling Extremal Events for Insurance and Finance*, Springer, Berlin.

Embrechts, P., Resnick, S. and Samorodnitsky, G.: 1999, Extreme value theory as a risk management tool, *North American Actuarial Journal* **3**, 30–41.

Fama, E.: 1965, The behaviour of stock market prices, *Journal of Business* **38**, 34–105.

Gemmill, G. and Saflekos, A.: 2000, How useful are implied distributions? Evidence from stock-index options, *Journal of Derivatives* **7**, 83–98.

Harrison, J.M. and Pliska, S.R.: 1981, Martingales and stochastic integrals in the theory of continuous trading, *Stochastic Processes Applications* **11**, 215–260.

Hauksson, H.A., Dacorogna, M., Domenig, T., Müller, U. and Samorodnitsky, G.: 2001, Multivariate extremes, aggregation, and risk estimation, *Quantitative Finance* **1**, 79–95.

Jackwerth, J.C.: 1999, Option-implied risk-neutral distributions and implied binomial trees: A literature review, *Quantitative Finance* **7**, 66–82.

Kaufmann, R. and Patie, P.: 2003, Strategic long-term financial risks: The one-dimensional case, *Risklab report*, ETH Zurich.

Mandelbrot, B.: 1963, The variation of certain speculative prices, *The Journal of Business* **36**(4), 394–419.

Mandelbrot, B.: 1967, The variation of some other speculative prices, *Journal of Business* **40**, 393–413.

Markose, S. and Alentorn, A.: 2005, The generalized extreme value (GEV) distribution, implied tail index and option pricing, *Discussion Paper 594*, University of Essex, Department of Economics.

McNeil, A.J. and Frey, R.: 2000, Estimation of tail-related risk measures for heteroscedastic financial time series: an extreme value approach, *Journal of Empirical Finance* **7**, 271–300.

Melick, W.R.: 1999, Results of the estimation of implied PDFs from a common data set, *in* Bank for International Settlements (ed.), *Estimating and Interpreting Probability Density Functions: Proceedings of the workshop held at the BIS on 14 June 1999*, Bank for International Settlement, Basel, pp. 21–30.

Melick, W.R. and Thomas, C.P.: 1996, Using option prices to infer PDFs for asset prices: An application to oil prices during the Gulf war crisis, *International Finance Discussion Paper 541*, Board of Governors of the Federal Reserve System.

Melick, W.R. and Thomas, C.P.: 1998, Confidence intervals and constant maturity series for probability measures extracted from option prices, *Paper presented at the conference Information contained in Prices of Financial Assets*, Bank of Canada.

Menkens, O.: 2004, Value-at-risk and self similarity, *Mimeo*, Centre for Computational Finance and Economic Agents, University of Essex, UK.

Ncube, M.: 1996, Modelling implied volatility with OLS and panel data models, *Journal of Banking and Finance* **20**, 71–84.

Newey, W.K. and West, K.D.: 1987, A simple, positive semi-definite, heteroskedasticity and autocorrelation consistent matrix, *Econometrica* **55**, 703–708.

Panigirtzoglou, N. and Skiadopoulos, G.: 2004, A new approach to modelling the dynamics of implied distributions: Theory and evidence from the S&P 500 options, *Journal of Banking and Finance* **28**, 1499–1520.

Provizionatou, V., Markose, S. and Menkens, O.: 2005, Empirical scaling laws in VaR, *Mimeo*, Centre for Computational Finance and Economic Agents, University of Essex, UK.

Reiss, R.D. and Thomas, M.: 2001, *Statistical Analysis of Extreme Values*, Birkhäuser, Basel. 2nd edition.

Ritchey, R.J.: 1990, Call option valuation for discrete normal mixtures, *Journal of Financial Research* **13**, 285–296.

Shiratsuka, S.: 2001, Information content of implied probability distributions: Empirical studies on japanese stock price index options, *IMES Discussion Paper 2001-E-1*, Institute for Monetary and Economic Studies, Bank of Japan.

von Mises, R.: 1936, La distribution de la plus grande de *n* valeurs, *in* P. Frank (ed.), *Selected papers of Richard von Mises*, American Mathematical Society, Providence, RI, pp. 271–294. Reprint 1954 (first published in Rev. Math. Union. Interbalcanique, **1**, 141–160, 1936).

Zigrand, J.-P. and Danielsson, J.: 2003, On time-scaling of risk and the square-root-of-time rule, *Discussion Papers 439*, FMG, London School of Economics, London.

Portfolio Optimization under VaR Constraints Based on Dynamic Estimates of the Variance-Covariance Matrix

Katja Specht[1] and Peter Winker[2]

[1] Pforzheim University of Applied Sciences, katja.specht@hs-pforzheim.de
[2] University of Giessen, peter.winker@wirtschaft.uni-giessen.de

Summary. The concept of Value at Risk (VaR) is eminently intuitive and easy to interpret. However, there are two major problems in its practical application: reliable estimations are difficult to perform and using the VaR as a constraint in portfolio optimization causes computational problems. Both problems are taken into account in the present application. First, the VaR based on estimates of the conditional covariance matrix with the "Principal Components GARCH model" (PC GARCH model) is calculated. Second, recent advances in heuristic optimization (a modified version of Memetic Algorithms) are applied in order to deal with the computational problems. The computational study indicates that the proposed method for estimating the conditional covariance matrix, the PC GARCH model, results in much better approximations of the conditional volatility structure than a simple historical estimate. This advantage in modelling is exploited by the optimization algorithm to identify portfolios with higher expected return given a fixed VaR constraint. However, adjusting the portfolio to the dynamic approximations of the conditional volatility structure also results in some overconfidence with regard to the risk constraint.

Key words: Portfolio optimization, value at risk, heuristic optimization, memetic algorithms, dynamic variance-covariance matrix.

1 Introduction

Within the framework of portfolio optimization in the traditional mean-variance world, one is interested in either a weighting system that maximizes the portfolio return under given variance or the portfolio that minimizes variance under given return. In this work, we are interested in the portfolio with the highest expected return. However, instead of considering the portfolio variance, the Value-at-Risk (VaR) is taken into account as risk measure. The concept of VaR is popular among investors because it is eminently intuitive and easy to interpret. Furthermore, VaR has been imposed on banks and other financial institutions by the Basel II Accord about banks' equity requirements (Basel Committee on Banking Supervision, 2003). Nevertheless, it is well known that VaR is not a coherent risk measure. For a comparison of VaR and conditional VaR (CVaR) see, e.g., Alexander and Baptista (2004).

Nevertheless, there are two major problems in the practical application of VaR in portfolio optimization: first, reliable estimations are difficult to perform and second, using the VaR as a constraint in portfolio optimization causes computational problems. In particular, if non-Gaussian distributions and additional constraints (maximum number of assets, minimum lot size etc.) are considered, portfolio optimization under VaR cannot be formulated as a standard quadratic programming problem as in the mean-variance approach.[3] In our approach we take account of both problems. First, we calculate the VaR based on estimates of the conditional covariance matrix with the 'Principal Components GARCH model' (PC GARCH model, Specht and Gohout (2003)). Second, we apply recent advances in heuristic optimization (a modified version of Memetic Algorithms) to deal with the computational problems (Maringer, 2005; Winker and Maringer, 2007).

Financial markets are characterized by time-dependent volatilities as well as covariances.[4] The empirical phenomenon of time-varying covariances between the returns has been demonstrated, e.g., in the empirical analyses from Longin and Solnik (2001) and de Goeij and Marquering (2004). Pojarliev and Polasek (2003) analysed some volatility models from the GARCH family in the context of portfolio optimization. They concluded from an out-of-sample comparison that multivariate GARCH models are superior to univariate ones. Of course, this is not surprising, because only multivariate models can adequately account for the relationships between the individual assets. Pojarliev and Polasek (2003) applied the BEKK model of Engle and Kroner (1995) to a low dimensional sample portfolio. Thus, the simultaneous estimation of the numerous parameters of the BEKK model has been feasible. More relevant in practice are portfolios with a large number of assets, but for these cases, the simultaneous multivariate estimation of the BEKK and similar models gives rise to numerical problems.

Instead of the simultaneous estimation of all variances and covariances of each single asset return, in the PC GARCH model just a few variance equations are estimated independently. Therefore, the PC GARCH model is suitable for dealing with high-dimensional and correlated portfolios as well. This model calculates orthogonal components from the standardized return series and applies a univariate volatility model, in this paper the GARCH and the GARCH-t model, to each principal component. Apart from the ARCH ef-

---

[3] For an application of linear programming on portfolio optimization under coherent risk measures such as CVaR without further constraints besides nonnegativity see Benati (2003).

[4] For implications on portfolio selection in a two asset model see Giannopoulos et al. (2005).

fects returns of financial data are often characterized by leverage or other asymmetry effects. Nonetheless, we have not used a more general GARCH specification as we did not model the features of the original returns themselves but of the orthogonal components. Future research will focus on the impact of applying more refined volatility models in this context.

The volatility estimates from the GARCH-models might be utilized to calculate the (volatile) covariance matrix of the original returns. The VaR is calculated based on these dynamic estimates of the covariance matrix. This is called the variance-covariance approach, i.e., using parametric models by assuming certain distributional properties of the asset returns. Alternatively, the VaR can be estimated by historical simulation and Monte Carlo approaches. According to the Basel II Accord, banks are free to choose one of these approaches for assessing the VaR of their asset portfolios.

Independent of the VaR estimation method, the fact that the VaR is given as a constraint on risk, combined with a non-negativity constraint on asset weights and an integer constraint on the number of assets traded, results in an optimization problem that cannot be solved analytically. Local search heuristics like Threshold Accepting have been applied successfully to such problems, e.g., by Dueck and Winker (1992) and Gilli and Këllezi (2002a). However, for the optimization of a high dimensional real vector, as the asset weights in a portfolio, population based methods might result in an improved performance. Thus, we use a modified version of Memetic Algorithms as proposed by Winker and Maringer (2007) in the context of portfolio optimization. This heuristic combines principles of population based algorithms with refined local search methods. In particular, for a set of candidate solutions, on each element a heuristic local search as in Threshold Accepting is performed for several iterations. Then, the resulting set is updated based on competition and cooperation between pairs of elements. The switching between individual local search and interaction between candidate solutions helps to obtain a wide coverage of the search space and accurate approximations to local and global optima.

The computational study is based on daily returns of twenty DAX stocks for the period January 1996 through July 2006. In a moving window we perform historical and conditional variance-covariance estimations. Descriptive analyses show that the conditional estimation of the covariance structure is superior to the historical estimation. Based on these estimation approaches we perform heuristic portfolio optimization under VaR.

The remainder of the paper is organized as follows. Section 2 presents our model, i.e., the optimization model and the procedure for estimating the VaR. In Section 3, the optimization heuristic to solve the complex optimization problem under VaR is described. Section 4 introduces the data used for the empirical analysis and presents the empirical results and Section 5 offers concluding remarks.

2 Model

2.1 Optimization Model

Let us assume that a portfolio manager has an initial amount of V_0, that can be invested in stocks. Given that the losses up to time τ must not exceed a fixed value of VaR $= \delta^{VaR} \cdot V_0$ with a given probability of $(1 - \alpha)$, the investor searches for the portfolio P with the maximum expected portfolio return, $E(r_P)$, that does not violate this VaR constraint.[5]

The described optimization model, combined with the constraints that the number of traded shares, n_i, has to be nonnegative (no short positions) and an integer for all assets i ($i = 1, 2, ..., N$), and the number of different assets actually included in the portfolio with positive weight, N_P, has to be less or equal to a maximum, $N_{P_{max}}$, can be formalized as

$$\max_{n_i} E(r_P) = \sum_i \frac{n_i \cdot S_{i,0}}{V_0} \cdot E(r_i) \tag{1}$$

subject to

$$n_i \in \mathbf{N}_0^+ \quad \forall i \,, \quad N_P \leq N_{P_{max}} \,, \quad \sum_i n_i \cdot S_{i,0} \leq V_0 \,,$$
$$prob(V_\tau \leq V_0(1 - \delta^{VaR})) = \alpha \,,$$

where $S_{i,0}$ and $E(r_i)$ are current closing prices (in EUR) and expected returns, respectively, of stock i ($i = 1, 2, ..., N$). In the empirical application, $E(r_i)$ will be estimated by the mean return over some historical periods. V_τ is the value of the portfolio at time τ.

The portfolio manager will be endowed with $V_0 =$ EUR 1,000,000 and the VaR constraint is specified such that the value of the portfolio five days ahead should not fall below EUR 995,000 (i.e., $\delta^{VaR} = 0.005$) with a probability of $(1 - \alpha) = 0.95$. The maximum number of assets in the portfolios is set to $N_{P_{max}} = 5$. In our computational study the weights of the portfolio will be calculated every five trading days, based on the most recent $T = 999$ observed daily returns as a moving window. This shall simulate the approach of a portfolio management team rebalancing the portfolio (approximately) once a week.

2.2 Value at Risk Estimation

The optimization problem is well described by equation (1) above introducing VaR as the objective function. However, the estimation and prediction of VaR for a given portfolio represents itself a major challenge. Kuester et al. (2006) provide a recent overview on some techniques employed to this end.

---

[5] For an extensive introduction to VaR methodology see, e.g., Jorion (2000) and Saunders and Allen (2002).

Winker and Maringer (2007) provide evidence that in an optimization approach, the portfolio selection process is markedly influenced by the method chosen to estimate risk. In particular, historical simulation is compared with parametric approaches. Although the normal distribution does not provide a good approximation to the historical data, portfolios generated based on this approximation outperformed those based on historical simulation with regard to meeting the risk constraint ex post. Winker and Maringer (2007) argue that the optimization routine might exploit specific multivariate correlation structures of the historical data which are unlikely to persist out of sample.

Consequently, we concentrate on parametric methods in this contribution. The VaR of a portfolio expresses the expected maximum loss of the portfolio over a target horizon within a given confidence level α. This expected maximum loss can be calculated for single positions as well as for portfolios. In connection with portfolios, besides the risk of each single position, the correlation and the diversification effects, respectively, have to be taken into account. Therefore the VaR of a portfolio is not the sum of the single VaRs but is defined as follows:

$$\text{VaR}_t = \sqrt{\mathbf{M}'_t \cdot \mathbf{H}_t \cdot \mathbf{M}_t} \cdot z_\alpha \, , \tag{2}$$

where $\mathbf{M}'_t = (m_{t,1}, m_{t,2}, ..., m_{t,N})$ is the vector of the values of each single position $(m_{t,i} = n_i \cdot S_{i,t})$ in the portfolio. Thereby $\mathbf{H}_t$ denotes the estimated covariance matrix and, using a (conditional) normal approximation, z_α is the respective percentile of the standard normal distribution. Consequently, the VaR estimation and prediction is very sensitive to the estimates of the covariance matrix.

Besides the historical estimate we are estimating the covariance matrix forecasts with the PC GARCH model in connection with the GARCH and the GARCH-t model which is suitable for high-dimensional portfolios.[6] The motivation for this procedure is the dimension of our model. The common volatility of twenty time series could hardly be estimated by a multivariate volatility model. Instead, the orthogonal principal components allow for independent and numerically stable estimation.[7] Additionally, we are utilizing this decomposition approach to exclude some 'noise' from the covariance by concentrating on the most volatile principal components. Thereby we do not only reduce the computational load as compared to the simultaneous estimation of all variances and covariances, but also reduce the variance of estimation due to the much smaller number of parameters to be estimated.

---

[6] The PC GARCH is also known under the label O-GARCH proposed by Alexander and Chibumba (1996).

[7] Alexander (2001) reports some identification problems in the case of weak correlation of the data. Given the substantial correlations between our series, this problem does not appear. Nevertheless, we might explore the generalization of the model proposed by van der Weide (2002) and Boswijk and van der Weide (2006) under the label GO-GARCH in future work.

The starting point for this method is the $(T \times N)$ matrix $\mathbf{R}$ of the stock returns. In the following, we consider the matrix $\widetilde{\mathbf{R}}$ of the standardized returns,

$$\widetilde{\mathbf{R}} = (\mathbf{R} - \mathbf{R}_m) \cdot \mathbf{S}^{-1}, \tag{3}$$

where $\mathbf{R}_m$ is the $(T \times N)$ matrix of the respective sample means and $\mathbf{S}$ is the diagonal matrix of the sample standard deviations of the returns. Let $\mathbf{W}$ denote the orthonormal matrix of the (column) eigenvectors of all eigenvalues of the sample correlation matrix

$$\mathbf{C} = \widetilde{\mathbf{R}}' \cdot \widetilde{\mathbf{R}}/T \tag{4}$$

in descending order of the eigenvalues. Then the matrix of all principal components may be calculated as

$$\mathbf{P} = \widetilde{\mathbf{R}} \cdot \mathbf{W}. \tag{5}$$

Since $\mathbf{W} \cdot \mathbf{W}' = \mathbf{I}$, we get

$$\widetilde{\mathbf{R}} = \mathbf{P} \cdot \mathbf{W}'. \tag{6}$$

If we are restricting ourselves to the m_r most volatile principal components in the $(T \times m_r)$ matrix $\mathbf{P}_r$ and the corresponding matrix $\mathbf{W}_r$ of eigenvectors, we get the approximation

$$\widetilde{\mathbf{R}} \approx \mathbf{P}_r \cdot \mathbf{W}'_r. \tag{7}$$

Resolving the standardization yields

$$\mathbf{R} \approx \mathbf{R}_m + \mathbf{P}_r \cdot \mathbf{W}'_r \cdot \mathbf{S}. \tag{8}$$

Finally, we derive the robust approximation of the conditional covariance matrix

$$\mathbf{H}_t \approx \mathbf{S} \cdot \mathbf{W}_r \cdot \mathbf{\Sigma}_{P_r,t} \cdot \mathbf{W}'_r \cdot \mathbf{S} \tag{9}$$

of the (non-standardized) returns, where $\mathbf{\Sigma}_{P_r,t}$ is the diagonal matrix of conditional variances of the selected principal components.[8] These variances are to be estimated by a univariate volatility model. Here we apply the GARCH(1,1) model (Bollerslev, 1986) and the GARCH-t(1,1) model (Bollerslev, 1987) for this purpose. We do not consider more general GARCH models to capture e.g. asymmetry effects like the well-known leverage effect because we did not model the features of the original returns themselves but of the principal components. This is left for future research.

---

[8] In our application we choose the principal components which explain 75 per cent of the total variance. Future research will focus on the impact of varying this percentage when dealing with the trade-off between fit and robustness.

3 Optimization Method

Given value-at-risk as risk constraint, the non-negativity constraint on port-folio weights, the integer constraint on the number of shares per stock and the constraint on the number of different assets in the portfolio, the optimiza-tion problem introduced in (1) cannot be solved analytically or by standard numerical optimization routines. Instead, we have to resort to heuristic opti-mization techniques. Although approaches for portfolio optimization by means of heuristic optimization date back to Dueck and Winker (1992), it is only re-cently that they become more widespread given the increasing availability of computational resources. Recent contributions include a collection of applica-tions to portfolio management in Maringer (2005), index tracking in Gilli and Këllezi (2002b) as well as optimization under VaR, e.g. in Gilli and Këllezi (2002a), Gilli et al. (2006) or Winker and Maringer (2007).

To tackle the optimization problem, we use the modified version of Memetic Algorithms (Moscato, 1989) introduced by Winker and Maringer (2007) (a more detailed description is provided in Maringer and Winker (2003)) which combines principles from the local search heuristic Thresh-old Accepting with population based evolutionary search strategies. The ba-sic idea of the Threshold Accepting heuristic first introduced by Dueck and Scheuer (1990) consists in starting with a random solution and to generate slight modifications for the current solution in each search step. The modified solutions are kept if they improve the current solution — or at least do not downgrade it beyond a certain threshold. While this limit is rather gener-ous for the early iterations, it becomes more and more strict later on. The intention is to allow the algorithm to escape out of bad local optima in the beginning, while it should converge eventually to a solution as close as possible to the global optimum. A more detailed description of the Threshold Accept-ing algorithm and implementation details can be found in Winker (2001) and Gilli and Winker (2007).

In contrast to the standard local search heuristics such as Threshold Ac-cepting, Memetic Algorithms use a whole population of agents working in parallel. Most of the time, each of these agents performs independent local search with the Threshold Accepting heuristic. However, at regular intervals, the agents "compete" and "cooperate". Competition aims at eliminating in-ferior solutions, while cooperation combines solutions obtained by different agents. Algorithm 1 provides the pseudo code of the Memetic Algorithm.[9] It is assumed that the agents are located on a circle. Starting with one agent, the competition is always with the next neighbour on the circle. If this neigh-bour corresponds to a better solution, it will replace the first agent. It will also replace the first agent, if the solution is only slightly worse. Thereby, the current threshold value indicates up to what degree a downgrade is accept-able. Cooperation is quite similar to the crossover step in genetic algorithms.

---

[9] Further details on the implementation are available on request from the authors.

In this step, the solutions of agents which are located on opposite sides of the circle are combined to generate new offsprings replacing their parents. Again, the decision on whether to replace the parents or not is based on the threshold criterion. For a more detailed description and discussion of modified versions of Memetic Algorithms in the context of portfolio optimization see also Maringer (2005), p. 152ff.

Algorithm 1: Pseudo code for Memetic Algorithm.

 1: Initialize population
 2: **while** stopping criteria not met **do**
 3:     **for** all agents **do**
 4:         Perform optimization with threshold method
 5:     **end for**
 6:     Compete
 7:     **for** all agents **do**
 8:         Perform optimization with threshold method
 9:     **end for**
10:     Cooperate
11:     Adjust acceptance criterion for threshold method
12: **end while**

Our implementation for the present application uses the following parameters. The population size is set to 12. Each agent is initialized by randomly selecting N_{max} assets and distributing the initial wealth uniformly on these assets. Due to the integer constraint, the initial wealth might not be invested completely. The rest is kept in cash. As stopping criterion, we use a preset maximum of iterations ranging from 50 000 to 200 000. In each iteration, a single optimization step with the threshold method is performed before competition and cooperation among agents. For all three components, a common threshold sequence is used which determines up to what extent a worsening in the portfolio quality is accepted in each step. This threshold sequence is geometrically decreasing with the number of iterations.

As an extension to the standard Memetic Algorithm, an elitist strategy is followed, i.e., the best solution obtained during the optimization process is stored separately and the corresponding portfolio is reintroduced in the population after the competition step with a certain probability (0.2 in the present application).[10] This extension guarantees that the algorithm will not eventually lose the information about promising parts of the search space. The final reported solution corresponds to the elitist, i.e., the best solution found during the complete optimization run.

---

[10] For a detailed description of the elitist principle see Maringer (2005, pp. 154ff). Typically, the quality of results does not depend heavily on the chosen (positive) probability of reintroducing the elitist. A detailed analysis of the present implementation will be subject of future research.

The optimizations were run on standard Pentium IV computers (3.2 GHz) using Matlab 7.3. A single run of the algorithm for one of the objective functions with 50 000 iterations takes about 100 seconds including the estimation of the PC GARCH model. Increasing the number of iterations or restarting the algorithm several times for different seeds leads to an almost linear scaling of computation time. Restarting the algorithm provides information on the distribution of the (stochastic) results obtained. As an example, Figure 1 exhibits the empirical distribution function for 100 restarts of the algorithm using 200 000 iterations in each replication. The values of the objective function refer to the first five days period considered in the application and the variance matrix forecast based on the PC GARCH model.

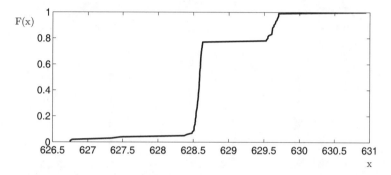

Fig. 1. Empirical distribution function $F(x)$ of expected portfolio return x for 100 restarts (200 000 iterations).

Given the high inherent complexity of the optimization problem and the stochastic nature of the optimization algorithm employed, a one point distribution cannot be expected. In fact, the results point towards the existence of two clusters of solutions, one with value of the objective function around 628.5 and the other one with values above 629.7. We expect that further increasing the number of iterations should result in a concentration of the results in the cluster with higher values of expected returns. Nevertheless, already for 200 000 iterations, we find that the difference between the best and worst solution found in 100 replications is smaller than 0.7%. The same qualitative findings are obtained when considering other time periods.

Figure 2 shows histograms describing the composition of the 100 optimized portfolios. First, it has to be noted that out of the 20 assets in the investment universe, only eight are included at least once in a portfolio. However, out of this eight assets, two have been included only once or twice with a number of stocks close to zero. Thus, only the distribution for the remaining six stocks are shown which are included a relevant number of times. It turns out that the two clusters found in Figure 1 correspond to the inclusion of stocks from MAN or Muenchner Rueck. and adjustments in the weights of the other assets. Nevertheless, the portfolio composition appears to be rather robust, in particular with regard to the selection of assets. Further tuning of the algorithm might be helpful to decrease the variance of asset weights.

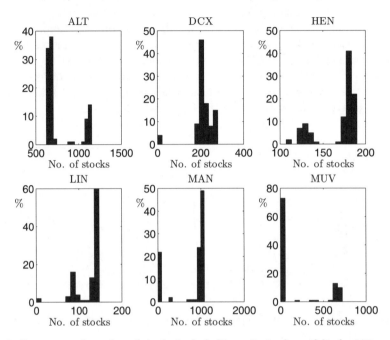

Fig. 2. Frequency of number of stocks included in optimized portfolio for 100 restarts (200 000 iterations). For the stock symbols please refer to Table 1 below.

4 Empirical Analysis

4.1 Data

The computational study is based on daily closing prices (in EUR) and returns of twenty DAX stocks, as listed in Table 1, for the period 01–01–1996 through 07–28–2006. This corresponds to a total number of $T = 2760$ trading days. The exclusion of other DAX stocks is due to a lack of historical data in the relevant period.

The portfolios will be calculated every five trading days based on the most recent 999 observed daily returns as a moving window. This shall simulate the approach of a portfolio management team rebalancing the portfolio (approximately) once a week.

The statistical properties of the five-days returns in the back-testing period are given in Table 1. The distributions are not symmetric, but mostly left skewed, and leptokurtic. These deviations from the normal distribution make it advisable to test for ARCH effects. We apply Engle's (1982) test for significant ARCH effects in the returns and the Ljung-Box (LB) test for the absence of significant correlation in the squared residuals. The results are given in Table 1.

Table 1. Description, ARCH and Ljung-Box (LB) statistics of the five-days returns (10–29–1999 to 07–28–2006).

stock	symbol	mean	standard deviation	skewness	kurtosis	ARCH statistic	LB statistic $(l = 5)$
Adidas-Salomon	ADS	0.0018	0.0415	-0.3707	5.76	28.67**	38.71**
Allianz	ALV	-0.0023	0.0526	0.0644	5.33	30.58**	83.41**
BASF	BAS	0.0010	0.0414	0.1135	7.96	63.44**	84.70**
Bayer	BAY	0.0001	0.0527	-0.049	8.56	12.45**	63.29**
BMW	BMW	0.0010	0.0444	-0.3860	6.76	7.04*	11.93*
Commerzbank	CBK	-0.0008	0.0543	-0.2819	5.44	98.05**	154.28**
Continental	CON	0.0038	0.0438	-0.1039	5.24	39.28**	43.43**
Deutsche Bank	DBK	0.0007	0.0470	0.0662	5.40	38.63**	58.82**
EON	EOA	0.0018	0.0369	-0.0835	4.28	53.29**	77.48**
Henkel	HEN	0.0010	0.0344	0.2989	6.51	19.96**	26.09**
Lufthansa	LHA	-0.0009	0.0549	-0.7057	8.74	51.33**	64.80**
Linde	LIN	0.0008	0.0388	-0.0217	4.99	16.05**	22.13**
MAN	MAN	0.0015	0.0517	-0.0435	4.82	34.92**	51.69**
Muenchner Rueck.	MUV	-0.0019	0.0561	-0.3364	7.40	23.11**	47.38**
RWE	RWE	0.0017	0.0385	0.0536	5.08	11.39**	21.37**
SAP	SAP	0.0006	0.0710	0.2234	5.58	33.01**	55.39**
Schering	SCH	0.0025	0.0415	-0.0531	7.78	3.04	6.90
Siemens	SIE	0.0001	0.0549	-0.0905	3.74	15.59**	55.78**
TUI	TUI	-0.0032	0.0588	-0.0374	7.78	56.77**	90.43**
Volkswagen	VOW	0.0004	0.0480	-0.0903	4.27	12.73**	14.88**

Under the null hypothesis of no ARCH effects, the ARCH statistic is asymptotically χ^2 distributed with $p+q = 2$ degrees of freedom as we consider the GARCH(1,1) model. The LB statistics (e.g. $l = 5$) are also asymptotically χ^2 distributed with $l = 5$ degrees of freedom, under the null hypothesis of no serial correlations. The statistics of the present data are almost all significant on the 1%–level (**) or the 5%–level (*). Therefore, it appears advisable to perform dynamic variance–covariance estimations.

4.2 Results

Variance–Covariance Estimation

In our computational study the covariance matrix and the VaR are estimated followed by the optimization of the weights of the portfolio every five trading days in a moving window of the most recent 999 returns. This procedure implies 351 five-day forecasts of the covariance matrix and 351 portfolio optimizations in the back-testing period using the resulting VaR.

For each of the 351 considered periods, we extract principal components from the standardized returns. Those principal components which explain 75 per cent of the total variance are utilized. It turns out that ten or eleven principal components are needed during each moving window in the back-testing period for this purpose. The less volatile principal components are assumed to represent mere 'noise' and are therefore neglected. After that, the conditional covariance matrix is approximated according to the procedure presented in Section 2.2 based on (i) the GARCH(1,1) model and (ii) the GARCH-t(1,1) model. Furthermore, we calculate (iii) the historical covariance matrix for the considered periods.

For a first impression, in Figure 3 the squared residuals as a 'proxy' for the unobservable volatility and the estimated volatilities of the five-day returns of the stock Allianz are shown exemplarily. Obviously, the historical estimation is not appropriate to model the time-dependent volatility. The two approaches for conditional estimates, PC GARCH(1,1) and PC GARCH-t(1,1), seem to provide very similar results. This is a finding which is representative for all stocks in our asset universe.

For assessing the properties of the covariance estimations the residuals have to be standardized using the conditional variance estimates obtained from the different approaches. In case of adequate modeling these standardized residuals should not show ARCH effects any more. In Table 2 the ARCH statistics and in Table 3 the LB statistics are listed for the returns, according to Table 1, and the standardized (by the conditional variance) returns. Both statistics show that the ARCH effects are not eliminated through the conditional estimates of the covariance matrix but markedly reduced. PC GARCH(1,1) and PC GARCH-t(1,1) provide very similar results, as already seen in Figure 3. Unlike the conditional estimates, the historical covariance matrix does hardly reduce the ARCH effects.

Besides the conditional volatilities the time-dependent correlations are crucial to the portfolio optimization. As an instance of this phenomenon, in Figure 4 the estimated correlations between the returns of Allianz (ALV) and Henkel (HEN) are presented. The dynamically estimated correlations between the considered stocks from the industries 'insurance' and 'trade' are characterized by a strong variation and a rather low level. Industry-specific factors seem to determine the development of the stocks to a large extent. The regarded stocks are only connected by the very volatile factor 'market'. Figure 4

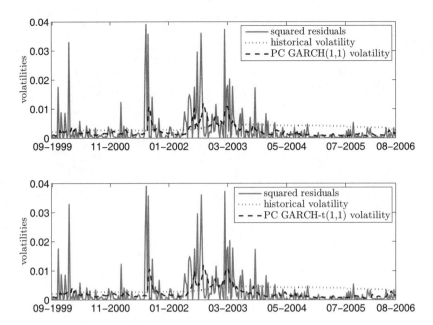

Fig. 3. Estimated volatilities of Allianz (ALV) returns on the 351 considered trading days.

(ALV–HEN) demonstrates that the historical correlation is not adequate to map this fact.

In contrast, the returns of two stocks from the same sector, e.g., BASF (BAS) and Bayer (BAY) from 'chemicals', are highly correlated with less variation, see Figure 4 (BAS–BAY). Both stocks are influenced by similar factors, i.e., diversification effects are limited. In this situation, the historical correlations seems to understate the interdependence between the stocks.

For testing the specification of the conditional correlations we use the Composite Lagged Covariance test (CLC test) introduced by Ding and Engle (2001), in which the orthogonality of the models' squared standardized residuals is tested.[11] The test statistic is T times the uncentered R^2 from the auxiliary regression of a constant on the sum of all k–th lagged sample covariances. Under the null of no covariances, the CLC statistic is asymptotically χ^2 distributed with k degrees of freedom. In Table 4 the CLC statistics up to the fifth lag before and after standardization are reported. The statistics of the non-standardized data are all significant on the 1%–level (**) or the 5%–level (*). The two models for conditional correlations provide very similar

---

[11] Engle and Marcucci (2006) report possible problems with other covariance tests suggested by Ding and Engle (2001) due to the high number of degrees of freedom for which the asymptotic theory might not work.

Table 2. ARCH statistics before and after standardization.

| stock | returns | ARCH statistic standardized returns | | |
		PC GARCH	PC GARCH-t	historical
ADS	28.6631**	18.0300**	17.095**	21.4516**
ALV	30.5764**	12.9034**	11.6609**	27.9545**
BAS	63.4379**	16.3595**	15.548**	69.3401**
BAY	12.4462**	4.7443	4.7682	13.5474**
BMW	7.0370*	0.2662	0.1593	5.3766*
CBK	98.0495**	101.3886**	97.6234**	118.8571**
CON	39.2758**	11.7981**	11.5663**	34.2906**
DBK	38.6314**	16.2601**	13.6977**	34.7676**
EOA	53.2875**	12.1481**	12.5117**	45.0737**
HEN	19.9551**	1.725	1.8105	11.5995**
LHA	51.3315**	29.1055**	27.8385**	50.9172**
LIN	16.0526**	0.4252	0.4183	15.4873**
MAN	34.9230**	27.6322**	27.871**	35.6667**
MUV	23.1066**	0.5445	0.5837	22.5782**
RWE	11.3939**	0.6378	0.6686	6.9185*
SAP	33.0092**	12.5375**	12.7876**	25.8742**
SCH	3.0360	7.5044*	7.6166*	3.2185
SIE	15.5858**	3.1978	2.9941	13.9617**
TUI	56.7686**	29.4747**	28.4413**	63.7981**
VOW	12.7253**	0.6607	0.6178	11.3075**

results again. Both appear to be well-specified in contrast to the historical correlations.

Portfolio Structure

Based on the above-discussed estimations of the covariance matrix and the resulting VaR, optimized weights of the portfolios are calculated for each of the 351 weeks under consideration. Looking at the resulting portfolio structures, we can see that the three estimates of the dynamic covariance matrix lead to mostly the same stocks being selected, but in different quantities. The portfolio structure of all three approaches is characterized by the fact that the upper bound imposed on the number of stocks in the portfolio ($N_{P_{max}} = 5$) is reached most of the time. Obviously, this finding does not come as a surprise given that a higher number of stocks in the portfolio increases diversification effects.

Figures 5 to 7 give three examples of stocks which were selected in a large number of weeks in the back-testing period. In the upper part of the figures the quantities of these stocks are shown for comparison with the prices in the lower part. It is visible that the optimization algorithm works in a plausible manner in the sense that in less volatile periods, the algorithm tends to pick out stocks with increasing prices (e.g. Adidas-Salomon (ADS) and

Table 3. LB statistics before and after standardization.

| stock | returns | LB statistic squared stand. residuals | | |
		PC GARCH	PC GARCH-t	historical
ADS	38.7103**	18.1099**	17.0805**	24.9471**
ALV	83.4056**	15.2399**	14.3252**	68.6384**
BAS	84.6931**	18.1058**	17.3978**	92.1541**
BAY	63.2875**	17.4852**	17.0647**	45.3193**
BMW	11.9328*	0.5305	0.3434	7.7612
CBK	154.2761**	102.2998**	99.5841**	171.6452**
CON	43.4279**	15.8777**	15.6268**	38.2483**
DBK	58.8162**	20.0541**	16.9019**	52.4925**
EOA	77.4780**	15.8378**	16.1776**	68.0791**
HEN	26.0942**	8.0070	9.2080	15.5803**
LHA	64.8000**	27.7552**	26.6667**	61.9173**
LIN	22.1262**	2.5928	3.2931	21.7298**
MAN	51.6875**	33.829**	34.0555**	51.9801**
MUV	47.3838**	3.6831	3.742	46.4367**
RWE	21.3689**	5.5258	5.8859	16.3444**
SAP	55.3862**	13.4903*	13.8535*	43.4762**
SCH	6.9021	8.8114	8.9455	7.7328
SIE	55.7809**	9.9787	9.5981	62.8768**
TUI	90.4273**	32.7650**	31.8031**	93.2321**
VOW	14.8778**	1.6982	1.4905	12.8478*

Table 4. CLC statistics before and after standardization.

| lag | returns | CLC statistic squared stand. residuals | | |
		PC GARCH	PC GARCH-t	historical
1	7.5546**	2.7120	2.8026	5.8063*
2	8.0158*	2.7201	2.8279	6.0794*
3	25.7545**	3.5790	3.4080	23.8718**
4	36.0365**	3.7653	3.5669	39.9692**
5	48.2055**	4.7113	4.3471	47.4708**

Contintental (CON)) and vice versa (e.g. BMW). In particular, the quantities of the selected stocks are very similar for the optimization under VaR based on both dynamic estimates of the covariance matrix. In contrast, in our examples, the historical covariance matrix induces optimized portfolios in which the quantities of stocks selected differ from those in the aforementioned portfolios.

Furthermore, it is an appreciable property that the selection process seems to be stable in the back-testing period, i.e., the selected stocks are not switched very often and the variance of quantities held is also limited. Of course, the approach might be extended by dynamic considerations, i.e., taking into account

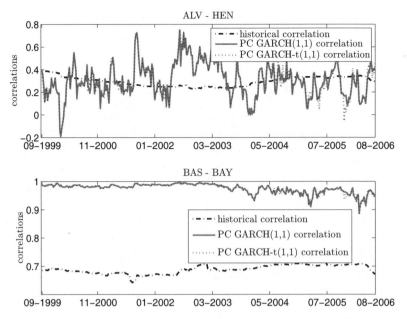

Fig. 4. Estimated correlations between the returns of Allianz (ALV) and Henkel (HEN) (upper plot) and BASF (BAS) and Bayer (BAY) (lower plot).

(expected) trading volumes induced by (expected) future portfolio adjustment. However, this extension of the model has to be left for future analysis.

Obviously, one of the most interesting questions is the performance of the portfolios. In Figure 8 the histograms of the 351 portfolio yields are plotted. It is noticeable that the number of extreme realizations in the case of portfolio optimization under VaR based on PC GARCH(1,1) and PC GARCH-t(1,1) estimates of the covariance matrix is higher than for the portfolios optimized under the historical estimates. Moreover, considering the means and the standard deviations of the portfolio weekly returns for all 351 weeks, given in Table 5, the conclusion is obvious that in our empirical work the 'PC GARCH portfolios' are characterized by higher returns in connection with higher risk than the 'historical portfolios'. The results of both 'PC GARCH approaches' are quite similar.

Table 5. Performance and risk of the resulting portfolios in the back-testing period.

covariance matrix	mean (weekly returns)	standard deviation (weekly returns)	VaR shortfalls
historical	EUR 88.22	EUR 2,865.30	13 (3.70%)
PC GARCH	EUR 215.15	EUR 3,981.80	25 (7.12%)
PC GARCH-t	EUR 217.00	EUR 3,964.70	25 (7.12%)

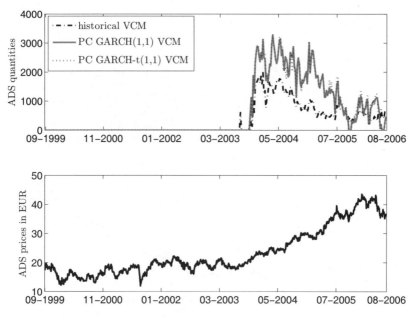

Fig. 5. Adidas-Salomon (ADS): quantities and prices.

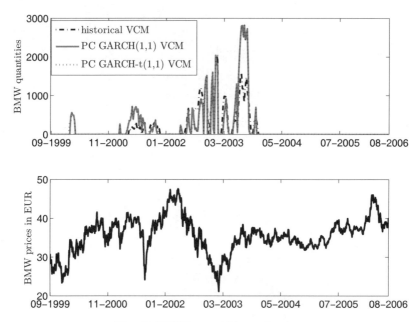

Fig. 6. BMW: quantities and prices.

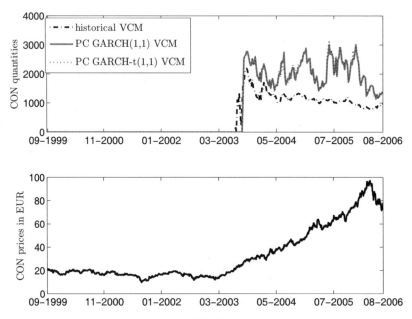

Fig. 7. Contintental (CON): quantities and prices.

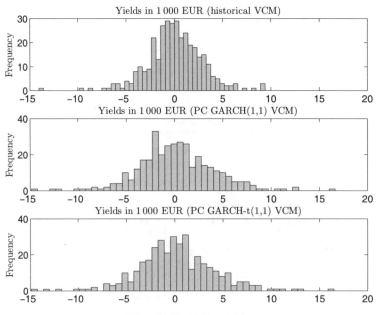

Fig. 8. Portfolio yields.

Besides the previous argumentation concerning the portfolio risk, we have to review the actual probability and magnitude of VaR shortfalls. The resulting shortfalls in our computational study, shown in Table 5, support the fact that the portfolios optimized under dynamic estimates of the covariance matrix are riskier than the portfolios optimized under the historical covariance matrix. For the 'PC GARCH portfolios' as well as for the 'PC GARCH-t portfolios' the probability of shortfalls exceeds its nominal level ($\alpha = 0.05$) to a small extent.[12] However, it is worth mentioning that there is no remarkable concentration of shortfalls for all three approaches over the 351 weeks, as can be seen from Figure 9.

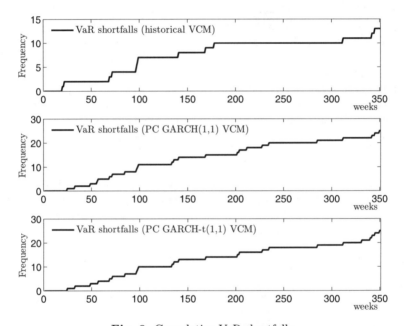

Fig. 9. Cumulative VaR shortfalls.

5 Conclusion

This paper describes an approach for portfolio optimization under VaR constraints, where the VaR forecasts are based on a dynamic estimate of the conditional covariance matrix. It turns out that the proposed method for estimating the conditional covariance matrix, the PC GARCH or PC GARCH-

---

[12] Here, we consider only shortfalls at the end of the holding period. Following Giannopoulos et al. (2005), one might also be interested in possible shortfalls at intermediate times, e.g., at a daily frequency.

t model results in much better approximations of the conditional volatility structure than a simple historical estimate. Nevertheless, the theoretical and finite properties of the PC GARCH procedure have not been analyzed in detail yet. Thus, future research will also have to consider the consistency of the estimates.

The improved modelling of the conditional volatility structure resulting from the PC GARCH model is exploited by the optimization algorithm to identify portfolios with higher expected returns given a fixed VaR constraint. The resulting portfolios exhibit some nice properties, e.g., small changes in assets from period to period, quite low volatility of quantities held. However, adjusting the portfolio to the dynamic approximations of the conditional volatility structure also results in some overconfidence with regard to the risk constraint. In fact, the VaR constraint is violated more often during the backtesting period as imposed by the constraint. We assume that this is not just due to bad luck for the period under study, but the result of a kind of overfitting the portfolio weights similar to the one described by Winker and Maringer (2007) for the historical simulation approach.

Future research will concentrate on three issues related to the current findings. First, we will improve the performance of the algorithm and study in more detail the impact of parameter settings on the optimization results. Second, the "overfitting" issue in the context of dynamic covariance estimates deserves further attention. In particular, we will analyse whether it can be reduced by modifying the length of the estimation window. Third, the modelling approach will be extended to an explicit multiperiod setting, i.e., taking into account transaction costs.

Acknowledgements

We are indebted to Manfred Gilli and three anonymous referees for helpful comments on an earlier version of this paper. Financial support from the EU Commission through MRTN-CT-2006-034270 COMISEF is gratefully acknowledged. The first author also benefited from funding by the Dr.-Herbert-Stolzenberg-foundation at the Justus-Liebig-University of Giessen.

References

Alexander, C.: 2001, Orthogonal GARCH, *in* C. Alexander (ed.), *Mastering Risk*, Vol. 2, FT-Prentice Hall, London, pp. 21–38.

Alexander, C. and Chibumba, A.: 1996, Multivariate orthogonal factor GARCH, *Discussion Papers in Mathematics*, University of Sussex.

Alexander, G.J. and Baptista, A.M.: 2004, A comparison of VaR and CVaR constraints on portfolio selection with the mean-variance model, *Management Science* **50**(9), 1261–1273.

Basel Committee on Banking Supervision: 2003, Consultative document: The new Basel capital accord, Bank for International Settlement, www.bis.org.

Benati, S.: 2003, The optimal portfolio problem with coherent risk measure constraints, *European Journal of Operational Research* **150**, 572–584.

Bollerslev, T.: 1986, Generalized autoregressive conditional heteroscedasticity, *Journal of Econometrics* **31**(3), 307–327.

Bollerslev, T.: 1987, A conditionally heteroskedastic time series model for speculative prices and rates of return, *The Review of Economics and Statistics* **69**, 542–547.

Boswijk, H.P. and van der Weide, R.: 2006, Wake me up before you GO-GARCH, *Discussion Papers 06-079/4*, Tinbergen Institute.

de Goeij, P. and Marquering, W.: 2004, Modeling the conditional covariance between stock and bond returns: A multivariate GARCH approach, *Journal of Financial Econometrics* **2**(4), 531–564.

Ding, Z. and Engle, R.: 2001, Large scale conditional covariance matrix modeling, estimation and testing, *Academia Economic Papers* **29**(2), 157–184.

Dueck, G. and Scheuer, T.: 1990, Threshold accepting: A general purpose algorithm appearing superior to simulated annealing, *Journal of Computational Physics* **90**, 161–175.

Dueck, G. and Winker, P.: 1992, New concepts and algorithms for portfolio choice, *Applied Stochastic Models and Data Analysis* **8**, 159–178.

Engle, R.: 1982, Autoregressive conditional heteroscedasticity with estimates of the variance of United Kingdom inflation, *Econometrica* **50**(1), 987–1007.

Engle, R. and Marcucci, J.: 2006, A long-run pure variance common features model for the common volatilities of the Dow Jones, *Journal of Econometrics* **132**(1), 7–42.

Engle, R.F. and Kroner, K.F.: 1995, Multivariate simultaneous generalized ARCH, *Econometric Theory* **11**(1), 122–150.

Giannopoulos, K., Clark, E. and Tunaru, R.: 2005, Portfolio selection under VaR constraints, *Computational Management Science* **2**(2), 123–138.

Gilli, M. and Këllezi, E.: 2002a, A global optimization heuristic for portfolio choice with VaR and Expected Shortfall, *in* E.J. Kontoghiorghes, Berç Rustem and S. Siokos (eds), *Computational Methods in Decision-making, Economics and Finance*, Kluwer, pp. 165–181.

Gilli, M. and Këllezi, E.: 2002b, The threshold accepting heuristic for index tracking, *in* P. Pardalos and V.K. Tsitsiringos (eds), *Financial Engineering, E-Commerce, and Supply Chain*, Applied Optimization Series, Kluwer, pp. 1–18.

Gilli, M., Këllezi, E. and Hysi, H.: 2006, A data-driven optimization heuristic for downside risk minimization, *Journal of Risk* **8**(3), 1–18.

Gilli, M. and Winker, P.: 2007, Heuristic optimization methods in econometrics, *in* E. Kontoghiorghes et al. (ed.), *Handbook on Computational Econometrics*, Elsevier, Amsterdam. forthcoming.

Jorion, P.: 2000, *Value at Risk*, 2nd edn, McGraw-Hill, Chicago et al.

Kuester, K., Mittnik, S. and Paolella, M.S.: 2006, Value-at-risk prediction: A comparison of alternative strategies, *Journal of Financial Econometrics* **4**(1), 53–89.

Longin, F. and Solnik, B.: 2001, Extreme correlation of international equity returns, *Journal of Finance* **56**(2), 649–676.

Maringer, D.: 2005, *Portfolio Management with Heuristic Optimization*, Springer, Dordrecht.

Maringer, D. and Winker, P.: 2003, Portfolio optimization under different risk constraints with modified memetic algorithms, *Discussion Paper No. 2003-005E*, Faculty of Economics, Law and Social Sciences, University of Erfurt.

Moscato, P.: 1989, On evolution, search, optimization, genetic algorithms and martial arts: Towards memetic algorithms, *Report 790*, Caltech Concurrent Computation Program.

Pojarliev, M. and Polasek, W.: 2003, Portfolio construction by volatility forecasts: does the covariance structure matter?, *Financial Markets and Portfolio Management* **17**(1), 103–116.

Saunders, A. and Allen, L. (eds): 2002, *Credit Risk Measurement: New Approaches to Value-at-Risk and Other Paradigms*, 2nd edn, Wiley, New York, NY.

Specht, K. and Gohout, W.: 2003, Portfolio selection using the principal components GARCH model, *Financial Markets and Portfolio Management* **17**(4), 450–458.

van der Weide, R.: 2002, GO-GARCH: A multivariate generalized orthogonal GARCH model, *Journal of Applied Econometrics* **17**, 549–564.

Winker, P.: 2001, *Optimization Heuristics in Econometrics. Applications of Threshold Accepting*, John Wiley & Sons, ltd., Chichester et al.

Winker, P. and Maringer, D.: 2007, The hidden risks of optimizing bond portfolios under VaR, *Journal of Risk* **9**(4), 1–19.

Optimal Execution of Time-Constrained Portfolio Transactions

Farid AitSahlia[1], Yuan-Chyuan Sheu[2], and Panos M. Pardalos[3]

[1] Department of Industrial and Systems Engineering, University of Florida, Gainesville, Florida, 32611, USA. `farid@ise.ufl.edu`
[2] Department of Industrial and Systems Engineering, University of Florida, Gainesville, Florida, 32611, USA. `ycsheu@ufl.edu`
[3] Department of Industrial and Systems Engineering, University of Florida, Gainesville, Florida, 32611, USA. `pardalos@ise.ufl.edu`

Summary. Time-constrained dynamic optimal portfolio transactions for institutional investors are investigated. The resulting constrained dynamic programming problem is solved approximately through a succession of quadratic programs. The ensuing strategies are then tested on real data. The model extends a recent one by accounting for liquidity differences between stocks.

Key words: Trade price impact, dynamic programming, quadratic approximation.

1 Introduction

Transactions on world-leading exchanges are conducted overwhelmingly by large institutional investors whose trades directly affect prices (see e.g., Schwartz and Shapiro (1992)). As a result, these investors have a preference for spreading their orders over a certain time interval (cf. Chan and Lakonishok (1995), Keim and Madhavan (1995, 1996)).

Optimal strategies for large institutional trades in a single stock were studied in Bertsimas and Lo (1998) and then extended to a portfolio of securities in Bertsimas et al. (1999). These authors show that in the presence of a single constraint, dynamic programming is impractical. They then resort to obtain approximate optimal strategies through a series of quadratic programs, which can be efficiently handled with commercially available packages such as CPLEX or MINOS.

The set-up in Bertsimas et al. (1999) (henceforth BHL) is such that the institutional trader wants to purchase (or sell) all stocks by the same date. However, stocks have different liquidities and therefore investors will likely set different liquidation horizons for different stocks (cf. Keim and Madhavan (1995)). It is this perspective that we adopt in this paper and which results in several additional constraints. As stated in BHL, the approximation of the dynamic programming problem by a series of static quadratic programs is not rigorously justified. It is simply a device that may ultimately lead to the identification of trading strategies that are better than the naive spreading of purchases (or sales) over the transaction interval. In this paper we study the impact of liquidity differences on this method. In what follows, section 2 introduces the model, section 3 describes the adaptation of the BHL approximation approach to account for liquidity differences, section 4 contains a summary of the model parameter estimates to be used in section 5. The results of our Monte Carlo study are presented in section 6 and section 7 concludes.

2 Problem Definition

Consider a portfolio of n stocks, with $\bar{s}_i$ shares to be purchased[4] by time T_i, $1 \leq i \leq n$, which are horizons that capture the liquidity difference between stocks. Due to its importance, a large (institutional) trader's transaction in a stock will impact the stock price through a specified price dynamics model. The objective of this investor is to find an optimal set of trading schedules for the different stocks so as to minimize their entire purchasing cost. Specifically, letting $T = \max_{1 \leq i \leq n} T_i$, we seek to obtain dynamically $\mathbf{s}_t = [s_1^t, ..., s_n^t]'$, $1 \leq t \leq T$, where s_i^t is the number of shares of stock i purchased in period t, that solve

$$\text{Minimize } \mathrm{E}\left[\sum_1^T \mathbf{p}_t' \mathbf{s}_t\right], \qquad (1)$$

subject to constraints resulting from the dates $T_1, T_2, \ldots, T_n$ by which these transactions should occur, namely:

$$\sum_{t=1}^{T_i} s_i^t = \bar{s}_i, \ \ 1 \leq i \leq n, \qquad (2)$$

and as in BHL, we also impose the no-short-sales constraints

$$0 \leq \mathbf{s}_t \ \text{ for } t = 1, \ldots, T. \qquad (3)$$

3 Price Dynamics

In this section, we look into the fine structure of the price dynamics of the prices of the securities. Following BHL, we assume that the execution price $\mathbf{p}_t$ is the sum of two components:

---

[4] For simplicity, we will only show net purchasing while in practice net selling or purchasing and selling shares simultaneously on different stocks could be considered.

$$\mathbf{p}_t = \tilde{\mathbf{p}}_t + \boldsymbol{\delta}_t, \tag{4}$$

where $\tilde{\mathbf{p}}_t$ may be viewed as a "no-impact" price, i.e., the price that would prevail without the impact of a transaction by a large trader, and the impact $\boldsymbol{\delta}_t$ that results from such transaction.

For $\tilde{\mathbf{p}}_t$ we use the standard vector-geometric Brownian motion model:

$$\tilde{\mathbf{p}}_t = \exp(\mathbf{Z}_t)\tilde{\mathbf{p}}_{t-1}. \tag{5}$$

Here $\mathbf{Z}_t = \text{diag}[\mathbf{z}_t]$, where $\mathbf{z}_t$ is a multivariate normal random variable with mean $\boldsymbol{\mu}_z$ and covariance matrix $\boldsymbol{\Sigma}_z$. Note that $\text{diag}[\mathbf{z}_t]$ maps the vector $\mathbf{z}_t$ into a diagonal matrix with diagonal $\mathbf{z}_t$, and the matrix exponential $\exp(\mathbf{Z}_t)$ reduces to the element-wise exponential of $\mathbf{z}_t$.

For the price impact $\boldsymbol{\delta}_t$ we assume a (perturbed) linear percentage form

$$\delta_t = \tilde{\mathbf{P}}_t(A\tilde{\mathbf{P}}_t\mathbf{s}_t + \mathbf{B}\mathbf{x}_t), \tag{6}$$

where $\tilde{\mathbf{P}}_t = \text{diag}[\tilde{\mathbf{p}}_t]$, A accounts for the cross-sectional impact of trading (in percent), and the random perturbation $\mathbf{x}_t$ is a state variable that captures the influence of overall market conditions, including private information. For simplicity, we assume as in BHL that $\mathbf{x}_t$ is specified by an AR(1) process:

$$\mathbf{x}_t = \mathbf{C}\mathbf{x}_{t-1} + \boldsymbol{\eta}_t, \tag{7}$$

where $\boldsymbol{\eta}_t$ is a vector white noise with mean 0 and covariance matrix $\boldsymbol{\Sigma}_\eta$.

The decomposition (4) expresses a decoupling of market-microstructure effects from price dynamics that is in line with the concept of implementation shortfall of Pérold (1988). In addition, the assumption of percentage price increasing linearly with trade size is empirically validated through Loeb (1983), Birinyi (1995), and Leinweber (1994) among others.

4 An Approximation Approach

When $T_i = T$ for all stocks, BHL show that the dynamic programming problem of minimizing (1) under state dynamics (6) and (7) with constraints (2) and (3) is impractical. They then propose to solve it approximately through a succession of quadratic programs, which we adopt as described next. The basic idea is to solve a series of linked static optimization problems that are solved T times. Specifically, at each iteration (time) k, $k = 1, \ldots, T$, "optimal controls" $\left(\mathbf{s}_t^{(k)}\right)_{1 \le t \le T} = \left([s_1^{t,k}, \ldots, s_n^{t,k}]'\right)_{1 \le t \le T}$, are obtained as solutions for the problem (8) – (9) below, where $s_i^{t,k}$ is the number of shares of stock i purchased in period t after the kth iteration. These controls are linked in the following manner: at every iteration k, for every $1 \le i \le n$, $s_i^{t,k}$, $1 \le t < k$, is set to its value obtained at iteration t (i.e., $s_i^{t,t}$) and $\left(s_1^{t,k}, \ldots, s_n^{t,k}\right)_{k \le t \le T}$ are the only variables for (8) – (9), from which $\left(s_i^{k,k}\right)_{1\le i\le n}$ will be preserved and implemented once determined. The rationale for this approach is as follows: at time t, $\mathbf{s}_t$ should depend on all observed prices (and states) up to time t. However, as stated, the static approximation will, for example, compute $\mathbf{s}_1,\ldots,\mathbf{s}_T$. There is therefore a need to update these values at every iteration on the basis of the most recent observations. In a numerical experiment, the latter are in fact generated through a Monte Carlo procedure. Formally, the static problem is expressed as

$$
\begin{aligned}
\text{Minimize}\quad & \mathbf{e}_n'\tilde{\mathbf{P}}_1\mathbf{s}_1 + \mathbf{s}_1'\tilde{\mathbf{P}}_1\mathbf{A}'\tilde{\mathbf{P}}_1\mathbf{s}_1 + \mathbf{x}_1'\mathbf{B}'\tilde{\mathbf{P}}_1\mathbf{s}_1 + \qquad\qquad (8)\\
& \mathbf{e}_n'\mathbf{Q}\tilde{\mathbf{P}}_1\mathbf{s}_2 + \mathbf{s}_2'\tilde{\mathbf{P}}_1(\mathbf{A}'\cdot\mathbf{R})\tilde{\mathbf{P}}_1\mathbf{s}_2 + \mathbf{x}_1'\mathbf{C}'\mathbf{B}'\mathbf{Q}\tilde{\mathbf{P}}_1\mathbf{s}_2 + \\
& \mathbf{e}_n'\mathbf{Q}^2\tilde{\mathbf{P}}_1\mathbf{s}_3 + \mathbf{s}_3'\tilde{\mathbf{P}}_1(\mathbf{A}'\cdot\mathbf{R}\cdot\mathbf{R})\tilde{\mathbf{P}}_1\mathbf{s}_3 + \mathbf{x}_1'(\mathbf{C}^2)'\mathbf{B}'\mathbf{Q}^2\tilde{\mathbf{P}}_1\mathbf{s}_3 + \\
& \cdots + \mathbf{e}_n'\mathbf{Q}^{T-1}\tilde{\mathbf{P}}_1\mathbf{s}_T + \mathbf{s}_T'\tilde{\mathbf{P}}_1(\mathbf{A}'\cdot\mathbf{R}\cdot\cdots\cdot\mathbf{R})\tilde{\mathbf{P}}_1\mathbf{s}_T + \\
& \mathbf{x}_1'(\mathbf{C}^{T-1})'\mathbf{B}'\mathbf{Q}^{T-1}\tilde{\mathbf{P}}_1\mathbf{s}_T,
\end{aligned}
$$

subject to

$$
\bar{s}_i = \sum_{t=1}^T s_i^t, \quad i = 1,\ldots,n, \qquad\qquad (9)
$$

$$
0 \le \mathbf{s}_t, \quad t = 1,\ldots,T,
$$

where $\tilde{\mathbf{P}}_1 = \mathrm{diag}[\mathbf{p}_1]$, $\mathbf{e}_n$ denotes the n-parameter column vector $[1,...,1]'$, $\mathbf{Q}$ is an $(n\times n)$ diagonal matrix with entries

$$
q_i = \exp(\mu_{z,i} + \frac{1}{2}\Sigma_{z,ii}),
$$

and $\mathbf{R}$ is an $(n\times n)$ symmetric matrix with elements:

$$
r_{i,j} = \exp[\mu_{z,i} + \mu_{z,j} + \frac{1}{2}(\Sigma_{z,ii} + \Sigma_{z,jj} + 2\Sigma_{z,ij})].
$$

The matrix dot operator '$\cdot$' denotes an element-wise matrix multiplication, i.e., $\mathbf{A}\cdot\mathbf{B} \equiv [a_{ij}b_{ij}]$.

When different stocks have potentially different liquidation horizons T_i, $1 \le i \le n$, the above approach needs to be adjusted as follows. With $T = \max_{1\le i\le n}T_i$, the constraints (9) need to be augmented with the following conditions:

$$
s_i^t = 0, \quad \text{for } 1 \le i \le n \text{ and } T_i < t \le T,
$$

where the last condition may be withdrawn (empty set) when it is not satisfied because of $T_i = T$.

5 Numerical Example

In this section, we summarize parameter estimates and data provided in BHL in order to conduct a Monte Carlo experiment in the next section. BHL provide complete and detailed information on only five stocks, which will form our illustrative portfolio. Their data were drawn from three different sources. The first is a proprietary record of trades performed over the NYSE DOT system by the trading desk at Investment Technology Group (ITG) on every trading day between January 2, 1996 and December 31, 1996. The second is the NYSE TAQ data that extract quotes prevailing at the time of ITG trades. The last is S&P 500 tick data provided by Tick Data Inc. to get intraday levels for the S&P 500 index during 1996. As in BHL, we use the S&P 500 index to capture the overall market conditions. The five stocks we use in this study are listed in Table 1. Their mean returns ($\boldsymbol{\mu}_z$) and associated covariance matrix ($\boldsymbol{\Sigma}_z$) are listed in Table 2.

Table 1. Ticker symbols, company names, and closing prices are on a randomly selected day in 1996 for the five stocks in the portfolio (Source: BHL).

TICKER	Company Name	Closing Price
AHP	AMER HOME PRODS	64.0625
AN	AMOCO	70.5000
BLS	BELLSOUTH	37.2500
CHV	CHEVRON	62.6250
DD	DUPON	88.9375

Table 2. Mean return and correlation estimates (in percent/year) for the no-impact price process $\tilde{\mathbf{p}}_t$ (Source: BHL).

	AHP	AN	BLS	CHV	DD
$\hat{\mu}$	0.200	-0.103	-0.387	-0.167	0.283
$\hat{\sigma}$	0.268	0.196	0.283	0.222	0.225
AHP	1.000	0.196	0.284	0.226	0.345
AN	0.196	1.000	0.173	0.408	0.283
BLS	0.284	0.173	1.000	0.259	0.328
CHV	0.226	0.408	0.259	1.000	0.314
DD	0.345	0.283	0.328	0.314	1.000

We also use price impact estimates for (6) and (7) as obtained by BHL. Namely, $\hat{C} = 0.0354$, $\hat{\sigma}_\eta = 0.999$, and the regression estimates of (6) as given in Table 3, where the first row corresponds to the elements of the diagonal matrix B and the remaining entries form A.

Table 3. Price-impact parameter estimates (eq. (6)). All coefficients have been scaled up by 10^{10}, except those for the SPX index (x_t), which have been multiplied by 10^5 (Source: BHL).

Variable	AHP	AN	BLS	CHV	DD
SPX	3.74	2.28	0.26	11.70	2.38
AHP	12.40	-1.69	-1.99	1.04	-1.07
AN	-1.32	10.10	-1.96	1.09	-1.63
BLS	3.49	0.83	14.40	2.26	-2.45
CHV	2.09	-0.17	2.34	21.20	2.84
DD	-0.93	0.66	6.18	1.55	11.70

6 Monte Carlo Simulation

With the parameter estimates given in the previous section, we now investigate the performance of the time-constrained best-execution strategy via Monte Carlo simulation. Summary results of 10,000 runs are given in Tables 4 and 5. In the former, all shares are to be purchased in equal number ($\bar{s}$) by the same date T as in the BHL set-up. The corresponding approximate optimal strategy resulting from the linked solutions of (8) – (9) are in the column labelled "Approximate Optimal Strategy". The column labelled "Even Trading" refers to the strategy of spreading evenly the share purchases in the amount of $\bar{s}/T$ over time. In the latter table, "Even Trading" refers to the even purchase of $\bar{s}_i/T_i$ shares of stock i by time T_i, and the results under "Approximate Optimal Strategy" refer to the approximately optimal strategy resulting from the linked solutions of (8) – (9).

Table 4 reflects the same overall set-up as in Table 6 of BHL, namely identical number of shares and horizons for all stocks. Clearly, one major reason behind the differences between the two tables is the significantly different portfolio constitutions. However, the two tables do share overall similar average patterns such as the drop of execution costs with longer horizons, the increase of these costs with smaller order sizes, and the generally better performance of the static approximation strategy. In addition, Table 5 indicates that accounting for liquidity differences between stocks may lead to significantly higher execution costs. We note that, in fact, the latter result is the only one that clearly emerges for certain configurations of $\bar{s}$ and $(T_i)_{1 \leq i \leq n}$ from our (limited) study. All the other conclusions are somewhat less clear, as pointed out in BHL, because of the resulting standard errors.

7 Concluding Remarks

This study provided a small-portfolio instance that also accounts for liquidity differences between stocks to evaluate the effectiveness of a static-optimization algorithm that purports to approximately solve a dynamic programming prob-

Table 4. Average execution cost per share (in cents). For each stock, $\bar{s}$ shares are to be purchased by T. Standard errors in parentheses.

T	$\bar{s}$	Even trading	Approximate optimal strategy
20	10,000	1.15	1.14
		(0.42)	(0.41)
20	20,000	1.14	1.09
		(0.41)	(0.40)
20	30,000	1.16	1.07
		(0.42)	(0.40)
20	50,000	1.14	1.04
		(0.41)	(0.38)
20	100,000	1.16	1.09
		(0.42)	(0.40)
20	200,000	1.19	1.17
		(0.42)	(0.41)
5	100,000	1.50	1.50
		(0.69)	(0.68)
10	100,000	1.15	0.81
		(0.56)	(0.55)
15	100,000	1.16	1.14
		(0.47)	(0.46)

Table 5. Average execution cost per share (in cents). For stock i, $\bar{s}$ shares are to be purchased by $(T * i)/5$, $1 \leq i \leq 5$, with stocks ordered as in Table 1. Standard errors in parentheses.

T	$\bar{s}$	Even trading	Approximate optimal strategy
20	10,000	1.14	1.08
		(0.50)	(0.25)
20	20,000	1.16	1.15
		(0.50)	(0.26)
20	30,000	1.19	1.00
		(0.51)	(0.45)
20	50,000	1.19	0.78
		(0.50)	(0.33)
20	100,000	1.25	1.09
		(0.51)	(0.32)
20	200,000	1.36	1.70
		(0.51)	(0.48)
5	100,000	2.68	2.11
		(0.74)	(0.53)
10	100,000	1.51	1.45
		(0.65)	(0.55)
15	100,000	1.30	1.82
		(0.56)	(0.56)

lem. As anticipated, this approximation approach appears to be sensitive to its context.

Acknowledgements

We dedicate this article to Manfred Gilli on the occasion of his 65th birthday.

References

Bertsimas, D., Hummel, P. and Lo, A.: 1999, Optiomal control of execution costs for portfolios, *Computing in Science and Engineering* **1**(6), 40–53.

Bertsimas, D. and Lo, A.: 1998, Optimal control of execution costs, *Journal of Financial Markets* **1**(1), 1–50.

Birinyi, L.: 1995, What does institutional trading cost?, Birinyi Associates.

Chan, L. and Lakonishok, J.: 1995, The behavior of stock prices around institutional trades, *Journal of Finance* **50**, 1147–1174.

Keim, D. and Madhavan, A.: 1995, Anatomy of the trading process: empirical evidence on the behavior of institutional traders, *Journal of Financial Economics* **37**(3), 371–398.

Keim, D. and Madhavan, A.: 1996, The upstairs market for large-block transactions: analysis and measurement of price effects, *Review of Financial Studies* **9**(1), 1–36.

Leinweber, D.: 1994, Careful structuring reins in transaction costs, *Pensions and Investments* **25**, 431–468.

Loeb, T.: 1983, Trading cost: the critical link between investment information and results, *Financial Analysts J.* **39**, 39–44.

Pérold, A.: 1988, The implementation shortfall: paper versus reality, *J. Portfolio Management* **14**(3), 4–9.

Schwartz, R. and Shapiro, J.: 1992, The challenge of institutionalization for the equity markets, *in* A. Saunders (ed.), *Recent Developments in Finance*, Business One Irwin, pp. 31–45.

Semidefinite Programming Approaches for Bounding Asian Option Prices

Georgios V. Dalakouras[1], Roy H. Kwon[2], and Panos M. Pardalos[3]

[1] Department of Industrial and Systems Engineering, University of Florida, Weil Hall, P.O. Box 116595, Gainesville, FL 32611-6595, USA, `gbdalako@ufl.edu`
[2] Department of Mechanical & Industrial Engineering, University of Toronto, 5 King's College Road, Toronto, Ontario, M5S3G8, Canada, `rkwon@mie.utoronto.ca`
[3] Department of Industrial and Systems Engineering, University of Florida, Weil Hall, P.O. Box 116595, Gainesville, FL 32611-6595, USA, `pardalos@ufl.edu`

Summary. Semidefinite programming (SDP) approaches are considered for obtaining bounds for the price of an arithmetic average Asian option. A method for computing the moments of the distribution of prices is developed which enables the method of Bertsimas and Popescu to be extended for the case of the Asian option. In particular, several SDP formulations for upper and lower bounds of the price of an Asian option are given based on different representations of the payoffs of the option. The formulations are amenable to standard SDP computational methods.

Key words: Asian options, semidefinite programming.

Preamble

We wish Manfred the best on his birthday and for the future.

1 Introduction

Financial derivatives such as options have been a fundamental component of modern finance and risk management for the last few decades (Hull, 2003) . Options are derivative contracts that enable holders (buyers) of the option the right but not obligation to purchase or sell an asset (e.g., common stock or foreign exchange) for a predetermined price at some future point in time. As such, options can be very useful instruments in hedging financial exposures. In the case of a European call or put option, the holder of an option will decide to purchase or sell at the end of a specified time period depending on the price of the underlying asset at the end of the time period (exercise time), i.e., the payoff of the option at exercise time is a function of the price of the underlying asset. If the payoff is positive, then the holder of the option will exercise the right to purchase or sell else the holder will not exercise this right. A crucial issue is the pricing of an option, i.e., how does one determine a fair price for an option? For European options one can use a variant of the famous Black and Scholes (1973) pricing formula to obtain a fair price. A key assumption that is necessary for the Black-Scholes formula is that the distribution of the price of the underlying asset is lognormal.

In this paper, we consider the pricing of arithmetic average European-type Asian options. We will refer to this variant as simply "Asian options" for the rest of the paper. Asian options are financial derivatives whose payoff depends on the average price of an underlying asset over a specified (fixed) time duration where the exercise time is at the end of the time period (Boyle and Emanuel, 1980). Asian options are of great practical and theoretical interest as they are useful instruments for underlying assets such as currencies with low trading volume. We consider the arithmetic average case where the underlying asset may be assumed to follow a diffusion process such as geometric Brownian motion. A major difficulty in pricing Asian options is that the distribution of the average price of an underlying asset over a time period may not be lognormal even when the distribution of the price of the underlying asset is lognormal. Thus, closed form solutions such as the Black-Scholes pricing formula for European type options are unavailable for Asian options. In general, it is a challenge to price Asian options and there have been many different approaches such as numerical techniques of solving the relevant PDE's as in Rogers and Shi (1995), analytic approximations based on Taylor expansions as in Turnbull and Wakeman (1991), and methods based on Monte Carlo simulation as in Glasserman et al. (1999). However, it is still a challenge to approximate the price of an Asian option and current extensions are still at the early stages of development.

We consider semidefinite programming (SDP) approaches for bounding Asian option prices. In particular, methods to compute moments of the distribution of arithmetic average Asian option prices at exercise time are developed which enables the use of several SDP formulations that represent bounds for the price of an Asian option. The approach taken is based on Bertsimas and Popescu (2002) where they present SDP formulations for obtaining tightest upper bounds on the price of a European call option. In our approach, both upper and lower bounds are developed for the Asian option that are analogously the tightest possible. Several equivalent SDP formulations are presented for the bounds where each formulation will admit different computational strategies for solution based on the particular structure of the formulation. The formulations are amenable to efficient solution by standard SDP computational methods and approaches. The main contributions of the paper involve the development of methods to compute the first N moments of the distribution of the price of the Asian option and development of the associated semidefinite programming formulations that provide the tightest lower and upper bounds for the price of the Asian option. It is noted that in Bertsimas and Popescu (2002) no discussion is given of how moments are computed.

2 SDP Strategy for Bounding Option Prices

We extend the approach of Bertsimas and Popescu (2002) for providing tight bounds on Asian option prices. Their strategy proceeds as follows: given the first N moments of a risk-neutral distribution of the price of an underlying asset at the exercise time a mathematical programming formulation is given that represents a tight upper bound on the price of a European call option on the asset. The mathematical programming formulation is essentially an infinite dimensional linear program. Then, it is shown that the dual of the infinite dimensional linear program can be converted to an equivalent semidefinite program. Thus, computing a tight upper bound on the price of the option is reduced to that of finding the optimal solution of a semidefinite program.

In our extension, we consider SDP formulations for *both* upper and lower bounds on the price of an Asian option that are tight. We present a method to compute the moments of the distribution of the price of the Asian option assuming that the distribution of the underlying asset follows a geometric Brownian motion. This will enable the SDP approach since the moments are used as part of the SDP formulations. We present two different SDP formulations for the upper bounds. Each will be amenable to different efficient computational strategies for solving SDPs.

2.1 Arithmetic Average Asian Options

The price of an arithmetic average (fixed-strike) Asian European-type call option at exercise time T is given by

$$V_T = E\left[\left(\frac{1}{T}\int_0^T S_t dt - K\right)^+\right],$$

where T is the exercise time, K is the strike price, and S_t is the price of the underlying at time t. We can also express the price of the Asian option at time $t = 0$ as

$$price = S_0 E[\max(0, Z_T^*)],$$

where Z_t^* is given by

$$Z_t^* = q_t^* + \frac{e^{-r(T-t)}}{S_t}\left(\frac{1}{T}\int_0^T S_t dt - K\right),$$

where r is the risk-free rate. We can choose q_t^* so that $q_T^* = 0$ and so

$$Z_T^* = \frac{1}{S_T} \left(\frac{1}{T} \int_0^T S_t dt - K \right).$$

We assume that S_t follows a geometric Brownian motion.

Moments for Asian Options

An important requirement for the SDP approach is the availability of the first N moments of the distribution of the price of the Asian option. It is not difficult to compute the first two moments of a continuously sampled arithmetic average. However, the use of more moments results in better approximations. Methods for computing higher order moments have been considered by Dufresne (1989) using Ito's lemma, time reversal, and recurrence arguments. Other methods involving Laplace transforms in time have been considered by Geman and Yor (1993). Most if not all of these methods represent the payoffs in the standard way. We present two results that allow access to all of the moments in principle. Each version represents the payoffs of the Asian option in two different ways (one of the payoff representations is the standard one).

The use of different payoff representations will allow several different (but mathematically equivalent) SDP formulations for the bounding problems. The first version obtains moments of Z_T^* and the second version uses $Z_T = \frac{1}{T} \int_0^T S_t dt - K$. In the following, Q^* is the probability measure under which all *stock*-discounted asset prices are martingales and Q is the measure under which all *dollar*-discounted asset prices become martingales. First, we need the following lemma.

Lemma 1 *The integral of* $\exp\left[\sum_{l=1}^L A_l x_l\right]$ *over the* $L-$ *dimensional rectangular region* $0 \le x_1 \le x_2, ..., \le x_{L-1} \le x_L \le T$ *is given by*

$$\sum_{k=1}^{2^L} \frac{e^{TB_{L,k}}}{C_{L,k}}$$

with the coefficients $B_{l,k}$ *and* $C_{l,k}$ *defined recursively by (1)* $B_{l,2k-1} = B_{l-1,k} + A_l$, *where* $B_{l,2k} = 0$ *and (2)* $C_{l,2k-1} = C_{l-1,k}B_{l,2k-1}$ *and* $C_{l,2k} = -C_{l,2k-1}$ *for* $1 \le l \le L$ *and* $1 \le k \le 2^{l-1}$ *with* $B_{01} = 0$ *and* $C_{01} = 1$.

Proof: Follows by induction on L. $\square$

Proposition 1 *The* Nth *moment* q_N^* *of* $Z_T^* = \frac{1}{S_T}(\frac{1}{T}\int_0^T S_t dt - K)$ *under* Q^* *is given by*

$$q_N^* = \sum_{L=0}^N \binom{N}{L} \left(\frac{(-K)}{S_0}\right)^{N-L} (e^{-NrT + \frac{1}{2}(N^2-N)T\sigma^2}) Q_{NL},$$

where

$$Q_{NL} = \sum_{k=1}^{2^L} \frac{e^{TB_{L,k}}}{C_{L,k}},$$

where the coefficients $B_{l,k}$ and $C_{l,k}$ for $1 \leq l \leq L$ and $1 \leq k \leq 2^l$ are defined recursively as in Lemma 1 with $A_l = r - (N - l + L - 1)\sigma^2$ for $1 \leq l \leq L$.

Proof: See appendix. □

Proposition 2 *The Nth moment q_N of $Z_T = \frac{1}{T}\int_0^T S_t dt - K$ under Q is given by*

$$q_N = \sum_{L=0}^{N} \binom{N}{L} \left(\frac{(-K)^{N-L}}{T^N}\right) Q_{0L},$$

where

$$Q_{0L} = \sum_{k=1}^{2^L} \frac{e^{TB_{L,k}}}{C_{L,k}},$$

where the coefficients $B_{l,k}$ and $C_{l,k}$ for $1 \leq l \leq L$ and $1 \leq k \leq 2^l$ are defined recursively as in Lemma 1 with $A_l = r - (L - k)\sigma^2$ for $1 \leq l \leq L$.

Proof: See appendix. □

The proofs of Propositions 1 and 2 proceed by direct integration with the use of Fubini's theorem and Lemma 1 to facilitate the integration see the appendix and Dalakouras (2004) for more details.

2.2 Linear Programming Formulations for Upper and Lower Bounds on the Price of an Asian Option

Assuming that the first N moments $q_j^*(q_j)$ for $j = 1, \ldots, N$ of the probability density $P^*(P)$ of the price of the Asian option at exercise time T corresponding to the measure $Q^*(Q)$ are available then tight upper bounds on the price of the Asian option can be obtained by solving the following two infinite dimensional linear programming problems:

Upper Bound Version 1 (UB1)

$$\text{maximize } e^{-rt} \int_0^\infty \max(0, (Z_T^*)) P^*(Z_T^*) dZ_T^*$$

subject to

$$E_{Q^*}[(Z_T^*)^j] = e^{-rtj} \int_0^\infty (Z_T^*)^j P^*(Z_T^*) dZ_T^* = q_j^*, \ j = 0, 1, \ldots, N,$$

$$P^*(Z_T^*) \geq 0.$$

Let $X_T = \frac{1}{T}\int_0^T S_t dt$ and $Z_T = X_T - K$, then the second formulation is:

Upper Bound Version 2 (UB2)

$$\text{maximize } e^{-rt} \int_0^\infty \max(0, (X_T - K)) P(Z_T) dZ_T$$

subject to

$$E_Q[(Z_T)^j] = e^{-rtj} \int_0^\infty (Z_T)^j P(Z_T) dZ_T = q_j, \; j = 0, 1, \ldots, N,$$

$$P(Z_T) \geq 0.$$

The second formulation *UB2* is equivalent to *UB1* since the distributions of Z_T^* and Z_T are the same (except that one distribution is a scaled version of the other), i.e., they represent identical payoffs based on the same random underlying asset. It will be shown that *UB2* will admit a different SDP formulation than the one that will be obtained for *UB1*.

Tight lower bounds for the price of the Asian option can be obtained by solving the following optimizations problems:

Lower Bound Version 1 (LB1)

$$\text{minimize } e^{-rt} \int_{-\infty}^\infty \max(0, (Z_T^*)) P^*(Z_T^*) dZ_T^*$$

subject to

$$E_{Q^*}[(Z_T^*)^j] = e^{-rtj} \int_{-\infty}^\infty (Z_T^*)^j P^*(Z_T^*) dZ_T^* = q_j^*, \; j = 0, 1, \ldots, N,$$

$$P^*(Z_T^*) \geq 0.$$

Lower Bound Version 2 (LB2)

$$\text{minimize } e^{-rt} \int_{-\infty}^\infty \max(0, (X_T - K)) P(Z_T) dZ_T$$

subject to

$$E_Q[(Z_T)^j] = e^{-rtj} \int_{-\infty}^\infty (Z_T)^j P(Z_T) dZ_T = q_j, \; j = 0, 1, \ldots, N,$$

$$P(Z_T) \geq 0.$$

The duals of *UB1* and *UB2* can be formed by associating a dual variable y_r within each of the N constraints in $UB1$ and $UB2$ as follows (without loss of generality we refer to Z_T as z and X_T as x):

Dual of UB1 (DUB1)

$$\text{minimize } \sum_{r=0}^N q_r^* y_r$$

subject to

$$\sum_{r=0}^{n} y_r z^r \geq \max(0, z) \quad \forall z \in R_+^1.$$

Dual of UB2 (DUB2)

minimize $\sum_{r=0}^{N} q_r y_r$

subject to

$$\sum_{r=0}^{N} y_r x^r \geq \max(0, x - e^{-rt}K) \quad \forall x \in R_+^1.$$

The duals of *LB1* and *LB2* can be written as follows:

Dual of LB1 (DLB1)

maximize $\sum_{r=0}^{N} q_r^* y_r$

subject to

$$\sum_{r=0}^{N} y_r z^r \leq \max(0, z) \quad \forall z \in R_+^1.$$

Dual of LB2 (DLB2)

maximize $\sum_{r=0}^{n} q_r y_r$

subject to

$$\sum_{r=0}^{N} y_r x^r \leq \max(0, x - e^{-rt}K) \quad \forall x \in R_+^1.$$

Ishii (1963) shows that strong duality will hold between the primal and dual pairs.

SDP Formulations

We now present some key lemmas that enable the conversion of the infinite linear programming bounding formulations above to equivalent semidefinite programs. The lemmas below are from Gotoh and Konno (2002) and are slightly modified versions from Bertsimas and Popescu (2002).

Lemma 1. *The polynomial* $g(x) = \sum_{r=0}^{N} y_r x^r \geq 0$ *for all* $x \geq 0$ *if and only if there exists a positive semidefinite matrix* $X = [x_{ij}]_{i,j=0,...,N}$ *such that (1)* $\sum_{i,j:i+j=2l-1} x_{ij} = 0$, $l = 1, ..., N$, *and (2)* $\sum_{i,j:i+j=2l} x_{ij} = y_l$, $l = 0, ..., N$.

Lemma 2. *The polynomial* $g(x) = \sum_{r=0}^{N} y_r x^r \geq 0$ *for all* $x \in [0, a)$ *if and only if there exists a positive semidefinite matrix* $X = [x_{ij}]_{i,j=0,...,N}$ *such that (1)* $\sum_{i,j:i+j=2l-1} x_{ij} = 0$, $l = 1, ..., N$, *and (2)* $\sum_{i,j:i+j=2l} x_{ij} = \sum_{r=0}^{l} y_r \binom{N-r}{l-r} a^r$, $l = 0, ..., N$.

Lemma 3. *The polynomial $g(x) = \sum_{r=0}^{N} y_r x^r \geq 0$ for all $x \in [a, \infty)$ if and only if there exists a positive semidefinite matrix $X = [x_{ij}]_{i,j=0,...,N}$ such that (1) $\sum_{i,j:i+j=2l-1} x_{ij} = 0$, $l = 1, ..., N$, and (2) $\sum_{i,j:i+j=2l} x_{ij} = \sum_{r=l}^{n} y_r \binom{r}{l} a^{r-l}$, $l = 0, ..., N$.*

It is not hard to see that if one defines 0^0 to be 1, then for the $a = 0$ case we get the following simplification of Lemma 3 in that (1) remains the same, but (2) can be written as $y_l = \sum_{i,j:i+j=2l} x_{ij}$, $l = 0, ..., N$.

Lemma 4. *The polynomial $g(x) = \sum_{r=0}^{N} y_r x^r \geq 0$ for all $x \in (-\infty, a]$ if and only if there exists a positive semidefinite matrix $X = [x_{ij}]_{i,j=0,...,N}$ such that (1) $\sum_{i,j:i+j=2l-1} x_{ij} = 0$, $l = 1, ..., N$, and (2) $\sum_{i,j:i+j=2l} x_{ij} = \sum_{r=l}^{n} (-1)^l y_r \binom{r}{l} a^{r-l}$, $l = 0, ..., N$.*

Similarly, when $a = 0$, (2) in Lemma 4 can simplify to

$$y_l = (-1)^l \sum_{i,j:i+j=2l} x_{ij}, \ l = 0, ..., N.$$

Theorem 1. *DUB1 is equivalent to the following SDP formulation:*

$$(SDP\_DUB1)$$

$$minimize \ \sum_{r=0}^{N} q_r^* y_r$$

subject to

$$\sum_{i,j:i+j=2l-1} x_{ij} = 0, \ l = 1, ..., N,$$

$$y_l = (-1)^l \sum_{i,j:i+j=2l} x_{ij}, \ l = 0, ..., N,$$

$$\sum_{i,j:i+j=2l-1} z_{ij} = 0, \ l = 1, ..., N,$$

$$y_0 = z_{00}, \ y_1 = 1 + \sum_{i,j:i+j=2} z_{ij},$$

$$y_l = \sum_{i,j:i+j=2l} z_{ij}, \ l = 1, ..., N,$$

$$X, Z \succeq 0.$$

Proof: The constraints of *DUB1* can be represented as (a) $\sum_{r=0}^{N} y_r z^r \geq 0$ for $z \in [0, \infty)$ and (b) $\sum_{r=0}^{N} y_r z^r \geq 0$ for $z \in (-\infty, 0]$ and so the formulation *SDP-DB1* follows immediately from Lemmas 3 and 4 for the case where $a = 0$. □

We have the following analogous result for LB1.

Theorem 2. *DLB1 is equivalent to the following SDP formulation:*

(SDP_DLB1)

$$maximize \ \sum_{r=0}^{N} q_r^* y_r$$

subject to

$$\sum_{i,j:i+j=2l-1} x_{ij} = 0, \ l = 1, ..., N,$$

$$y_l = (-1)^{l+1} \sum_{i,j:i+j=2l} x_{ij}, \ l = 0, ..., N,$$

$$\sum_{i,j:i+j=2l-1} z_{ij} = 0, \ l = 1, ..., N,$$

$$y_0 = -z_{00}, \ y_1 = 1 - \sum_{i,j:i+j=2} z_{ij},$$

$$y_l = - \sum_{i,j:i+j=2l} z_{ij}, \ l = 1, ..., N,$$

$$X, Z \succeq 0.$$

Proof: Similar to Theorem 1.   □

Standard SDP Form for *SDP_DUB1* **and** *SDP_DLB1*

Both of the formulations *SDP_DUB1* and *SDP_DLB1* can be put in standard semidefinite programming form which enables many powerful SDP algorithms such as the barrier/interior point methods and many others (see Pardalos and Wolkowicz (1998), Ramana and Pardalos (1996), and Vandenberghe and Boyd (1996)) to be applied to the formulations. The formulations from Bertsimas and Popescu (2002) are not guaranteed to be cast in standard SDP form, but are amenable to effective SDP cutting plane strategies. A standard primal form SDP takes on the form

$$\min \tilde{C} \bullet \tilde{X}$$

subject to $A_k \bullet \tilde{X} = b_k$ for $k = 1, ..., m$ and where $\tilde{X} \succeq 0$ and $\tilde{C} \bullet \tilde{X} = \sum_{ij} \tilde{X}_{ij} \tilde{Y}_{ij} = Trace(\tilde{C}\tilde{X})$.

The corresponding dual SDP form is max $b^t y$ subject to $\sum_{k=1}^{m} y_k A_k + S = C$ and $S \succeq 0$. For example, one can form the primal SDP version of *SDP_DUB1* by forming a matrix $\tilde{X}$ as a block diagonal with the matrices X and Z from the *SDP_DUB1* and *SDP_DLB1* formulations. For the objective one can express variables y_r in terms of the x_{ij} representation. Finally, one can readily place the remaining constraints in the form $A_k \bullet \tilde{X} = b_k$.

Other SDP Formulations

By using the other formulations of the bounding problems other equivalent SDP formulations can be derived as in the original approach of Bertsimas and Popescu (2002). For example, if we consider $UB2$ as the upper bound formulation, i.e., the expression of the payoff is disaggregated into $X_T - K$ as opposed to being represented by Z_T. Observe that by Proposition 2 we can obtain the first N moments $q_1, \ldots, q_n$ of the distribution of prices of $Z_T^* = X_T - K$. Then, we observe the constraints of $DUB2$ can represented as $\sum_{r=0}^{N} y_r x^r \geq 0$ for all $x \geq 0$ and $y_0 + \bar{K} + (y_1 - 1)x + \sum_{r=2}^{N} y_r x^r \geq 0$ for all $x \geq 0$ where $\bar{K} = e^{-rt} K$. So by Lemma 1, we have the following SDP equivalent to $UB2$:

$$(SDP\_UB2)$$

$$\text{minimize } \sum_{r=0}^{N} q_r y_r$$

subject to

$$\sum_{i,j:i+j=2l-1} x_{ij} = 0, \ l = 1, \ldots, N,$$

$$y_l - \sum_{i,j:i+j=2l} x_{ij} = 0, \ l = 0, \ldots, N,$$

$$\sum_{i,j:i+j=2l-1} z_{ij} = 0, \ l = 1, \ldots, N,$$

$$y_0 - z_{00} = -\bar{K}, \ y_1 = 1 + \sum_{i,j:i+j=2} z_{ij},$$

$$y_l = \sum_{i,j:i+j=2l} z_{ij}, \ l = 1, \ldots, N,$$

$$X, Z \succeq 0.$$

One can yet form other SDP formulations of the bounding problems by using alternate descriptions of the feasible region of the bounding formulations and by representing the payoff by $X_T - K$ using the various combinations of Lemmas in Section 2.2 (see Bertsimas and Popescu (2002) and Gotoh and Konno (2002)). In general, SDP formulations based on the payoff representation $X_T - K$, e.g., SDP_UB2 will be amenable to algorithms such as the cutting place method in Gotoh and Konno (2002) where the method is observed to be very effective at $N = 4$ moments.

3 Conclusion

We have considered a (non-parametric) SDP approach for bounding the price of arithmetic average Asian option. Methods were presented to compute moments for two different representations of the distributions of the price of the option which allows the development of several different but mathematically equivalent SDP formulations that represent tight upper and lower bounds. One class of formulations can be cast readily in standard SDP form which allows many existing powerful methods based on interior point methods (e.g., barrier) for solution. The other class of problems are amenable to efficient SDP cutting plane methods as in Gotoh and Konno (2002). It remains to compare the computational performance of the different formulations and methods and this is the subject of future research. The formulations are the tightest bounds possible and so will be useful. Results similar to that achieved for European option prices can be expected.

References

Bertsimas, D. and Popescu, I.: 2002, On the relation between option and stock prices: a convex optimization approach, *Operations Research* **50**(2), 358–374.

Black, F. and Scholes, M.: 1973, The pricing of options and corporate liabilities, *Journal of Political Economy* **81**, 637–654.

Boyle, P.P. and Emanuel, D.: 1980, The pricing of options on the generalized means, *Working paper*, University of British Columbia, Vancouver, B.C., Canada.

Dalakouras, G.V.: 2004, Computation of the moments of the continuously sampled arithmetic average, *Working paper*, Department of Mathematics, University of Michigan, Ann Arbor.

Dufresne, D.: 1989, Weak convergence of random growth processes with applications to insurance, *Insurance Mathematics & Economics* **8**, 187–201.

Geman, H. and Yor, M.: 1993, Bessel processes, Asian options and perpetuities, *Mathematical Finance* **3**(4), 349–375.

Glasserman, P., Heidelberger, P. and Shahabuddin, P.: 1999, Asymptotically optimal importance sampling and stratification for pricing path dependent options, *Mathematical Finance* **9**, 117–152.

Gotoh, J.Y. and Konno, H.: 2002, Bounding option prices by semidefinite programming: a cutting plane algorithm, *Management Science* **48**, 665–678.

Hull, J.C.: 2003, *Options, Futures, and Other Derivatives*, Prentice Hall, Englewood Cliffs, NJ. (5th ed.).

Ishii, K.: 1963, On sharpness of Chebyshev-type inequalities, *Ann Inst. Statist Meth.* **14**, 185–197.

Pardalos, P.M. and Wolkowicz, H. (eds): 1998, *Topics in Semidefinite and Interior-Point Methods*, Vol. 18 of *Fields Institute Communications Series*, American Mathematical Society.

Ramana, M. and Pardalos, P.M.: 1996, Semidefinite programming, *in* T. Terlaky (ed.), *Interior Point Methods of Mathematical Programming*, Kluwer Academic Publishers, Dordrecht, pp. 369–398.

Rogers, L.C. and Shi, Z.: 1995, The value of an Asian option, *Journal of Applied Probability* **32**, 1077–1088.

Turnbull, S.M. and Wakeman, L.M.: 1991, A quick algorithm for pricing European average options, *Journal of Financial and Quantitative Analysis* **26**, 377–389.

Vandenberghe, L. and Boyd, S.: 1996, Semidefinite programming, *SIAM Review* **38**, 49–95.

Appendix

A.1 We prove Propositions 1 and 2. **Note:** Proposition 1 (2) will involve a measure $Q^*(Q)$ under which the stock-discounted (dollar-discounted) prices of all assets become martingales. The price S_t under measure $Q^*(Q)$ takes on the form

$$S_t = S_0 e^{(r-\frac{\sigma^2}{2})t+\sigma W_t^*} \; (S_t = S_0 e^{(r-\frac{\sigma^2}{2})t+\sigma W_t}),$$

and the two Brownian motions W_t^* and W_t are related via $W_t^* = W_t - \sigma t$.

Proof of Proposition 1

Recall $Z_T^* = \frac{1}{S_T}(\frac{1}{T}\int_0^T S_t dt - K)$ and so then

$$Z_N^{T*} = \sum_{L=0}^{N} \binom{N}{L} \left(\int_0^T S_u du\right)^L \left(\frac{(-K)^{N-L}}{S_T^N}\right)$$

and so by the linearity of expectation we have

$$q_N^* = \sum_{L=0}^{N} \binom{N}{L}(-K)^{N-L} E_{Q^*}\left[\left(\int_0^T S_u du\right)^L \frac{1}{S_T^N}\right].$$

Now,

$$E_{Q^*}\left[\left(\int_0^T S_u du\right)^L \frac{1}{S_T^N}\right] = E_{Q^*}\left[\prod_{i=1}^{L}(\int_0^T S_u du)\frac{1}{S_T^N}\right]$$

or by Fubini's theorem as multiple integral

$$E_{Q^*}\left[\left(\int_0^T \int_0^T \cdots \int_0^T S_{u_1} S_{u_2}...S_{u_L} du_1 du_2...du_L\right)\frac{1}{S_T^N}\right],$$

and by ordering the $u_i's$ using symmetry (there are $L!$ arrangements) and by using Fubini's theorem again we get

$$L! \int_0^T \int_0^T \cdots \int_0^T E_{Q^*} \left[S_{u_1} S_{u_2} ... S_{u_L} \frac{1}{S_T^N} \right] du_1 du_2 ... du_L ,$$

where $0 = u_0 \leq u_1 \leq ... \leq u_L \leq u_{L+1} = T$.
Define

$$u_{NL} = S_{u_1} S_{u_2} ... S_{u_L} \frac{1}{S_T^N} .$$

The expectation of u_{NL} will be calculated under the risk-neutral measure Q^*. Then, this expectation will be integrated to arrive at the final result. The calculation of the iterated integral will be handled through Lemma 1.

Now

$$u_{NL} = S_{u_1} S_{u_2} ... S_{u_L} \frac{1}{S_T^N} = \frac{1}{S_T^N} S_{u_i} ,$$

and we can rewrite the product using the Brownian motion W_t^* (for simplicity we denote $W_i = W_{u_i}$) to get

$$S_0^{-N} e^{-N(r+\frac{\sigma^2}{2})T - N\sigma W_i^*} \prod_{i=1}^{L} S_o e^{(r+\frac{\sigma^2}{2})u_i + \sigma W_i^*} .$$

Separating the deterministic and stochastic parts we get

$$S_0^{L-N} \exp\left[-N\left(r + \frac{\sigma^2}{2}\right)T + \sum_{i=1}^{L}\left(r + \frac{\sigma^2}{2}\right)u_i \right] \exp\left[-N\sigma W_T^* + \sum_{i=1}^{L} \sigma W_T^* \right] .$$

For what follows, we use the assumption that $u_{l+1} = T$ and $u_0 = 0$. We need to take the expectation of $\exp[-N\sigma W_T^* + \sum_{i=1}^{L} \sigma W_i^*]$. The exponent of this quantity is

$$-\sigma\left(N W_T^* - \sum_{i=1}^{L} W_i^* \right) = -\sigma \sum_{i=1}^{L}(N - L + i)(W_{i+1}^* - W_i^*) .$$

Now $W_{i+1}^* - W_i^*$ are independent increments that have the same distribution as $W_{u_{i+1}}^* - W_{u_i}^*$ since W_t^* is a Brownian motion. For a Brownian motion W_t^* we have $aW_t^* \approx N(0, a^2 t)$ and $E[e^{aW_t^*}] = e^{a^2 t/2}$. Then, the expectation of the stochastic term is

$$\exp\left[\sum_{i=0}^{L} \frac{\sigma^2(N - L + i)^2}{2}(u_{i+1} - u_i) \right] =$$

$$\exp\left[\frac{\sigma^2}{2}\left(N^2 T - \sum_{i=0}^{L}[2(N - i + L) - 1]u_i \right) \right]$$

after grouping the $u_i's$ together. Therefore,

$$E[u_{NL}] = S_0^{L-N} \exp\left[-NT + \sum_{i=1}^{L}\left(r + \frac{\sigma^2}{2}\right)u_i\right.$$

$$\left. + \frac{\sigma^2}{2}\left(N^2T - \sum_{i=0}^{L}[2(N-i+L)-1]u_i\right)\right]$$

which simplifies to

$$S_0^{L-N}\exp\left[-N\left(r+\frac{\sigma^2}{2}\right)T + \frac{\sigma^2}{2}N^2T\right]\exp\left[\sum_{i=1}^{L}(r-\sigma^2(N-i+L-1))u_i\right].$$

The last factor is of the form $\prod_{i=1}^{L} e^{A_i u_i}$ with $A_i = r - \sigma^2(N-i+L-1)$. Then, by Lemma 1 (the integration is over triangular N-dimensional region defined by $0 \le u_1 \le \ldots \le u_L \le T$) the final result follows. □

Proof of Proposition 2

The argument follows the proof of Proposition 1 mutatis mutandis, except that $Z_T = \frac{1}{T}\int_0^T S_t dt - K$ is used instead of Z_T^* and that $S_t = S_0 e^{(r-\frac{\sigma^2}{2})t - \sigma W_t}$ is used instead of $S_t = S_0 e^{(r-\frac{\sigma^2}{2})t + \sigma W_t^*}$. □

The Evaluation of Discrete Barrier Options in a Path Integral Framework

Carl Chiarella[1], Nadima El–Hassan[1], and Adam Kucera[2]

[1] School of Finance and Economics,University of Technology, Sydney, PO Box 123, Broadway NSW 2007 Australia `carl.chiarella@uts.edu.au`, `nadima.el-hassan@uts.edu.au`
[2] Integral Energy, Australia `adam.kucera@integral.com.au`

Summary. The pricing of discretely monitored barrier options is a difficult problem. In general, there is no known closed form solution for pricing such options. A path integral approach to the evaluation of barrier options is developed. This leads to a backward recursion functional equation linking the pricing functions at successive barrier points. This functional equation is solved by expanding the pricing functions in Fourier-Hermite series. The backward recursion functional equation then becomes the backward recurrence relation for the coefficients in the Fourier-Hermite expansion of the pricing functions. A very efficient and accurate method for generating the pricing function at any barrier point is thus obtained. A number of numerical experiments with the method are performed in order to gain some understanding of the nature of convergence. Results for various volatility values and different numbers of basis functions in the Fourier-Hermite expansion are presented. Comparisons are given between pricing of discrete barrier option in the path integral framework and by use of finite difference methods.

Key words: Discrete barrier options, path integral, Fourier-Hermite polynomials, backward recursion method.

1 Introduction

Barrier options are derivative securities with values contingent on the relationship between the value of the underlying asset and one or more barrier levels. In this paper, we consider the pricing of barrier options which are monitored at particular points over the life of the contract, also known as discrete barrier options. These types of exotic options have become a prominent feature of modern financial markets with many variations heavily traded in the foreign exchange, equity and fixed income markets. The emergence of such securities has also provided challenging research problems in the area of efficient pricing and hedging of such securities. Most pricing models (Merton, 1973; Rubinstein and Reiner, 1991; Heynen and Kat, 1996) consider barrier options whose barrier level is continuously monitored at every instant in time over the life of the option, allowing the derivation of closed form solutions. However most traded barrier options are monitored discretely, rather than on a continuous basis. The application of the solutions derived assuming continuous monitoring to the pricing of discretely monitored barrier options results in substantial pricing errors (Chance, 1994; Kat, 1995; Levy and Mantion, 1997). Hence, a method is needed to accurately and efficiently evaluate discretely monitored barrier options. Traditional lattice and Monte Carlo methods have difficulties in incorporating discrete monitoring principally because of the misalignment of the monitoring points.

Several papers have appeared in the literature proposing various methods to handle discrete monitoring. Variations to the traditional binomial and trinomial methods were proposed by Figlewski and Gao (1997), Tian (1996) and Boyle and Tian (1997). These include Broadie et al. (1997b) who propose a method based on the price of a continuous barrier options with a continuity correction for discrete monitoring. Broadie et al. (1997a) incorporate the correction term in developing a lattice method for determining accurate prices of discrete and continuous path-dependent options. As far as the pricing of discretely monitored barrier options is concerned, Wei (1998) proposes an interpolation method whereas Sullivan (2000) proposes a method that reduces the discrete time multidimensional integration required to a sequential numerical integration.

In this paper we examine a potentially powerful alternative to existing pricing methods. Using knowledge of the conditional transition density function we repeatedly apply the Chapman-Kolmogorov equation to relate the pricing function at successive monitoring points. We then expand the pricing function in a Fourier-Hermite series in terms of the price of the underlying asset. We derive recurrence relations involving orthogonal polynomials under a given measure. The proposed method works well for various discrete barrier structures, including single and double barriers, constant and time varying barrier levels, and for a number of payoff structures such as vanillas, digital and powers. The method can be made arbitrarily accurate by taking a sufficient number of terms in the expansion.

The techniques used extend the work of Chiarella et al. (1999) where Fourier-Hermite series expansions were applied to the valuation of European and American options. The novel aspect of our contribution is the expansion of the derivative security price at each time step in a Fourier-Hermite series expansion so that it is obtained as a continuous and differentiable function of the price of the underlying asset. The actual implementation of our method then becomes a question of determining the coefficients of the Fourier-Hermite series expansion at each time step. Using the orthogonalisation condition, it turns out that these can be generated recursively by working backwards from one time step to the next, by use of the recurrence relations which generate the Hermite polynomials. The implementation of these recurrence relations is in fact very efficient. Here we apply the method to pricing discretely monitored barrier options. Since we obtain a continuous and differentiable representation of the price, the hedge ratios delta and gamma can be obtained to a high level of accuracy and also very cheaply in terms of calculation time.

Our approach may be regarded as one way to implement the path integral techniques in option pricing problems. In recent years it has become appreciated that the path integral technique of statistical physics can be applied to derivative security valuation. We refer in particular to Linetsky (1997) who provides an overview of the path integral concept and its application to financial problems. The wide application of path integral techniques to financial modelling and in particular to the pricing and hedging of options, including path-dependent and exotic options, was first studied by Dash (1988). Dash's contribution to the area was largely in the formulation of many derivative security pricing problems in the path integral framework, including standard equity options, exotic options, path dependent options and the pricing of bond options under a number of popular term structure models. The framework provides an intuitive description of the value of derivative securities using relatively simple mathematics.

However, the application and implementation of solution techniques of path integrals to finance problems has been limited, with the cited authors focusing on a general framework and pointing to the potential for the application of these techniques to financial pricing problems. Path integrals can be evaluated in a number of ways including analytic approximations by means of moment expansions in a perturbation series, deterministic discretisation schemes of the path integral and Monte-Carlo simulation (Makici, 1995) methods. The method chosen will in general depend on the problem at hand and the stability of the solution technique used. We also refer to the contribution of Eydeland (1994) who provides a computational algorithm based on a Toeplitz matrix structure and fast Fourier transforms for evaluating financial securities in a path integral framework. This technique was successfully applied by Chiarella and El-Hassan (1997) to evaluate European and American bond options in an HJM framework.

The organisation of the paper is as follows: In section 2, we define the general barrier structure and outline the backward recursion procedure. In section 3, we show how the path integral formulation for the value of a discretely monitored barrier defined in section 2 can be represented as an expansion of the value function in a series of orthogonal polynomials and reduced to a backward recursion procedure. In section 4, we report some numerical results and in section 5 we make some concluding remarks.

2 The General Barrier Structure

We denote the underlying asset price by S and assume that under the risk-neutral measure it follows a geometric Brownian motion given by

$$dS = (r - q)Sdt + \sigma SdW,$$

where σ is the volatility, r is the risk-free rate of interest, q is the continuous dividend yield on the underlying asset and $W(t)$ is a standard Wiener process. By allowing for a continuous dividend yield q our framework may be applied to barrier options on indices, or to options on foreign exchange by setting $q = r_f$ where r_f is the risk free rate of interest in the foreign economy.

The implementation of the path integral method described below follows the framework laid out in Chiarella et al. (1999), which requires expansion of pricing functions in terms of Hermite polynomials. These are defined on an infinite interval, for this reason we need to transform the asset price to a variable defined on an infinite interval. This is most conveniently done by introducing the change of variable

$$\xi = \frac{1}{\sigma} \ln(S). \tag{1}$$

A straight forward application of Ito's lemma reveals that ξ satisfies the stochastic differential equation

$$d\xi = \frac{1}{\sigma} \left((r - q) - \frac{1}{2}\sigma^2 \right) dt + dW(t). \tag{2}$$

We recall that (2) implies that the transition probability density for ξ between two times $t', t(t' < t)$, denoted $\pi(\xi_t, t|\xi_{t'}, t')$, is normally distributed and is in fact given by

$$\pi(\xi_t, t|\xi_{t'}, t') = \frac{1}{\sqrt{2\pi(t - t')}} \exp \left[\frac{-\left[\xi_t - \sqrt{2(t - t')}\mu(\xi_{t'}, t - t') \right]^2}{2(t - t')} \right], \tag{3}$$

where it is convenient to define

$$\mu(\xi, t) = \frac{1}{\sqrt{2t}} \left(\xi + \frac{1}{\sigma} \left(r - q - \frac{1}{2}\sigma^2 \right) t \right). \tag{4}$$

We divide the time interval from initial time to option maturity into K subintervals $(t^{k-1}, t^k), (k = 0, 1, \ldots, K)$. The spacings between the barrier observation points t^k need not be constant. We set $\Delta t^k = t^k - t^{k-1}$ for $k = K, \ldots, 1$ with the implied notation that $t^0 = 0$ and $t^K = T$ so that $T = \sum_{k=1}^{K} \Delta t^k$.

We allow for barrier levels that can be time dependent. Thus at each time t^k there will be an upper barrier level, b_u^k, and a lower barrier level b_l^k, for $k = K - 1, \ldots, 1$ Here we note that we do not include barriers at expiry since they are part of the pay-off definition. With this notation we may handle the case of no lower barrier by setting $b_l^k = 0$, and the case of no upper barrier by letting $b_u^k \to \infty$. Figure 1 illustrates in the S, t plane a typical discretisation with a variety of possible barriers at the discretisation points.

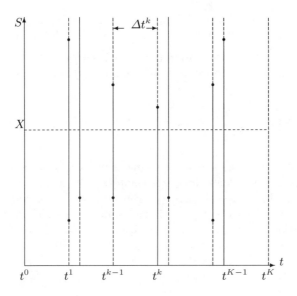

Fig. 1. The discretisation scheme.

3 The Barrier Option as a Functional Recurrence Relation Equation

Under the risk neutral measure the price of a derivative security at any point in time is the discounted expected payoff at the next point at which it may be exercised. The transition probability density function (3) is the one that is required to calculate the expected payoff. We use $F^k(\xi^k)$ to denote the value function of the barrier option at monitoring point t^k as a function of ξ^k, the volatility scaled log price. Then, the discounted expected value relation between the value functions at two successive monitoring points t^{k-1}, t^k is given by

$$F^{k-1}\left(\xi^{k-1}\right) = e^{-r\Delta t^k} \int_{\ln(b_l^k)/\sigma}^{\ln(b_u^k)/\sigma} \pi(\xi^k, t^k | \xi^{k-1}, t^{k-1}) F^k(\xi^k) d\xi^k. \qquad (5)$$

Substituting equation (3) and making a change of integration variable we obtain

$$F^{k-1}(\xi^{k-1}) = \frac{e^{-r\Delta t^k}}{\sqrt{\pi}} \int_{z_l^k}^{z_u^k} e^{-(\xi^k - \mu(\xi^{k-1}, \Delta t^k))^2} F^k(\sqrt{2\Delta t^k}\xi^k) d\xi^k, \qquad (6)$$

where, due to the the scaling factor $\sqrt{2\Delta t^k}$ in F^k, the limits of integration z_l^k and z_u^k are given by

$$z_l^k = \frac{\ln(b_l^k)}{\sigma\sqrt{2\Delta t^k}} \quad \text{and} \quad z_u^k = \frac{\ln(b_u^k)}{\sigma\sqrt{2\Delta t^k}}, \tag{7}$$

for $k = K-1, \ldots, 1$. Thus the notation implicitly carries the time dependence of the problem. With the above notation, equation (6) can be written as

$$F^{k-1}(\xi^{k-1}) = \frac{e^{-r\Delta t^k}}{\sqrt{\pi}} \int_{z_l^k}^{z_u^k} e^{-(\xi^k - \mu(\xi^{k-1}, \Delta t^k))^2} F^k(\sqrt{2\Delta t^k}\xi^k)d\xi^k. \tag{8}$$

Equation (8) is the functional recurrence equation that we need to solve. We note that in stepping back the range of integration, from $t = t^k$ to $t = t^{k-1}$ (i.e., propagating back the solution) is given by $0 \leq \xi^k < \infty$ for a call option and for a put option by $-\infty < \xi^k \leq 0$. In this way, the max function is handled by the integration limits. For ease of clarity and notation, when performing the path integrations across all the time intervals, we will denote ξ^k as x, ξ^{k-1} as ξ, z_l^k as z_l, z_u^k as z_u, Δt^k as Δt.

With the above notation, equation (1.7) can be written as

$$F^{k-1}(\xi) = e^{-r\Delta t}\frac{1}{\sqrt{\pi}} \int_{z_l}^{z_u} e^{-(x-\mu(\xi,\Delta t))^2} F^k(\sqrt{2\Delta t}x)dx, \quad k = K, \ldots, 1. \tag{9}$$

Figure 2 illustrates the region of integration and the concept of the backward propagation of the price function. The task at hand involves successive iteration of equation (9) from the given pay-off function, $F^k(\xi^k)$, through the barrier points back to $t = 0$. On completing the iterations, it is a simple matter to invert the log transformation and return to the price variable S and evaluate F^0 at the required spot value.

Because of our geometric Brownian motion assumption for the underlying asset price the option pricing function is homogeneous in $\frac{S}{X}$. Hence, for a call option we need only consider a pay-off function of the form $\max(S-1, 0)$ and for a put option a pay-off function $\max(1-S, 0)$. Under the log transformation the pay-off for a call option becomes $\max[0, e^{\sigma^k}\xi^k - 1]$. Similarly the pay-off function for a put option is $\max[0, 1 - e^{\sigma^k}\xi^k - 1]$. We use $F^K(\xi^K)$ to denote the payoff function at the final time t^K.

4 The Fourier-Hermite Series Expansion

The basic problem in implementing (9) is to obtain a convenient way to evaluate the integral on the right-hand side and to then build up successively the value functions $F^k(\xi)$ $(k = K-1, \ldots, 1)$. In this paper we solve this problem by expanding both $F^k(x)$ and $F^{k-1}(\xi)$ in Fourier-Hermite series. Thus we set

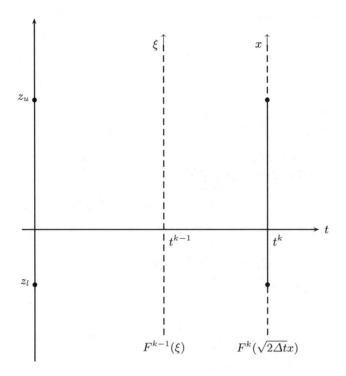

Fig. 2. Propagating the price function back from t^k to t^{k-1}.

$$
\left.
\begin{aligned}
F^k(x) &\approx \sum_{n=0}^{N} \alpha_n^k H_n(x) \\
\text{and} \\
F^{k-1}(\xi) &\approx \sum_{m=0}^{N} \alpha_m^{k-1} H_m(\xi)
\end{aligned}
\right\}
\tag{10}
$$

for N sufficiently large to ensure convergence of the series. By substituting the Fourier-Hermite expansions (10) into the backward recurrence (9) and making use of the orthogonality property of Hermite polynomials, it is possible to obtain a backward recurrence relation for the coefficients in the expansions (10). In this way, we are able to construct the value functions $F^k(\xi)$. We state this key result in proposition I, but before stating this proposition it is useful to introduce some special notation for quantities which occur frequently in the calculations below.

Notation[3]

First introduce the functions

$$L_m(x) = \frac{H_m(x)}{2^m m! v^m},$$ (11)

which are easily shown to satisfy the recurrence equation

$$L_m(x) = \frac{x L_{m-1}(x)}{mv} - \frac{L_{m-2}(x)}{2(m-1)v^2},$$ (12)

with $L_0(x) = 1$ and $L_1(x) = x/v$.

Next we set (recall $\phi(x) := e^{-x^2/2}/\sqrt{2\pi}$)

$$R_{m,n}(x) = \sqrt{2} L_m(x) H_n(vx + \beta) \phi(\sqrt{2}x),$$ (13)

from which we define

$$Q_{m,n}(x, y) = R_{m,n}(x) - R_{m,n}(y).$$ (14)

Finally we define

$$P(x, y) = \Phi(\sqrt{2}x) - \Phi(\sqrt{2}y),$$ (15)

where

$$\Phi(x) = \frac{1}{\sqrt{2\pi}} \int_{-\infty}^{x} e^{-\frac{u^2}{2}} \, du$$

is the cumulative normal density function.

Proposition I

The known coefficients α_n^k and the to be calculated coefficients α_m^{k-1} at the time t^{k-1} are connected by the recurrence relation

$$\alpha_m^{k-1} = e^{-r\Delta t} \sum_{n=0}^{N} a_{m,n}^k \alpha_n^k, \qquad \text{for} \qquad k = K, \dots, 1$$ (16)

or in matrix notation:

$$\boldsymbol{\alpha}^{k-1} = e^{-r\Delta t} A^k \boldsymbol{\alpha}^k, \qquad \text{for} \qquad k = K, \dots, 1$$ (17)

where

$$A^k = [a_{m,n}^k] \qquad \text{for} \qquad m = 0, 1, \dots, N \qquad \text{and} \qquad n = 0, 1, \dots, N \quad (18)$$

The coefficients $a_{m,n}^k$ are generated by the equations outlined in the proof below.

---

[3] See Appendix H for a summary of the notation.

Proof. Substituting the two series expansions (10) into the functional equation (9), using the orthogonality properties of the Hermite polynomials we find after some algebraic manipulations that

$$\alpha_m^{k-1} = e^{-r\Delta t} \sum_{n=0}^{N} \frac{n}{2^m m!} \frac{\alpha_n^k}{\sqrt{\pi}} \int_{z_l} 1 z_u H_n(\sqrt{2\Delta t}\, x) I_m(x) dx, \qquad (19)$$

where

$$I_m(x) = \frac{\sqrt{2\Delta t}}{v^{m+1}} H_m\left(\frac{\sqrt{2\Delta t}\, x - \beta}{v}\right) \exp\left[-\left(\frac{\sqrt{2\Delta t}x - \beta}{v}\right)^2\right], \qquad (20)$$

with

$$v = \sqrt{1 + 2\Delta t} \quad \text{and} \quad \beta = \frac{1}{\sigma}\left(r - q - \frac{1}{2}\sigma^2\right)\Delta t. \qquad (21)$$

Now, introducing the transformation

$$z = \frac{\sqrt{2\Delta t}x - \beta}{v}, \qquad (22)$$

equation (19) can be written as

$$\alpha_m^{k-1} = e^{-r\Delta t} \sum_{n=0}^{N} a_{m,n}^k \alpha_n^k, \qquad (23)$$

where the elements $a_{m,n}^k$ are given by

$$a_{m,n}^k = \frac{1}{2^m m! v^m} \frac{1}{\sqrt{\pi}} \int_{x_l}^{x_u} e^{-x^2} H_m(x) H_n(vx + \beta) dx. \qquad (24)$$

The new limits of integration x_l and x_u are given by

$$x_l = \frac{\sqrt{2\Delta t}z_l - \beta}{v} \quad \text{and} \quad x_u = \frac{\sqrt{2\Delta t}z_u - \beta}{v}. \qquad (25)$$

We note that at final time t^K for a call, the integration limits become

$$x_l = 0 \quad \text{and} \quad x_u \to \infty, \qquad (26)$$

whilst for a put, when $k = K$,

$$x_l \to -\infty \quad \text{and} \quad x_u = 0. \qquad (27)$$

At final time t^K, it is convenient to define w such that $w = 1(-1)$ applies to a call(put).

Thus taking into consideration the foregoing limits the first four elements are given by

$$a_{0,0}^K = \frac{w}{\sqrt{\pi}} \int_0^{w\infty} e^{-x^2} dx = \frac{1}{2}$$

$$a_{0,1}^K = \frac{w}{\sqrt{\pi}} \int_0^{w\infty} e^{-x^2} 2(vx + \beta) dx = \frac{w}{2\sqrt{\pi}} + \beta$$

$$a_{1,0}^K = \frac{w}{\sqrt{\pi}} \int_0^{w\infty} e^{-x^2} 2x dx = \frac{w}{2v\sqrt{\pi}}$$

$$a_{1,1}^K = \frac{w}{\sqrt{\pi}} \int_0^{w\infty} e^{-x^2} (2x)2(vx + \beta) dx = \frac{1}{2} + \frac{w\beta}{v\sqrt{\pi}}. \tag{28}$$

For $m = 2, 3, \ldots, N$ with $n = 0$

$$a_{m,0}^K = \frac{w}{2^m m! v^m \sqrt{\pi}} \int_0^{w\infty} e^{-x^2} H_m(x) dx = \frac{w}{2mv\sqrt{\pi}} L_{m-1}(0). \tag{29}$$

For $n = 2, 3, \ldots, N$ with $m = 0$

$$a_{0,n}^K = \frac{w}{\sqrt{\pi}} \int_0^{w\infty} e^{-x^2} H_n(vx + \beta) dx, \tag{30}$$

which upon use of the recurrence equation for Hermite polynomials yields the recurrence equation

$$a_{0,n}^K = \frac{wv}{\sqrt{\pi}} H_{n-1}(\beta) + 2\beta a_{0,n-1}^K + 2(v^2 - 1)(n-1)a_{0,n-2}. \tag{31}$$

For $m = 2, 3, \ldots, N$ and $n = 2, 3, \ldots, N$ application of the recurrence equation for Hermite polynomials yields

$$a_{m,n}^K = \frac{w\sqrt{2}}{2mv\sqrt{\pi}} L_{m-1}(0)H_n(\beta) + \frac{n}{m} a_{n-1,m-1}^K. \tag{32}$$

Next we consider the generation of the coefficients $a_{m,n}^k$ at any general time step k.

Then the first four elements are given by

$$a_{0,0}^k = \frac{1}{\sqrt{\pi}} \int_{x_l}^{x_u} e^{-x^2} dx = P(x_u, x_l), \qquad \text{(see Appendix A)}$$

$$a_{1,0}^k = \frac{1}{2v\sqrt{\pi}} \int_{x_l}^{x_u} e^{-x^2} 2x dx = \frac{1}{2v} Q_{0,0}(x_l, x_u), \qquad \text{(see Appendix B)}$$

$$a_{0,1}^k = \frac{1}{\sqrt{\pi}} \int_{x_l}^{x_u} e^{-x^2} 2(vx + \beta) dx = v Q_{0,0}(x_l, x_u) + 2\beta P(x_u, x_l),$$

$$\text{(see Appendix C)}$$

$$a_{1,1}^k = \frac{1}{2v\sqrt{\pi}} \int_{x_l}^{x_u} e^{-x^2} (2x)2(vx + \beta) dx = \frac{1}{2v} Q_{0,1}(x_l, x_u) + P(x_u, x_l).$$

$$\text{(see Appendix D)}$$

To alleviate the notation we shall henceforth set

$$P = P(x_u, x_l) \qquad \text{and} \qquad Q_{m,n} = Q_{m,n}(x_l, x_u).$$

For $m = 2, 3, \ldots, N$ with $n = 0$ (see Appendix E)

$$a_{m,0}^k = \frac{1}{2mv} Q_{m-1,0}. \tag{33}$$

For $n = 2, 3, \ldots, N$ with $m = 0$ (see Appendix F)

$$a_{0,n}^k = vQ_{0,n-1} + 2\beta a_{0,n-1}^k + 2(v^2 - 1)(n-1)a_{0,n-2}^k. \tag{34}$$

For $m = 2, 3, \ldots, N$ and $n = 2, 3, \ldots, N$ (see Appendix G)

$$a_{m,n}^k = \frac{1}{2mv} Q_{m-1,n} + \frac{n}{m} a_{m-1,n-1}^k. \tag{35}$$

Alternatively, if we introduce a new function

$$Z_{m,n} = \frac{1}{2(m+1)v} Q_{m,n}(x_l, x_u), \tag{36}$$

then the first four elements are given by

$$a_{0,0}^k = P, \qquad a_{0,1}^k = 2v^2 Z_{0,0} + 2\beta P,$$
$$a_{1,0}^k = Z_{0,0}, \quad a_{1,1}^k = Z_{0,1} + P. \tag{37}$$

For $m = 2, 3, \ldots, N$ and $n = 0$

$$a_{m,0}^k = Z_{m-1,0}. \tag{38}$$

For $n = 2, 3, \ldots, N$ and $m = 0$

$$a_{0,n}^k = 2v^2 Z_{0,n-1} + 2\beta a_{0,n-1}^k + 2(v^2 - 1)(n-1)a_{0,n-2}^k. \tag{39}$$

For $m = 2, 3, \ldots, N$ and $n = 2, 3, \ldots, N$

$$a_{m,n}^k = Z_{m-1,n} + \frac{n}{m} a_{m-1,n-1}^k. \tag{40}$$

All of the elements of the matrix A of the proposition are now defined.

Proposition I has specified the coefficients of the matrix A which determine the backward transition of the $\boldsymbol{\alpha}^k$ coefficients from final time back to initial time. The quantity that we need to initialise the entire backward propagation process is $\boldsymbol{\alpha}^K$, the set of α coefficients at final time. These are determined by the payoff function at final time. In fact they are computed by expanding the payoff function itself in a Fourier-Hermite series.

Recalling the payoff functions for calls and puts (see Figure 2) and the notation $w = 1(-1)$ to indicate call (put), the backward recursion from final payoff at t^K to the time step t^{K-1} may be written

$$F^{K-1}(\xi) = \frac{e^{-r\Delta t}}{\sqrt{\pi}} \int_{z_l}^{z_u} e^{-(x-\mu(\xi,\Delta t))^2} w(e^{\sigma x} - 1)dx. \tag{41}$$

In the case of a call $z_u = \infty$ and $z_l = 0$. In the case of a put $z_u = 0$ and $z_l = -\infty$. We note however that the notation employed for the limits of integration in (41) also allows us to cater for the situation when there are discrete barriers at final time. In this case z_u and z_l would be determined by the barrier points.

Thus in order to calculate the coefficients α_n^K we first need to expand $e^{\sigma x}$ in a Fourier-Hermite series.

First we note the result that

$$\frac{1}{\sqrt{\pi}} \int_{-\infty}^{\infty} e^{-x^2} H_n(x) e^{\sigma x} dx = \sigma^n e^{\frac{\sigma^2}{4}}. \tag{42}$$

Then, forming the Fourier-Hermite series

$$e^{\sigma x} = \sum_{n=0}^{\infty} \beta_n H_n(x), \tag{43}$$

we apply the orthogonality condition to obtain

$$\beta_n = \frac{1}{2^n n!} \frac{1}{\sqrt{\pi}} \int_{-\infty}^{\infty} e^{-x^2} H_n(x) e^{\sigma x} dx, \tag{44}$$

which by use of (41) reduces to

$$\beta_n = \frac{\sigma^n}{2^n n!} e^{\frac{\sigma^2}{4}}. \tag{45}$$

Therefore, for a call,

$$\alpha_n^K = \beta_n \quad \text{for} \quad n = 1, 2, \ldots, N \tag{46}$$

and $\alpha_0^K = e^{\sigma^2} - 1$, while for a put, one finds

$$\alpha_n^K = -\beta_n \quad \text{for} \quad n = 1, 2, \ldots, N \tag{47}$$

and $\alpha_0^K = 1 - e^{\sigma^2}$.

5 Results

In this section we present some preliminary results obtained from implementing the path integral framework for pricing discretely monitored barrier options. The method was tested for parameter values similar to those used by

Broadie et al. (1997b) and Wei (1998). The method was also tested on barrier puts and calls with digital type payoff functions. To ascertain the accuracy of the method, a Crank-Nicholson method was used to generate *true* prices for the options investigated.[4] Hence, the Crank-Nicholson scheme was used with very fine discretisations, with the stock price variable taken out to five times the value of the strike (scaled to unity) with 2000 steps *per unit*. The time variable was discretised to 100 steps *per day*.

Table 1. Discretely monitored down-and-out call option.

	Path Integral $m = 40$	Path Integral $m = 100$	CN
$\sigma = 0.60$	9.4895	9.4904	9.4905
$\sigma = 0.40$	7.0388	7.0939	7.0394
$\sigma = 0.20$	4.4336	4.4342	4.4344

Parameters: $S = K = 100$, $T = 0.2\,year$, $r = 0.1$
$q = 0$, $barrier = 95$, $monitoring frequency = 4$.

Table 1 gives results of the path integral method for a set of parameter values which differ in their volatility value. The results are presented using 40 and 100 basis functions and are compared with the Crank-Nicholson (CN) true solution using the high order discretisation mentioned earlier. It is evident that even with 40 basis functions the method is relatively accurate when compared with the true values, with the results approaching the true values when using 100 basis functions. Note that the parameter values used in this test include high volatility as well as the barrier level close to the current asset price which typically present problems in other pricing methods for discretely monitored barrier options.

Tables 2–4 show results of the method for at-the-money, out-of-the-money and in-the-money down-and-out options under two monitoring frequencies – monthly and weekly. Results are presented for various barrier levels including values approaching the current asset values. Also presented are percentage-pricing errors relative to the true prices obtained using the Crank-Nicholson method. These give the percentage error of the option prices calculated using the path integral method relative to the true value. Following Wei (1998), we also give the percentage errors of prices calculated using the Broadie-Glasserman-Kou (BGK) continuity correction (Broadie et al., 1997b). Comparing the percentage errors of the prices derived under the path integral method with those derived under the BGK methods, show that relative errors are of comparable magnitude when the barrier level lies away from the current asset price. However, as the barrier level approaches the current asset

---

[4] Our implementation of the Crank-Nicholson scheme for discrete barrier options follows the approach outlined in Tavella (2002).

Table 2. At-the-money down-and-out call option.

Barrier Level	Path Integral Solution	CN Solution	Percentage Price Error	
			Path Integral	BGK Correction
Monthly Monitoring				
85	8.1859	8.1861	-0.003%	-0.011%
90	7.8406	7.8403	0.004%	-0.008%
95	6.7450	6.7463	-0.019%	0.036%
99.5	4.9323	4.9338	-0.031%	-10.926%
99.9	4.7460	4.7474	-0.030%	-16.023%
Weekly Monitoring				
85	8.1248	8.1250	-0.003%	0.006%
90	7.5761	7.5763	-0.003%	0.080%
95	5.8936	5.8946	-0.016%	-0.870%
99.5	3.0078	3.0093	-0.050%	-13.570%
99.9	2.7599	2.7354	0.895%	-16.527%

Parameters: $S = K = 100$, $T = 0.5\,year$, $r = 0.05$
$q = 0$, $\sigma = 0.25$, 100 $basis\ functions$.

Table 3. In-the-money down-and-out call option.

Barrier Level	Path Integral Solution	CN Solution	Percentage Price Error	
			Path Integral	BGK Correction
Monthly Monitoring				
85	10.8029	10.8052	-0.022%	-0.010%
90	9.8831	9.8865	-0.035%	-0.007%
95	7.4328	7.4381	-0.071%	0.043%
99.5	3.6611	3.6623	-0.034%	-11.061%
99.9	3.3172	3.3181	-0.027%	-16.160%
Weekly Monitoring				
85	10.9187	10.9210	-0.021%	0.009%
90	10.3118	10.3139	-0.021%	0.088%
95	8.6362	8.6381	-0.022%	-0.970%
99.5	6.1197	6.1213	-0.027%	-14.001%
99.9	5.8716	5.8732	-0.0027%	-16.960%

Parameters: $S = 100$, $K = 105$, $T = 0.5\,year$, $r = 0.05$
$q = 0$, $\sigma = 0.25$, 100 $basis\ functions$.

Table 4. Out-of-the-money down-and-out call option.

Barrier Level	Path Integral Solution	CN Solution	Percentage Price Error	
			Path Integral	BGK Correction
Monthly Monitoring				
85	5.9226	5.9237	-0.018%	-0.014%
90	5.6068	5.6081	-0.022%	-0.005%
95	4.4945	4.4979	-0.075%	0.058%
99.5	2.3851	2.3847	0.018%	-10.882%
99.9	2.1751	2.1751	0.001%	-15.962%
Weekly Monitoring				
85	5.9545	5.9548	-0.005%	0.000%
90	5.7630	5.7642	-0.021%	0.080%
95	5.0803	4.0814	-0.022%	-0.825%
99.5	3.8346	3.8256	-0.026%	-13.343%
99.9	3.7011	3.7021	-0.026%	-16.258%

Parameters: $S = 100$, $K = 95$, $T = 0.5\,year$, $r = 0.05$
$q = 0$, $\sigma = 0.25$, 100 $basis\ functions$.

price, the relative errors using the path integral method are much smaller than those using the BGK method. Also, the size of the relative error under the path integral method is relatively stable across barrier levels.

Although not reported in this version of the paper, the computation time required for the path integral method is one of its strengths. For a 1-year-down-and-out barrier option with weekly monitoring, with 100 basis functions, the algorithm in its current form runs about three times faster than the Crank-Nicholson method to give the same level of accuracy. Of course the computation times for the path integral method increase with the number of basis functions and with the monitoring frequency. However we should point out that the computer codes for both the Hermite expansion method and the Crank-Nicholson method have not been optimised as much as they could be. We leave the task of doing a thorough comparison of the relative efficiency of these and other methods to future research.

6 Conclusion

In this paper we have presented a pricing method for the valuation of discretely monitored barrier options in a path integral framework. We show how the backward recursion algorithm of such derivative securities in this framework may be efficiently evaluated by expanding the price in a Fourier-Hermite series as a function of the underlying asset price. The method has the advantage of giving the price as a continuous function of the underlying asset price, hence the hedge ratios can be calculated to a high degree of accuracy with minimal additional computational effort. The method can handle various barrier structures with constant and time varying barrier levels for a variety of option payoffs. The method can be made arbitrarily accurate by increasing the number of basis functions in the expansions. Preliminary numerical results show that the method presented is relatively accurate and efficient.

There are several paths for future research. First, it would be of interest to calculate the deltas and delta hedging costs of an option in this framework. Second, the approach here could be compound for speed and accuracy with the recent approach of Fusai et al. (2006)[5] which solves numerically a Wiener-Hopf equation. Third, a natural and simple extension of the work presented in this article would be to apply this method to discretely monitored lookback and Parisian options. American versions of these options could also be conveniently handled in this framework.

Acknowledgements

The authors thank the Editors for the invitation to contribute to this special volume dedicated to the 65th birthday of their good friend and colleague Manfred Gilli. The authors would also like to thank the referees for comments on the first draft which led to an improved final version.

References

Boyle, P. and Tian, Y.: 1997, An explicit finite difference approach to the pricing of barrier options, *Working paper*, University of Waterloo and University of Cinicinnati.

Broadie, M., Glasserman, P. and Kou, S.: 1997a, Connecting discrete and continuous path-dependent options, *Working paper*, Columbia Business School and University of Michigan.

Broadie, M., Glasserman, P. and Kou, S.: 1997b, A continuity correction for discrete barrier options, *Mathematical Finance* **7**(4), 328–349.

Chance, D.M.: 1994, The pricing and hedging of limited exercise caps and spreads, *Journal of Financial Res.* **17**, 561–584.

Chiarella, C. and El-Hassan, N.: 1997, Evaluation of derivative security prices in the Heath-Jarrow-Morton framework as path integrals using Fast Fourier Transform techniques, *Journal of Financial Engineering* **6**(2), 121–147.

Chiarella, C., El-Hassan, N. and Kucera, A.: 1999, Evaluation of American option prices in a path integral framework using Fourier-Hermite series expansions, *Journal of Economic and Dynamic Control* **23**, 1387–1424.

Dash, J.: 1988, *Path Integrals and Options – Part I*, CRNS Preprints.

---

[5] We are grateful to a referee for drawing this paper to our attention.

Eydeland, A.: 1994, A fast algorithm for computing integrals in function spaces: Financial applications, *Computational Economics* **7**(4), 277–285.

Figlewski, S. and Gao, B.: 1997, The adaptive mesh model: A new approach to efficient option pricing, *Working Paper S-97-5*, Salomon Centre, NYU.

Fusai, G., Abrahams, I.D. and Sgarra, C.: 2006, An exact analytical solution for discrete barrier options, *Finance and Stochastic* **10**, 1–26.

Heynen, R. and Kat, H.M.: 1996, Discrete partial barrier options with a moving barrier, *Journal of Financial Engineering* **5**(3), 199–209.

Kat, H.M.: 1995, Pricing lookback options using binomial trees: An evaluation, *Journal of Financial Engineering* **4**, 375–397.

Levy, E. and Mantion, F.: 1997, Discrete by nature, *Risk* **10**(1), 8–22.

Linetsky, V.: 1997, The path integral approach to financial modeling and option pricing, *Computational Economics* **11**(1), 129–163.

Makici, M.S.: 1995, Path integral Monte Carlo method for valuation of derivative securities: Algorithm and parallel implementation, *Technical Report SCCS 650*, NPAC.

Merton, R.: 1973, The theory of rational option pricing, *Bell Journal of Economics and Management Science* **4**, 141–183.

Rubinstein, M. and Reiner, E.: 1991, Breaking down the barriers, *Risk* **4**(4), 28–35.

Sullivan, M.A.: 2000, Pricing discretely monitored barrier options, *Journal of Computational Finance* **3**(4), 35–54.

Tavella, D.: 2002, *Quantitative Methods in Derivatives Pricing*, Wiley Finance.

Tian, Y.: 1996, Breaking the barrier with a two-stage binomial model, *Working paper*, University of Cincicinnati.

Wei, J.Z.: 1998, Valuation of discrete barrier options by interpolations, *Journal of Derivatives* **6**(1), 51–73.

Appendix

A The Coefficient $a_{0,0}^k$

By definition,

$$a_{0,0}^k = \frac{1}{\sqrt{\pi}} \int_{x_l}^{x_u} e^{-x^2} dx, \tag{48}$$

which by change of variable becomes

$$a_{0,0}^k = \frac{1}{\sqrt{2\pi}} \int_{\sqrt{2}x_l}^{\sqrt{2}x_u} e^{-z^2/2} dz,$$

$$= \frac{1}{\sqrt{2\pi}} \int_{-\infty}^{\sqrt{2}x_u} e^{-z^2/2} dz - \frac{1}{\sqrt{2\pi}} \int_{-\infty}^{\sqrt{2}x_l} e^{-z^2/2} dz. \tag{49}$$

This can be expressed as

$$a_{0,0}^K = \Phi(\sqrt{2}x_u) - \Phi(\sqrt{2}x_l), \tag{50}$$

or using the notation as given by equation (15) we have that

$$a_{0,0}^k = P(x_l, x_u). \tag{51}$$

It is worth noting *some special values* for this coefficient:

At the pay-off, $k = K$, we have in the case of a <u>call</u>, that $x_l = 0$ and $x_u \to \infty$ so that

$$a_{0,0}^k = P(0, \infty) = \frac{1}{2}. \tag{52}$$

In the case of a <u>put</u>, we have that $x_l \to -\infty$ and $x_u = 0$ so that

$$a_{0,0}^k = P(-\infty, 0) = \frac{1}{2}. \tag{53}$$

Consider *the general time step*, $k = (K - 1, \ldots, 1)$, there are two special cases:

If we have <u>only an upper barrier</u>, then $x_l \to -\infty$ and

$$a_{0,0}^k = P(-\infty, x_u) = \Phi(\sqrt{2}x_u). \tag{54}$$

If we have <u>only a lower barrier</u>, then $x_u \to \infty$ and

$$a_{0,0}^k = P(x_l, \infty) = 1 - \Phi(\sqrt{2}x_l). \tag{55}$$

B The Coefficient $a_{1,0}^k$

By definition,

$$a_{1,0}^k = \frac{1}{2v} \frac{1}{\sqrt{\pi}} \int_{x_l}^{x_u} 2x e^{-x^2} dx \tag{56}$$

which easily evaluates to

$$a_{1,0}^k = \frac{1}{2v}\sqrt{2} \left[\frac{1}{\sqrt{2\pi}} e^{-(\sqrt{2}x_l)^2/2} - \frac{1}{\sqrt{2\pi}} e^{-(\sqrt{2}x_u)^2/2} \right], \tag{57}$$

which upon use of the notation $\phi(x) = e^{-x^2/2}/\sqrt{2\pi}$ becomes

$$a_{1,0}^k = \frac{1}{2v} \left[\phi(\sqrt{2}x_l) - \phi(\sqrt{2}x_u) \right]. \tag{58}$$

Using the notation of equation (14) we can also write

$$a_{1,0}^k = \frac{1}{2v}Q_{0,0}(x_l, x_u). \tag{59}$$

We note **some special values**.
At the pay-off, $k = K$, we have in the case of a <u>call</u> that $x_l = 0$ and $x_u \to \infty$, then

$$a_{1,0}^K = Q_{0,0}(0, \infty) = \frac{1}{2v\sqrt{\pi}}. \tag{60}$$

In the case of a <u>put</u>, $x_l \to -\infty$ and $x_u = 0$, then

$$a_{1,0}^k = Q_{0,0}(-\infty, 0) = \frac{-1}{2v\sqrt{\pi}}. \tag{61}$$

Here, we see that the difference between the two cases is a simple sign change, that is

$$a_{1,0}^k(call) = -a_{1,0}^K(put). \tag{62}$$

At the general time step k, there are two special cases of interest:
If we have only an <u>upper barrier</u> so that $x_l \to -\infty$ then

$$
\begin{aligned}
a_{1,0}^k &= \frac{1}{2v}Q_{0,0}(-\infty, x_u) \\
&= \frac{1}{2v}\left[R_{0,0}(-\infty) - R_{0,0}(x_u)\right] \\
&= \frac{-\sqrt{2}}{2v}\phi(\sqrt{2}x_u).
\end{aligned} \tag{63}
$$

If we have only a lower barrier so that $x_u \to \infty$ then

$$
\begin{aligned}
a_{1,0}^k &= \frac{1}{2v}Q_{0,0}(x_l, \infty) \\
&= \frac{1}{2v}\left[R_{0,0}(x_l) - R_{0,0}(\infty)\right] \\
&= \frac{\sqrt{2}}{2v}\phi(\sqrt{2}x_l).
\end{aligned} \tag{64}
$$

C The Coefficient $a_{0,1}^k$

By definition

$$a_{0,1}^k = \frac{1}{\sqrt{\pi}}\int_{x_l}^{x_u} e^{-x^2}2(vx + \beta)dx, \tag{65}$$

which can be written

$$a_{0,1}^k = vI_1 + 2\beta I_2, \tag{66}$$

where

$$I_1 = \frac{1}{\sqrt{\pi}} \int_{x_l}^{x_u} 2x e^{-x^2} dx, \qquad I_2 = \frac{1}{\sqrt{\pi}} \int_{x_l}^{x_u} e^{-x^2} dx. \qquad (67)$$

Using results from Appendices A and B, we find that

$$I_1 = \sqrt{2} \left[\phi(\sqrt{2}x_l) - \phi(\sqrt{2}x_u) \right],$$
$$I_2 = \sqrt{2} \left[\phi(\sqrt{2}x_u) - \phi(\sqrt{2}x_l) \right]. \qquad (68)$$

Hence, on using equation (14) and (15)

$$a_{0,1}^k = v Q_{0,0}(x_l, x_u) + 2\beta P(x_l, x_u). \qquad (69)$$

We consider some **special values**. *At the pay-off, $k = K$.* In the case of a call we have $x_l = 0$ and $x_u \to \infty$ so that

$$a_{0,1}^k = v Q_{0,0}(0, \infty) + 2bP(0, \infty) = \frac{1}{2v\sqrt{\pi}} + \beta. \qquad (70)$$

In the case of a put we have $x_l \to -\infty$ and $x_u = 0$ so that

$$a_{0,1}^K = v Q_{0,0}(-\infty, 0) + 2\beta P(-\infty, 0) = \frac{-1}{2v\sqrt{\pi}} + \beta. \qquad (71)$$

At the general time step k, there are two special cases of interest. The case of only an upper barrier $(x_l \to -\infty)$ so that

$$\begin{aligned} a_{0,1}^k &= v Q_{0,0}(-\infty, x_u) + 2\beta P(-\infty, x_u) \\ &= v \left[R_{0,0}(-\infty) - R_{0,0}(x_u) \right] + 2\beta P(-\infty, x_u) \\ &= -v\sqrt{2}\phi(\sqrt{2}x_u) + 2b\Phi(\sqrt{2}x_u). \end{aligned} \qquad (72)$$

In the case of only a lower barrier $(x_u \to \infty)$ we have

$$\begin{aligned} a_{0,1}^k &= v Q_{0,0}(x_l, \infty) + 2\beta P(x_l, \infty) \\ &= v \left[R_{0,0}(x_l) - R_{0,0}(\infty) \right] + 2\beta P(x_l, \infty) \\ &= v\sqrt{2}\phi(\sqrt{2}x_l) + 2\beta \left[1 - \Phi(\sqrt{2}x_l) \right]. \end{aligned} \qquad (73)$$

D The Coefficient $a_{1,1}^k$

By definition

$$a_{1,1}^k = \frac{1}{2v\sqrt{\pi}} \int_{x_l}^{x_u} e^{-x^2} (2x) 2(vx + \beta) dx, \qquad (74)$$

which when integrating by parts and using (14) and (15) can be written as

$$a_{1,1}^k = \frac{1}{2v}Q_{0,1}(x_l, x_u) + P(x_u, x_l). \tag{75}$$

*At the pay-off, $k = K$, we have **two special cases**.* The call for which $x_l = 0$ and $x_u \to \infty$ so that

$$a_{1,1}^K = \frac{1}{2v}Q_{0,1}(0, \infty) + P(0, \infty),$$

$$= \frac{\beta}{2v\sqrt{\pi}} + \frac{1}{2}, \tag{76}$$

and the put for which $x_l \to -\infty$ and $x_u = 0$ so that

$$a_{1,1}^K = \frac{1}{2v}Q_{0,1}(-\infty, 0) + P(-\infty, 0),$$

$$= \frac{-\beta}{v\sqrt{\pi}} + \frac{1}{2}. \tag{77}$$

*At the general time step k, there are **two special cases** of interest.* Only upper barrier ($x_l \to -\infty$) so that

$$a_{1,1}^k = \frac{1}{2v}Q_{0,1}(-\infty, x_u) + P(-\infty, x_u)$$

$$= \frac{1}{2v}[R_{0,1}(-\infty) - R_{0,1}(x_u)] + P(-\infty, x_u)$$

$$= \frac{-\sqrt{2}}{2v}H_1(vx_u + b) + \Phi(\sqrt{2}x_u). \tag{78}$$

The case of only lower barrier ($x_u \to \infty$) so that

$$a_{1,1}^k = \frac{1}{2v}Q_{0,1}(x_l, \infty) + P(x_l, \infty)$$

$$= \frac{1}{2v}[R_{0,1}(x_l) - R_{0,1}(\infty)] + P(x_l, \infty)$$

$$= \frac{\sqrt{2}}{2v}H_1(vx_l + \beta) + \left[1 - \Phi(\sqrt{2}x_l)\right]. \tag{79}$$

E The Coefficient $a_{m,0}^k$

By definition

$$a_{m,0}^k = \frac{1}{2^m m! v^m}\frac{1}{\sqrt{\pi}}\int_{x_l}^{x_u} e^{-x^2}H_m(x)dx. \tag{80}$$

Using the three-term recurrence relation,

$$H_m(x) = 2xH_{m-1}(x) - 2(m-1)H_{m-2}(x), \tag{81}$$

equation (80) can be written as

$$a^k_{m,0} = W - \frac{1}{2v^2} a^k_{m-2,0}, \tag{82}$$

where we set

$$W = \frac{1}{2^m m! v^m} \frac{1}{\sqrt{\pi}} \int_{x_l}^{x_u} 2x e^{-x^2} H_{m-1}(x) dx. \tag{83}$$

Integrating by parts we find that

$$W = \frac{\sqrt{2}}{2^m m! v^m} \left[H_{m-1}(x_l)\phi(\sqrt{2}x_l) - H_{m-1}(x_u)\phi(\sqrt{2}x_u) \right] + \frac{1}{2mv^2} a^k_{m,0}, \tag{84}$$

which when substituted back into (82) gives

$$a^k_{m,0} = \frac{\sqrt{2}}{2^m m! v^m} \left[H_{m-1}(x_l)\phi(\sqrt{2}x_l) - H_{m-1}(x_u)\phi(\sqrt{2}x_u) \right]. \tag{85}$$

Use of (14) finally allows us to write

$$a^k_{m,0} = \frac{1}{2mv} Q_{m-1,0}(x_l, x_u). \tag{86}$$

At the pay-off, $k = K$, we have **two special cases.** The <u>call</u> for which $x_l = 0$ and $x_u \to \infty$ so that

$$
\begin{aligned}
a^K_{m,0} &= \frac{1}{2mv} Q_{m-1,0}(0, \infty) \\
&= \frac{1}{2mv} [R_{m-1,0}(0) - R_{m-1,0}(0)] \\
&= \frac{1}{2^m m! v^m} \frac{H_{m-1}(0)}{\sqrt{\pi}} \quad \text{(use of (13))} \\
&= \frac{1}{2mv\sqrt{\pi}} L_{m-1}(0) \quad \text{(use of (11))}.
\end{aligned}
\tag{87}
$$

The <u>put</u> for which $x_l \to -\infty$ and $x_u = 0$ so that

$$
\begin{aligned}
a^K_{m,0} &= \frac{1}{2mv} Q_{m-1,0}(-\infty, 0) \\
&= \frac{1}{2mv} [R_{m-1,0}(-\infty) - R_{m-1,0}(0)] \\
&= \frac{-1}{2^m m! v^m} \frac{H_{m-1}(0)}{\sqrt{\pi}} \quad \text{(use of (13))} \\
&= \frac{-1}{2mv\sqrt{\pi}} L_{m-1}(0) \quad \text{(use of (11))}.
\end{aligned}
\tag{88}
$$

At the *general time step* k, there are **two special cases** of interest. Only <u>upper barrier</u> $(x_l \to -\infty)$ when,

$$a^k_{m,0} = \frac{1}{2mv} Q_{m-1,0}(-\infty, x_u)$$

$$= \frac{1}{2mv} [R_{m-1,o}(-\infty) - R_{m-1,o}(x_u)]$$

$$= \frac{-\sqrt{2}}{2^m m! v^m} H_{m-1}(x_u)\phi(\sqrt{2}x_u) \qquad \text{(use of (13))}$$

$$= \frac{-\sqrt{2}}{2mv} L_{m-1}(x_u)\phi(\sqrt{2}x_u) \qquad \text{(use of (11))}. \qquad (89)$$

Only <u>lower barrier</u> $(x_u \to \infty)$ when

$$a^k_{m,0} = \frac{1}{2mv} Q_{m-1,0}(x_l, \infty)$$

$$= \frac{1}{2mv} [R_{m-1,o}(x_l) - R_{m-1,o}(\infty)]$$

$$= \frac{\sqrt{2}}{2^m m! v^m} H_{m-1}(x_l)\phi(\sqrt{2}x_l) \qquad \text{(use of (13))}$$

so finally

$$= \frac{\sqrt{2}}{2mv} L_{m-1}(x_l)\phi(\sqrt{2}x_l) \qquad \text{(use of (11))}. \qquad (90)$$

F The Coefficient $a^k_{0,n}$

By definition

$$a^k_{0,n} = \frac{1}{\sqrt{\pi}} \int_{x_l}^{x_u} e^{-x^2} H_n(vx + b)dx. \qquad (91)$$

Using the three-term recurrence relation,

$$H_n(vx + b) = 2(vx + \beta)H_{n-1}(vx + b) - 2(n - 1)H_{n-2}(vx + b), \qquad (92)$$

it follows that

$$a^k_{0,n} = vW + 2\beta a^k_{0,n-1} - 2(n - 1)a^k_{0,n-2}, \qquad (93)$$

where we set

$$W = \frac{1}{\sqrt{\pi}} \int_{x_l}^{x_u} 2xe^{-x^2} H_{n-1}(vx + \beta)dx. \qquad (94)$$

Integrating by parts, we find that

$$W = \left[-H_{n-1}(vx + \beta)\frac{e^{-x^2}}{\sqrt{\pi}} \right]_{x_l}^{x_u} + 2v(n - 1)a^k_{0,n-2}, \qquad (95)$$

which reduces (93) to

$$a_{0,n}^k = v\sqrt{2}\left[H_{n-1}(vx_l + \beta)\phi(\sqrt{2}x_l) - H_{n-1}(vx_u + \beta)\phi(\sqrt{2}x_u)\right]$$
$$+ 2\beta a_{0,n-1}^k + 2(v^2 - 1)(n - 1)a_{0,n-2}^k, \tag{96}$$

which by use of (14) can be written as

$$a_{0,n}^k = vQ_{0,n-1}(x_l, x_u) + 2\beta a_{0,n-1}^k + 2(v^2 - 1)(n - 1)a_{0,n-2}^k. \tag{97}$$

*At the pay-off, $k = K$, we have **two special cases.*** The call for which $x_l = 0$ and $x_u \to \infty$, so that

$$a_{0,n}^K = vQ_{0,n-1}(0, \infty) + 2\beta a_{0,n-1}^K + 2(v^2 - 1)(n - 1)a_{0,n-2}^K$$
$$= v\left[R_{0,n-1}(0) - R_{0,n-1}(\infty)\right] + 2\beta a_{0,n-1}^K + 2(v^2 - 1)(n - 1)a_{0,n-2}^K, \tag{98}$$

which by use of (14) finally reduces to

$$a_{0,n}^K = \frac{v}{\sqrt{\pi}}H_{n-1}(\beta) + 2\beta a_{0,n-1}^K + 2(v^2 - 1)(n - 1)a_{0,n-2}^K. \tag{99}$$

The put for which $x_l \to -\infty$ and $x_u = 0$ so that

$$a_{0,n}^k = vQ_{0,n-1}(-\infty, 0) + 2\beta a_{0,n-2}^K + 2(v^2 - 1)(n - 1)a_{0,n-2}^K$$
$$= v\left[R_{0,n-1}(-\infty) - R_{0,n-1}(0)\right] + 2ba_{0,n-1}^K + 2(v^2 - 1)(n - 1)a_{0,n-2}^K, \tag{100}$$

which by use of (14) finally reduces to

$$a_{0,n}^K = \frac{-v}{\sqrt{\pi}}H_{n-1}(b) + 2\beta a_{0,n-1}^K + 2(v^2 - 1)(n - 1)a_{0,n-2}^K. \tag{101}$$

*At the general time step k, there are **two special cases** of interest.* The case of only upper barrier $(x_l \to -\infty)$, when

$$a_{0,n}^k = vQ_{0,n-1}(-\infty, x_u) + 2\beta a_{0,n-1}^k + 2(v^2 - 1)(n - 1)a_{0,n-2}^k$$
$$= v\left[R_{0,n-1}(-\infty) - R_{0,n-1}(x_u)\right] + 2ba_{0,n-1}^k + 2(v^2 - 1)(n - 1)a_{0,n-2}^k, \tag{102}$$

which by use of (14) reduces to

$$a_{0,n}^k = -v\sqrt{2}H_{n-1}(vx_u + \beta)\phi(\sqrt{2}x_u) + 2\beta a_{0,n-1}^k + 2(v^2 - 1)(n - 1)a_{0,n-2}^k. \tag{103}$$

The case of only lower barrier $(x_u \to \infty)$, when

$$a_{0,n}^k = vQ_{0,n-1}(x_l, \infty) + 2\beta a_{0,n-1}^k + 2(v^2 - 1)(n - 1)a_{0,n-2}^k$$
$$= v\left[R_{0,n-1}(x_l) - R_{0,n-1}(\infty)\right] + 2\beta a_{0,n-1}^k + 2(v^2 - 1)(n - 1)a_{0,n-2}^k, \tag{104}$$

which by use of (14) reduces to

$$a_{0,n}^k = v\sqrt{2}H_{n-1}(vx_l + \beta)\phi(\sqrt{2}x_l) + 2\beta a_{0,n-1}^k + 2(v^2 - 1)(n - 1)a_{0,n-2}^k.$$

(105)

G The Coefficient $a_{m,n}^k$

By definition

$$a_{m,n}^k = \frac{1}{2^m m! v^m} \frac{1}{\sqrt{\pi}} \int_{x_l}^{x_u} e^{-x^2} H_m(x) H_n(vx + b)dx.$$

(106)

Using the three-term recurrence relation,

$$H_m(x) = 2xH_{m-1}(x) - 2(m - 1)H_{m-2}(x),$$

(107)

equation (106) can be written as

$$a_{m,n}^k = W - \frac{1}{2mv^2}a_{m-2,n}^k,$$

(108)

where we set

$$W = \frac{1}{2^m m! v^m} \frac{1}{\sqrt{\pi}} \int_{x_l}^{x_u} 2xe^{-x^2} H_{m-1}(x) H_n(vx + b)dx.$$

(109)

Integrating by parts we find that

$$W = \frac{\sqrt{2}}{2^m m! v^m} \cdot$$
$$\left[H_{m-1}(x_l)H_n(vx_l + \beta)\phi(\sqrt{2}x_l) - H_{m-1}(x_u)H_n(vx + \beta)\phi(\sqrt{2}x_u) \right]$$
$$+ \frac{1}{2mv^2}a_{m-2,n}^k + \frac{n}{m}a_{m-1,n-1}^k,$$

(110)

which by use of (15) can be written as

$$W = \frac{1}{2mv}Q_{m-1,n} + \frac{1}{2mv^2}a_{m-2,n}^k + \frac{n}{m}a_{m-1,n-1}^k.$$

(111)

Finally, substituting (111) back into (108) yields

$$a_{m,n}^k = \frac{1}{2mv}Q_{m-1,n} + \frac{n}{m}a_{m-1,n-1}^k.$$

(112)

At the pay-off, $k = K$, we have **two special cases.** The <u>call</u> for which $x_l = 0$ and $x_u \to \infty$, so that

$$a_{m,n}^K = \frac{1}{2mv}Q_{m-1,n}(0,\infty) + \frac{n}{m}a_{m-1,n-1}^K$$

$$= \frac{1}{2mv}[R_{m-1,n}(0) - R_{m-1,n}(\infty)] \qquad \text{(use of (14))}$$

$$+ \frac{n}{m}a_{m-1,n-1}^k, \qquad (113)$$

which by use of (13) reduces to

$$a_{m,n}^K = \frac{\sqrt{2}}{2mv\sqrt{\pi}}L_{m-1}(0)H_m(\beta) + \frac{n}{m}a_{m-1,n-1}^K. \qquad (114)$$

The $\underline{\text{put}}$ for which $x_l \to -\infty$ and $x_u = 0$, so that

$$a_{m,n}^K = \frac{1}{2mv}Q_{m-1,n}(-\infty,0) + \frac{n}{m}a_{m-1,n-1}^K$$

$$= \frac{1}{2mv}[R_{m-1,n}(-\infty) - R_{m-1,n}(0)] + \frac{n}{m}a_{m-1,n-1}^K, \qquad (115)$$

which by use of (13) reduces to

$$a_{m,n}^K = \frac{-\sqrt{2}}{2mv\sqrt{\pi}}L_{m-1}(0)H_m(\beta) + \frac{n}{m}a_{m-1,n-1}^K. \qquad (116)$$

*At the general time step k, there are **two special cases.** The case of only upper barrier ($x_l \to -\infty$), when*

$$a_{m,n}^k = \frac{1}{2mv}Q_{m-1,n}(-\infty,x_u) + \frac{n}{m}a_{m-1,n-1}^k$$

$$= \frac{1}{2mv}[R_{m-1,n}(-\infty) - R_{m-1,n}(x_u)] + \frac{n}{m}a_{m-1,n-1}^k, \qquad (117)$$

which by use of (13) finally reduces to

$$a_{m,n}^k = \frac{-\sqrt{2}}{2mv}L_m(x_u)H_n(vx_u + \beta)\phi(\sqrt{2}x_u) + \frac{n}{m}a_{m-1,n-1}^k. \qquad (118)$$

The case of only $\underline{\text{lower barrier}}$ ($x_u \to \infty$), when

$$a_{m,n}^k = \frac{1}{2mv}Q_{m-1,n}(x_l,\infty) + \frac{n}{m}a_{m-1,n-1}^k \qquad (119)$$

$$= \frac{1}{2mv}[R_{m-1,n}(x_l) - R_{m-1,n}(\infty)] + \frac{n}{m}a_{m-1,n-1}^k, \qquad (120)$$

which by use of (13) finally reduces to

$$a_{m,n}^k = \frac{\sqrt{2}}{2mv}L_{m-1}(x)H_n(vx_l + b) + \frac{n}{m}a_{m-1,n-1}^k. \qquad (121)$$

H Useful Notation

$$P(x_l, x_u) = \Phi(\sqrt{2}x_u) = \Phi(\sqrt{2}x_l). \tag{122}$$

$$Q_{m,n}(x_l, x_u) = R_{m,n}(x_l) - R_{m,n}x_u, \tag{123}$$

where

$$R_{m,n}(x) = \sqrt{2}L_m(x)H_n(vx + b)\phi(\sqrt{2}x) \tag{124}$$

with

$$L_m(x) = \frac{1}{2^m m! V^m} H_m(x) \tag{125}$$

and

$$\phi(x) = \frac{1}{\sqrt{2\pi}} e^{\frac{-x^2}{2}}, \qquad \Phi(x) = \frac{1}{\sqrt{2\pi}} \int_0^x e^{\frac{-\xi^2}{2}} d\xi. \tag{126}$$

Furthermore,

$$H_n(x) = 2xH_{n-1}(x) - 2(n-1)H_{n-2}(x) \tag{127}$$

with

$$H_o(x) = 1 \qquad \text{and} \qquad H_1(x) = 2x, \tag{128}$$

$$L_m(x) = \frac{x}{mV} L_{m-1}(x) - \frac{1}{2mV^2} L_{m-2}(x) \tag{129}$$

with

$$L_o(x) = 1 \qquad \text{and} \qquad L_1(x) = \frac{x}{V}. \tag{130}$$

Let us define

$$L_m(x) = \frac{1}{2^m m! V^m} H_m(x). \tag{131}$$

Using the three term recurrence relation

$$H_m(x) = 2xH_{m-1}(x) - 2(m-1)H_{m-2}(x) \tag{132}$$

and nothing that

$$\frac{2}{2^m m! V^m} = \frac{1}{mV} \left[\frac{1}{2^{m-1}(m-1)! V^{m-1}} \right], \tag{133}$$

$$\frac{2(m-1)}{2^m m! V^m} = \frac{1}{2mV^2} \left[\frac{1}{2^{m-2}(m-2)! V^{m-2}} \right], \tag{134}$$

we instantly have that

$$L_m(x) = \frac{x}{mV} L_{m-1}(x) - \frac{1}{2mV^2} L_{m-2}(x) \tag{135}$$

with

$$L_0(x) = 1, \tag{136}$$

$$L_1(x) = \frac{x}{v}. \tag{137}$$

Thus, the L_m's can be generated easily using (135), (136), (137).

Now, let us define

$$R_{m,n}(x) = \sqrt{2}L_m(x)H_n(vx + b)\phi(\sqrt{2}x). \tag{138}$$

Special values ($x = 0$):

$$R_{m,n}(0) = \frac{1}{\sqrt{\pi}}L_m(0)H_n(b) \tag{139}$$

and in particular,

$$R_{m,0}(0) = \frac{1}{\sqrt{\pi}}L_m(0), \tag{140}$$

$$R_{0,n}(0) = \frac{1}{\sqrt{\pi}}H_n(b). \tag{141}$$

Special values ($x \to \infty, x \to -\infty$):

$$\lim_{x \to -\infty} R_{m,n}(x) = 0, \tag{142}$$

$$\lim_{x \to \infty} R_{m,n}(x) = 0. \tag{143}$$

Furthermore, using the above notation,

$$Q_{m,n}(x_l, x_u) = R_{m,n}(x_l) - R_{m,n}(x_u). \tag{144}$$

In general,

$$P(x_l, x_u) = \left[\Phi\left(\sqrt{2}x_u\right) - \Phi\left(\sqrt{2}x_l\right) \right]. \tag{145}$$

Special cases:

$$P(-\infty, \infty) = 1, \tag{146}$$

$$P(-\infty, 0) = \frac{1}{2}, \tag{147}$$

$$P(0, \infty) = \frac{1}{2}, \tag{148}$$

$$P(-\infty, x_u) = \Phi(\sqrt{2}x_u), \tag{149}$$

$$P(x_l, \infty) = 1 - \Phi(\sqrt{2}x_l), \tag{150}$$

$$P(x_l, 0) = \frac{1}{2} - \Phi(\sqrt{2}x_l), \tag{151}$$

$$P(0, x_u) = \Phi(\sqrt{2}x_u) - \frac{1}{2}, \tag{152}$$

Part II

Estimation and Classification

Robust Prediction of Beta

Marc G. Genton[1] and Elvezio Ronchetti[2]

[1] Department of Econometrics, University of Geneva, Bd du Pont-d'Arve 40, CH-1211 Geneva 4, Switzerland. Marc.Genton@metri.unige.ch
[2] Department of Econometrics, University of Geneva, Bd du Pont-d'Arve 40, CH-1211 Geneva 4, Switzerland. Elvezio.Ronchetti@metri.unige.ch

Summary. The estimation of β plays a basic role in the evaluation of expected return and market risk. Typically this is performed by ordinary least squares (OLS). To cope with the high sensitivity of OLS to outlying observations and to deviations from the normality assumptions, several methods suggest to use robust estimators. It is argued that, from a predictive point of view, the simple use of either OLS or robust estimators is not sufficient but that some shrinking of the robust estimators toward OLS is necessary to reduce the mean squared error. The performance of the proposed shrinkage robust estimator is shown by means of a small simulation study and on a real data set.

Key words: Beta, CAPM, outliers, robustness, shrinkage.

Preamble

In the past 20 years Manfred Gilli has played an important role in the development of optimization techniques and computational methods in a variety of disciplines, including economics and finance, econometrics and statistics. He has played a leading role both in teaching and research in the Department of Econometrics of the University of Geneva, where he has always been a fixed point and a reference in this field and a constant source of advice and intellectual stimulation. It is therefore a great pleasure to contribute to this Festschrift in his honor and to wish him many more years of productive activity.

1 Introduction

In spite of some of its recognized shortcomings, the Capital Asset Pricing Model (CAPM) continues to be an important and widely used model for the estimation of expected return and the evaluation of market risk. Typically, the rate of return r_t of a given security with respect to that of the risk-free asset r_{ft} is regressed on the rate of return of the market index r_{mt} and the following single factor model is fitted by ordinary least squares (OLS)

$$r_t - r_{ft} = \alpha + \beta(r_{mt} - r_{ft}) + \sigma \cdot u_t, \qquad t = 1, \ldots, n, \tag{1}$$

where α and β are unknown regression parameters and σ is the standard deviation of the errors. It is well known that from a statistical point of view, standard OLS technology presents several drawbacks. In particular, its high sensitivity in the presence of outliers and its dramatic loss of efficiency in the presence of small deviations from the normality assumption have been pointed out in the statistics literature (see, for instance, the books by Huber (1981), Hampel et al. (1986), Rousseeuw and Leroy (1987), Maronna et al. (2006), in the econometric literature (see, for instance, Koenker and Bassett (1978), Chow (1983), Krasker and Welsch (1985), Peracchi (1990), Krishnakumar and Ronchetti (1997), Ronchetti and Trojani (2001)), and in the finance literature (see, for instance, Sharpe (1971), Chan and Lakonishok (1992), Knez and Ready (1997), Martin and Simin (2003)).

Robust statistics was developed to cope with the problem arising from the approximate nature of standard parametric models; see references above. Indeed robust statistics deals with deviations from the stochastic assumptions on the model and develops statistical procedures which are still reliable and reasonably efficient in a small neighborhood of the model. In particular, several well known robust regression estimators were proposed in the finance literature as alternatives to OLS to estimate β in (1). Sharpe (1971) suggested to use least absolute deviations (or the L_1-estimator) instead of OLS. Chan and Lakonishok (1992) used regression quantiles, linear combinations of regression quantiles, and trimmed regression quantiles. In a study to analyze the risk premia on size and book-to-market, Knez and Ready (1997) show that the risk premium on size estimated by Fama and French (1992) is an artefact of using OLS on data with outliers and it disappears when using robust estimators such as least trimmed squares. Finally, Martin and Simin (2003) proposed to estimate β by means of so-called redescending M-estimators.

These robust estimators produce values of β which are more reliable than those obtained by OLS in that they reflect the majority of the historical data and they are not unduly influenced by abnormal returns. In fact, robust estimators downweight (by construction) outlying observations by means of weights which are not fixed in an arbitrary way by the analyst but which are automatically determined by the data. This is a good practice if one wants to have a measure of risk which reflects the structure of the underlying process as revealed by the bulk of the data. However, a familiar criticism of this approach in finance is that "abnormal returns *are the important observations*," that "the analyst is precisely interested in these data and therefore they *should not be downweighted*." We believe that this criticism does not hold if the main goal of the analysis is to produce an estimate of β which reflects the pattern of historical data, but it has some foundation from the point of view of *prediction*.

Indeed if abnormal returns are not errors but legitimate outlying observations, they will likely appear again in the future and downweighting them by using robust estimators will result in a potentially severe bias in the prediction of β. On the other hand, it is true that OLS will produce in this case unbiased estimators of β but this is achieved by paying a potentially important price of a large variability in the prediction of β. Therefore, we are in a typical situation of a trade-off between bias and variance and we can improve upon a simple use of either OLS or a robust estimator. This calls for some form of shrinkage of the robust estimator toward OLS to achieve the minimization of the mean squared error. Similar ideas have been used in the framework of sample surveys containing outliers to estimate totals and quantiles of a given variable in a finite population; see the original paper by Chambers (1986), Welsh and Ronchetti (1998), and Kuk and Welsh (2001).

The paper is organized as follows. In Section 2 we derive shrinkage robust estimators of β by using a Bayesian argument in the framework of the standard linear model. This leads to a particular form for the shrinkage estimator of β. In Section 3 we provide some empirical evidence to illustrate the behavior of different prediction methods on two data sets. Section 4 presents the results of a small Monte Carlo simulation study and shows the potential gains in prediction power that can be obtained by using shrinkage robust estimators. Finally, some conclusions and a brief outlook are provided in Section 5.

2 Shrinkage Robust Estimators of Beta

In this section we derive our shrinkage robust estimator of β in the general setup of the standard linear model

$$y_i = x_i^T \beta + \sigma \cdot u_i, \qquad i = 1, \ldots, n, \tag{2}$$

where $\beta \in \mathcal{R}^p$, $\sigma > 0$, and the u_i's are iid with density $g(\cdot)$.

Our shrinkage robust (SR$_c$) estimator of β is defined by

$$\tilde{\beta}_c = \hat{\beta}_R + \left(\sum_{i=1}^n x_i x_i^T \right)^{-1} \sum_{i=1}^n \hat{\sigma} \psi_c \left(\frac{y_i - x_i^T \hat{\beta}_R}{\hat{\sigma}} \right) x_i, \tag{3}$$

where $\hat{\beta}_R$ is a robust estimator of β, $\hat{\sigma}$ is a robust estimator of scale, and $\psi_c(\cdot)$ is the Huber function defined by $\psi_c(r) = \min(c, \max(-c, r))$. Typically, $\hat{\beta}_R$ is one of the available robust estimators for regression such as, e.g., an M-estimator (Huber, 1981), a bounded-influence estimator (Hampel et al., 1986), the least trimmed squares estimator (Rousseeuw, 1984) or the MM estimator (Yohai, 1987), and $\hat{\sigma}$ is a robust estimator of scale such as the median absolute deviation (Hampel et al., 1986). The tuning constant c of the Huber function in (3) is typically larger than the standard value 1.345; see below.

The estimator $\tilde{\beta}_c$ defined by (3) can be viewed as a least squares estimator with respect to the "pseudovalues" $\tilde{y}_i = x_i^T \beta + \sigma \psi_c \left(\frac{y_i - x_i^T \beta}{\sigma} \right)$.

The structure of $\tilde{\beta}_c$ is as follows. The parameter β is first estimated by means of a robust estimator $\hat{\beta}_R$ and scaled residuals $r_i = (y_i - x_i^T \hat{\beta}_R)/\hat{\sigma}$ are obtained. Next the Huber function $\psi_c(r_i)$ is applied to each residual. When r_i is larger in absolute value than c, the residual is brought in to the value $c \operatorname{sign}(r_i)$, whereas it is not modified otherwise. By using a moderate value of c, i.e., a value larger than the standard value of 1.345 typically used in robust statistics, only very large residuals are trimmed thereby insuring both robustness (low variability) and small bias. The limiting case $c \to \infty$ gives back OLS (unbiasedness but high variability) while a very small value of c leads to $\hat{\beta}_R$ (low variability but large bias).

Further insight can be gained from a predictive Bayesian argument. Assume a prior density $h(\cdot)$ for β and a quadratic loss function. Then, the Bayes estimator of β is given by

$$\beta_B(y_1, \ldots, y_n) = E[\beta \mid y_1, \ldots, y_n]$$

$$= \frac{\int \beta \Pi_{i=1}^n \sigma^{-1} g((y_i - x_i^T \beta)/\sigma) h(\beta) d\beta}{\int \Pi_{i=1}^n \sigma^{-1} g((y_i - x_i^T \beta)/\sigma) h(\beta) d\beta}$$

$$= \frac{\int \beta \exp\{-nk(\beta; y_1, \ldots, y_n)\} d\beta}{\int \exp\{-nk(\beta; y_1, \ldots, y_n)\} d\beta}, \tag{4}$$

where $k(\beta; y_1, \ldots, y_n) = -\frac{1}{n} \left[\sum_{i=1}^n \log \left(\sigma^{-1} g((y_i - x_i^T \beta)/\sigma) \right) \right] - \frac{1}{n} \log h(\beta)$.

Using Laplace's method to approximate the integrals in (4), we can write

$$\beta_B(y_1, \ldots, y_n) = \hat{\beta} \left[1 + O(n^{-1}) \right], \tag{5}$$

where $\hat{\beta} = \arg \max_{\beta} k(\beta; y_1, \ldots, y_n)$.

If we choose $g(r) \propto \exp(-\rho(r))$ for some positive, symmetric function ρ, and the prior $h(\beta)$ as $N(\mu, \delta^2 I)$, where I is the identity matrix, then $\hat{\beta}$ satisfies the estimating equation

$$\sum_{i=1}^n \frac{1}{\sigma} \psi \left((y_i - x_i^T \hat{\beta})/\sigma \right) x_i - \frac{\hat{\beta} - \mu}{\delta^2} = 0, \tag{6}$$

where $\psi(r) = \frac{d\rho(r)}{dr}$.

Expanding the first term on the left hand side of (6) about β_0 and ignoring terms of order $\|\hat{\beta} - \beta_0\|^2$ or smaller we obtain

$$0 \approx \sum_{i=1}^{n} \frac{1}{\sigma} \psi \left((y_i - x_i^T \beta_0)/\sigma \right) x_i \tag{7}$$

$$- \left(\sum_{i=1}^{n} \frac{1}{\sigma^2} \psi' \left((y_i - x_i^T \beta_0)/\sigma \right) x_i x_i^T \right) (\hat{\beta} - \beta_0) - \frac{\hat{\beta} - \mu}{\delta^2}, \tag{8}$$

and

$$(\Omega + I) \hat{\beta} \approx \delta^2 \sum_{i=1}^{n} \frac{1}{\sigma} \psi \left((y_i - x_i^T \beta_0)/\sigma \right) x_i + \Omega \beta_0 + \mu,$$

where $\Omega = \sum_{i=1}^{n} \frac{\delta^2}{\sigma^2} \psi' \left((y_i - x_i^T \beta_0)/\sigma \right) x_i x_i^T$.

Therefore,

$$\hat{\beta} = (\Omega + I)^{-1} \Omega \hat{\beta}_R + (\Omega + I)^{-1} \mu \tag{9}$$
$$= (\Omega + I)^{-1} \Omega \left[\hat{\beta}_R + \Omega^{-1} \mu \right],$$

where $\hat{\beta}_R = \beta_0 + \Omega^{-1} \delta^2 \sum_{i=1}^{n} \frac{1}{\sigma} \psi \left((y_i - x_i^T \beta_0)/\sigma \right) x_i$.

Notice that $\hat{\beta}_R$ is a one-step M-estimator defined by ψ (Hampel et al., 1986, p. 106). Also it is natural to let Ω be based on $\hat{\beta}_R$. For large δ^2 (which corresponds to a vague prior for β) we have from (9) :

$$\hat{\beta} \approx \hat{\beta}_R + \hat{\Omega}^{-1} \mu, \tag{10}$$

where $\hat{\Omega} = \frac{\delta^2}{\sigma^2} \sum_{i=1}^{n} \psi' \left((y_i - x_i^T \hat{\beta}_R)/\sigma \right) x_i x_i^T$, which corresponds to (3) when we replace $\hat{\Omega}$ by the slightly simpler $\frac{\delta^2}{\sigma^2} \sum_{i=1}^{n} x_i x_i^T$ and we estimate the prior mean μ by $\sum_{i=1}^{n} \sigma \psi_c \left((y_i - x_i^T \hat{\beta}_R)/\sigma \right) x_i$.

Therefore, $\tilde{\beta}_c$ can be viewed as the Bayes estimator in this setup. This argument provides a justification for the use of an additive adjustment in (3). However, it allows for a wide class of such adjustments, interpreted as alternative estimates for the prior mean μ.

The tuning constant c controls the amount of shrinking and is typically larger than standard values used for robust Huber estimators. It can be chosen by cross-validation to minimize the mean squared error or an alternative measure of prediction error; see also Chambers (1986).

3 Empirical Evidence

In this section, we investigate the prediction of beta from two stock returns (Shared Technologies Inc. and Environmental Tectonics Corp.) versus market returns. These data, also analyzed by Martin and Simin (2003), were obtained from the Center for Research on Stock Prices (CRSP) database and are displayed in the top row of Figure 1. Some outlying values can be observed in both plots, especially for the stocks of Environmental Tectonics Corp. In the bottom row of Figure 1, we plot OLS (solid curve) and robust (dotted curve) betas computed using a 104-week moving window. There are dramatic differences between the two curves (solid and dotted) occurring when the outlying value(s) enters the moving window. We also computed a shrinkage robust estimator (dashed curve) of beta obtained from Equation (3) with $c = 1.5$. This estimator shrinks the robust estimator of beta toward the OLS beta in order to account for possible future outliers.

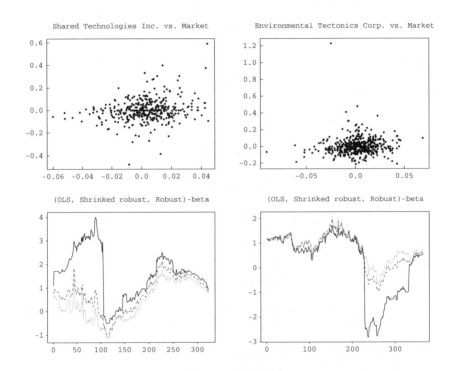

Fig. 1. Top row: plots of stock returns versus market returns. Bottom row: plots of one-week ahead predictions of betas using a 104-week moving window and: OLS-beta (solid curve); shrinkage robust-beta (dashed curve); robust-beta (dotted curve).

In order to illustrate the effect of outliers on betas, the top panel of Figure 2 depicts the CAPM model fitted by OLS on three weeks of Shared Technologies Inc. vs market data without outlier (solid line) and with one additional outlier (short-dashed line) represented by an open circle. The effect of this single

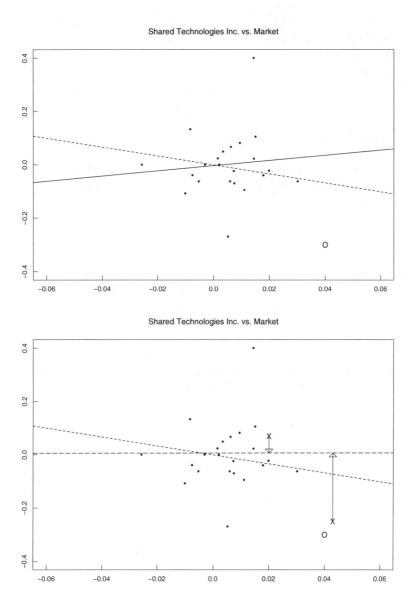

Fig. 2. Top panel: fitted CAPM by OLS on three weeks of Shared Technologies Inc. vs market data: without outlier (solid line), with one additional outlier (short-dashed line). Bottom panel: fitted CAPM by OLS (short-dashed line) and robust procedure (long-dashed line). The prediction of future new points (X) could be inaccurate if outliers occur again.

outlier is a leverage that changes the positive value of beta to a negative one. The bottom panel of Figure 2 depicts the previous contaminated situation with the CAPM model fitted by a robust procedure (long-dashed line). The prediction of future new points represented by an X could be inaccurate if outliers occur again. This points to the need of shrinking the robust estimator toward the OLS estimator.

4 Monte Carlo Simulations

We investigate the robust prediction of beta through Monte Carlo simulations. For the sake of illustration, we consider the model (2) with $p = 1$, that is,

$$y_i = \beta x_i + \sigma \cdot u_i, \qquad i = 1, \ldots, n, \tag{11}$$

with $\sigma = 1$, u_i's iid $N(0, 1)$, and x_i's iid $N(0, 4)$. We choose values of beta that are representative of situations found in real data sets, $\beta = 0.5$ and $\beta = 1.5$, and construct two simulated testing samples from model (11) of size $m = 400$ each. These two testing samples represent future observations that are not available at the modeling stage. In order to reproduce possible outliers found in practice, we include 10% of additive outliers to the x_i's and to the y_i's from a $N(0, \tau^2)$ distribution with $\tau = 5$. Figure 3 depicts these two testing samples where some outliers, similar to the ones found in real data, can be seen.

We use the least trimmed squares (LTS) regression estimator, $\hat{\beta}_{LTS}$, to obtain robust estimates of beta. It has a bounded influence function and a breakdown-point of nearly 50%. The LTS estimator is defined as the value of β that minimizes the sum of the smallest $\lfloor n/2 \rfloor + 1$ of the squared residuals, where $\lfloor . \rfloor$ denotes rounding down to the next smallest integer. Alternative robust estimators could be used instead of the LTS. In particular, we ran our simulation with the redescending M-estimator used by Martin and Simin (2003) and obtained similar results.

The LTS and OLS estimators of beta on the two testing samples are $\hat{\beta}_{LTS} = 0.379$ and $\hat{\beta}_{OLS} = 0.327$ for the case $\beta = 0.5$, and $\hat{\beta}_{LTS} = 1.492$ and $\hat{\beta}_{OLS} = 0.982$ for the case $\beta = 1.5$. The corresponding regression lines are drawn in Figure 3. They indicate that the effect of outliers on the estimation of beta is not too strong in the first testing sample, whereas it is much stronger in the second one.

The shrinkage robust (SR_c) estimator (3) of β in our context becomes

$$\tilde{\beta}_c = \hat{\beta}_{LTS} + \left(\sum_{i=1}^{n} x_i^2 \right)^{-1} \sum_{i=1}^{n} \hat{\sigma} \psi_c \left(\frac{y_i - \hat{\beta}_{LTS} x_i}{\hat{\sigma}} \right) x_i, \tag{12}$$

where $\hat{\sigma}$ is the median absolute deviation (MAD), a robust estimator of scale. (Again, other robust estimators of scale could be used instead.) In order to analyze the effect of outliers on the estimation of beta, we simulate 1,000

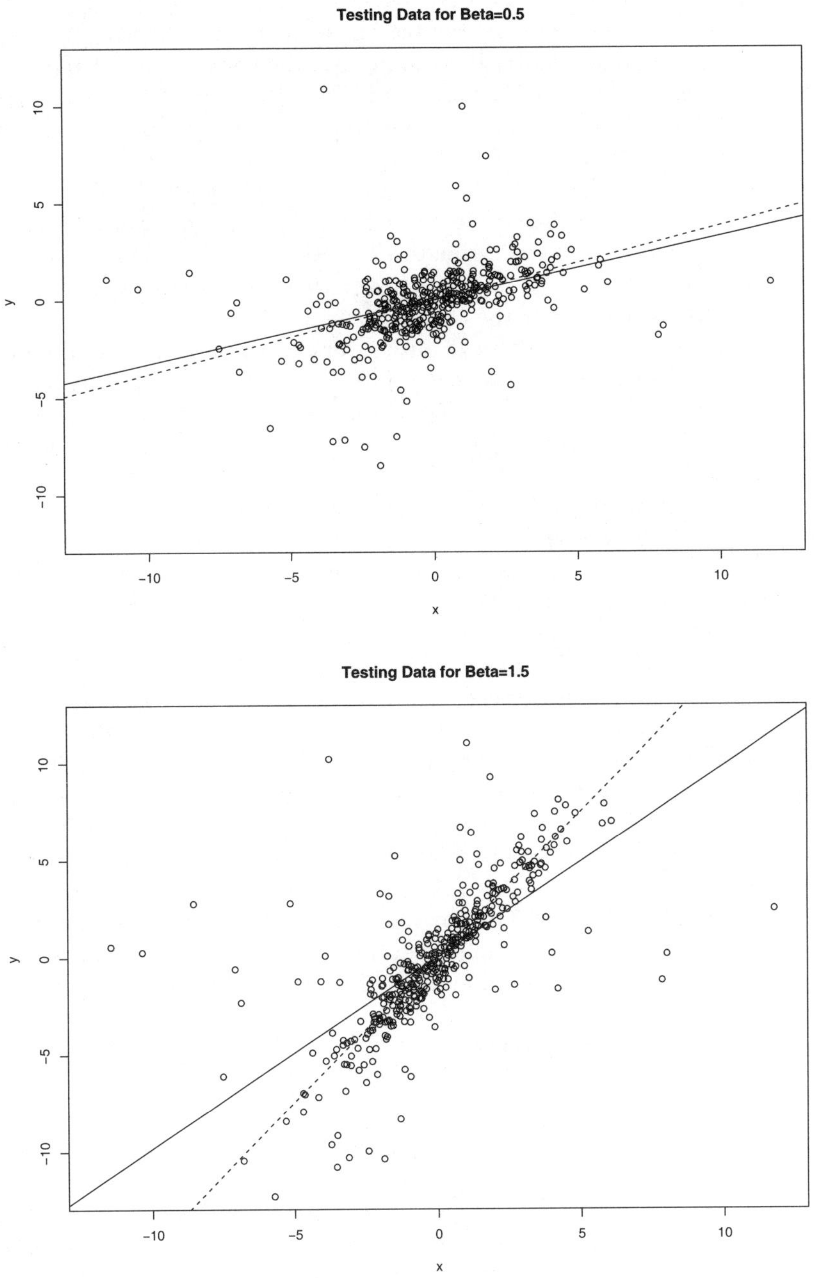

Fig. 3. Scatter plot of $m = 400$ testing data containing 10% outliers with fitted regression line by OLS (solid line) and LTS (dashed line). Top panel: $\beta = 0.5$. Bottom panel: $\beta = 1.5$.

training data sets of size $n = 100$ each, containing outliers, following the same procedure as described above for the testing samples. Hence the structure of the outliers is the same in both the training and the testing samples. For each sample, we estimate β by LTS, OLS, and SR_c with $c = 1, \ldots, 10$. Figure 4 depicts boxplots of these estimates over the 1,000 simulated training data containing outliers for $\beta = 0.5$ (top panel) and $\beta = 1.5$ (bottom panel). The horizontal solid line represents the true value of beta. The robust estimator LTS has a much smaller bias than the OLS estimator, but usually a larger variance. The shrinkage robust estimators behave between the boundary cases SR_0 =LTS and SR_∞ =OLS. Note however that for some values of c, the variance of SR_c is reduced at the cost of a small increase in bias.

Next, we investigate the effect of outliers and robust estimates of β on the prediction of future observations represented by the two testing data plotted in Figure 3. Specifically, for each of the 1,000 estimates $\hat{\beta}$, we compute the predicted values $\hat{y}_i = \hat{\beta}x_i$, $i = 1, \ldots, m$ with $m = 400$, on the two testing samples.

A measure of quality of prediction should depend on the given problem and there is a large amount of literature in econometrics and economics on the choice of the loss function for prediction purposes. Some authors (see, e.g., Leitch and Tanner (1991) and West et al. (1993)) argue that the loss function should reflect an economically meaningful criterion, while others consider statistical measures of accuracy which include, e.g., mean squared error, absolute error loss, linex loss among others. While we agree that these discussions are important, it is not our purpose in this paper to compare the different proposals. Instead we argue that the robustness issue (and therefore the possible deviations from the stochastic assumptions of the model) should also be taken into account when choosing the loss function.

In the absence of specific information, a typical choice is the root mean squared error. Note, however, that this measure is related directly or indirectly to the Gaussian case. In non-Gaussian cases there are no standard choices and a good recommendation is to compute several such measures. To illustrate this point, we focus here on a particular class of measures defined by

$$Q(p) = \left(\frac{1}{m} \sum_{i=1}^{m} \mid y_i - \hat{y}_i \mid^p \right)^{1/p}. \tag{13}$$

Three choices of p give the following quantities: $p = 2$ yields the root mean squared error (RMSE); $p = 1$ the mean absolute error (MAE); and $p = 1/2$ the square root absolute error (STAE). We use these measures to evaluate the quality of the prediction. For each of them, Table 1 reports the percentage of selection of a minimum measure of prediction for a range of values of $c = 0, 1, 2, \ldots, 10, \infty$ over the 1,000 simulation replicates for the scenarios with $\beta = 0.5$ and $\beta = 1.5$. The highest percentage for each measure is identified in bold fonts. As can be seen, the optimal c which minimizes a certain measure of quality of prediction is not exclusively concentrated at the boundary esti-

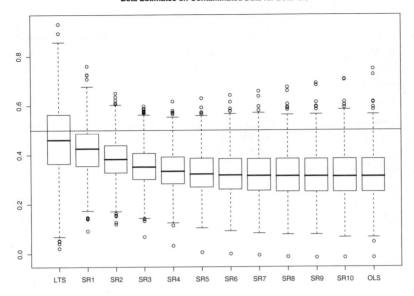

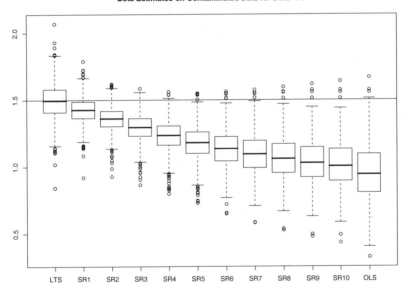

Fig. 4. Boxplots of estimates of beta on 1,000 simulated training data containing 10% outliers. The horizontal solid line represents the true value of beta. Top panel: $\beta = 0.5$. Bottom panel: $\beta = 1.5$.

mators LTS and OLS. Although a more rigorous comparison would involve formal tests of predictive accuracy such as the standard Diebold and Mariano (1995) test or its more general extension by Giacomini and White (2006), the effects seem to be clear: it is fair to say that better predictions of future beta can be obtained by making use of shrinkage robust estimators, that is, by shrinking the robust estimator toward OLS to minimize the trade-off between bias and variance.

Table 1. Percentage of selection of a minimum measure of prediction (RMSE, MAE, STAE) for a range of values of c $(0, 1, 2, \ldots, 10, \infty)$ over 1,000 simulation replicates for the scenarios with $\beta = 0.5$ and $\beta = 1.5$. The highest percentage for each measure is identified in bold fonts.

	LTS	SR_1	SR_2	SR_3	SR_4	SR_5	SR_6	SR_7	SR_8	SR_9	SR_{10}	OLS
$\beta = 0.5$												
RMSE	10.2	10.6	14.2	16.5	11.6	9.3	5.3	2.6	1.2	0.9	0.6	**17.0**
MAE	18.0	18.2	**22.5**	13.1	6.5	4.6	3.3	1.2	1.3	0.4	0.5	10.4
STAE	**22.8**	21.5	21.2	11.1	5.2	3.9	2.6	2.1	1.2	0.7	0.7	7.0
$\beta = 1.5$												
RMSE	1.3	0.0	0.3	0.9	2.4	6.5	10.2	8.1	9.6	7.5	11.4	**41.8**
MAE	15.9	25.2	**31.2**	14.0	6.5	2.8	1.8	0.8	0.6	0.2	0.2	0.8
STAE	32.5	**35.0**	20.6	7.6	2.9	0.7	0.4	0.1	0.1	0.1	0.0	0.0

The relative gains in reduction of the measure of quality of prediction obtained with shrinkage robust estimators compared to LTS and OLS are investigated next. Specifically, we define the relative gain by

$$RG_{c,d}(p) = \frac{Q_d(p) - Q_c(p)}{Q_d(p)}, \tag{14}$$

where $Q_c(p)$ denotes a measure of quality of prediction obtained with a shrinkage robust estimator SR_c. Denote by c^* the optimal value of c, that is, the value of c minimizing $Q_c(p)$ for a fixed p. Figure 5 depicts boxplots over the 1,000 simulation replicates of $RG_{c^*,0}(p)$ and $RG_{c^*,\infty}(p)$ for $p = 2, 1, 1/2$. In terms of RMSE, the gains of the SR_{c^*} compared to the OLS are fairly small, although they can reach 10-20%. In terms of MAE and STAE, the gains are more important and can reach up to 40% in the case $\beta = 1.5$. The gains compared to LTS go in the other direction.

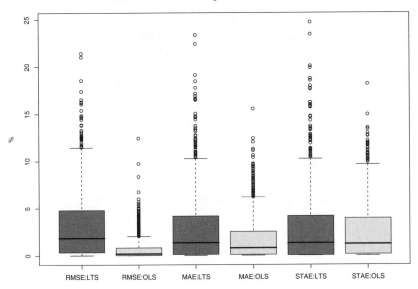

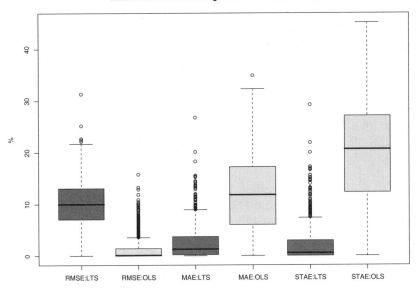

Fig. 5. Relative gains obtained with shrinkage robust estimators compared to LTS and OLS on various measures of prediction (RMSE, MAE, STAE). Top panel: $\beta = 0.5$. Bottom panel: $\beta = 1.5$.

5 Conclusion

In this paper we presented a shrinkage robust estimator which improves upon OLS and standard robust estimators of β from a predictive point of view in the presence of deviations from the normality assumption. The ideas developed here can be used beyond the simple framework of the CAPM and the standard linear model and are basic building blocks to introduce robustness in any prediction problem. This is particularly useful in financial applications but also in many other fields in economics and natural sciences.

Acknowledgements

The authors would like to thank the referees for helpful comments and relevant references.

References

Chambers, R.L.: 1986, Outlier robust finite population estimation, *Journal of the American Statistical Association* **81**, 1063–1069.

Chan, L.K.C. and Lakonishok, J.: 1992, Robust measurement of beta risk, *Journal of Financial and Quantitative Analysis* **27**, 265–282.

Chow, G.: 1983, *Econometrics*, McGraw-Hill, New York.

Diebold, F.X. and Mariano, R.S.: 1995, Comparing predictive accuracy, *Journal of Business & Economic Statistics* **13**, 253–263.

Fama, E.F. and French, K.R.: 1992, The cross-section of expected stock returns, *The Journal of Finance* **47**, 427–466.

Giacomini, R. and White, H.: 2006, Tests of conditional predictive ability, *Econometrica* **74**, 1545–1578.

Hampel, F.R., Ronchetti, E.M., Rousseeuw, P.J. and Stahel, W.A.: 1986, *Robust Statistics: The Approach Based on Influence Functions*, Wiley, New York.

Huber, P.J.: 1981, *Robust Statistics*, Wiley, New York.

Knez, P.J. and Ready, M.J.: 1997, On the robustness of size and book-to-market in cross-sectional regressions, *The Journal of Finance* **52**, 1355–1382.

Koenker, R. and Bassett, G.: 1978, Regression quantiles, *Econometrica* **46**, 33–50.

Krasker, W.S. and Welsch, R.E.: 1985, Resistant estimation for simultaneous equations models using weighted instrumental variables, *Econometrica* **53**, 1475–1488.

Krishnakumar, J. and Ronchetti, E.: 1997, Robust estimators for simultaneous equations models, *Journal of Econometrics* **78**, 295–314.

Kuk, A.Y.C. and Welsh, A.H.: 2001, Robust estimation for finite populations based on a working model, *Journal of the Royal Statistical Society Series B* **63**, 277–292.

Leitch, G. and Tanner, J.E.: 1991, Economic forecast evaluation: Profits versus the conventional error measures, *American Economic Review* **81**, 580–590.

Maronna, R.A., Martin, R.D. and Yohai, V.J.: 2006, *Robust Statistics: Theory and Methods*, Wiley, New York.

Martin, R.D. and Simin, T.: 2003, Outlier resistant estimates of beta, *Financial Analysts Journal* **59**, 56–69.

Peracchi, F.: 1990, Robust M-estimators, *Econometric Reviews* **9**, 1–30.

Ronchetti, E. and Trojani, F.: 2001, Robust inference with GMM estimators, *Journal of Econometrics* **101**, 37–69.

Rousseeuw, P.J.: 1984, Least median of squares regression, *Journal of the American Statistical Association* **79**, 871–880.

Rousseeuw, P.J. and Leroy, A.M.: 1987, *Robust Regression and Outlier Detection*, Wiley, New York.

Sharpe, W.F.: 1971, Mean-absolute-deviation characteristic lines for securities and portfolios, *Management Science* **18**, B1–B13.

Welsh, A.H. and Ronchetti, E.: 1998, Bias-calibrated estimation from sample surveys containing outliers, *Journal of the Royal Statistical Society Series B* **60**, 413–428.

West, K.D., Edison, H.J. and Cho, D.: 1993, A utility-based comparison of some models of exchange rate volatility, *Journal of International Economics* **35**, 23–45.

Yohai, V.: 1987, High breakdown-point and high efficiency robust estimates for regression, *The Annals of Statistics* **15**, 642–656.

Neural Network Modelling with Applications to Euro Exchange Rates

Michele La Rocca and Cira Perna

Department of Economics and Statistics, University of Salerno, Via Ponte Don Melillo, 84084 Fisciano (SA), Italy [larocca,perna]@unisa.it

Summary. Neural networks have shown considerable success when used to model financial data series. However, a major weakness of this class of models is the lack of established procedures for misspecification testing and tests of statistical significance for the various estimated parameters. These issues are particularly important in the case of financial engineering where data generating processes are very complex and dominantly stochastic. After a brief review of neural network models, an input selection algorithm is proposed and discussed. It is based on a multistep multiple testing procedure calibrated by using subsampling. The simulation results show that the proposed testing procedure is an effective criterion for selecting a proper set of relevant inputs for the network. When applied to Euro exchange rates, the selected network models show that information contained in past percentage changes can be relevant to the prediction of future percentage changes of certain time series. The apparent predictability for some countries which we analysed does not seem to be an artifact of data snooping. Rather, it is the result of a testing procedure constructed to keep the family wise error rate under control. The results also remain stable while changing the subseries length.

Key words: Feedforward neural networks, subsampling, variable selection, Euro exchange rates.

1 Introduction

Nowadays, it is widely accepted that non-linearity exists in financial markets and that non-linear models, both parametric and non-parametric, can be effectively used to reveal this pattern in the field of financial engineering (Qi, 1996). The increase of interest in non-linear techniques was made possible by advances in processing power of personal computers and the availability of large data sets with long time series. This interest was further enhanced by the development of fast numerical optimization methods which are in some cases already able to take advantage of parallel hardware structures of modern central processing units.

A class of very flexible non-linear models, designed to mimic biological neural systems, are artificial neural networks (ANN). They have become the focus of considerable attention as a possible tool for modelling complex non-linear systems by using highly interconnected non-linear memoryless computing elements. Artificial neural networks can be considered as parallel distributed models made up of simple data processing units, organized on multiple layers, with one or more hidden (latent) layers which add flexibility to the model. This class of models offers some clear advantages over classical techniques. Because of their massively parallel structure, they can perform very fast computations if implemented on dedicated hardware; due to their adaptive nature, they can learn the characteristics of input signals and adapt to changes in data; given their non-linear nature they can perform functional approximations which are beyond optimal linear techniques (Cybenko, 1989, inter alia).

For all these reasons, neural networks have shown considerable success in modelling financial data series and can be seen as an effective alternative tool to classical parametric modelling, especially when the underlying data generating process is not fully understood or when the nature of the relationships being modelled may display a complex structure (Refenes et al., 1997; Lam, 2004; Thawornwong and Enke, 2004; West et al., 2005; Alvarez-Diaz and Alvarez, 2007, inter alia).

However, the "atheoretical" nature of the tool and the intrinsic misspecification of the model, in the sense of White (1989), make the choice of a suitable neural network model a hard task. The most widely used approaches, such as pruning and regularization, information criteria and cross-validation, may lead to satisfactory results even if they have no inferential statistical perspective.

The lack of generally accepted procedures for performing tests for misspecified models as well as tests of statistical significance for the estimated parameters is felt to be one of the main drawbacks of neural modelling. This is a serious disadvantage in fields where there is a strong culture for testing the predictive power of a model, for assessing the sensitivity of the dependent variable to changes in the inputs or, more generally, for testing the statistical significance of a hypothesis of interest at a given level of confidence. Those issues are particularly important in financial engineering, where data generating processes are very complex and dominantly stochastic.

In the case of the single hidden feedforward neural network class, model selection basically involves the choice of the number and type of input neurons and the number of neurons in the hidden layer. In this framework, standard neural network modelling strategies might contain a strong judgemental component and in many cases they do not give, explicitly, any information on the most "significant" variables, with the exception being the procedures suggested by Qi and Maddala (1996) and Refenes and Zapranis (1999). Following these latter approaches, we emphasize that a better model selection strategy should stress the role of the explanatory variables (useful for the identification and interpretation of the model) and it should treat the hidden layer size as a smoothing parameter, taking into account the trade-off between estimation bias and variability (Perna and Giordano, 2001).

Clearly, variable selection is the key issue for knowledge extraction from the model especially when attention is devoted to the identification and interpretation of the relationship among output and input variables. Therefore, in our opinion, while the hidden layer size could be chosen according to one of the many methods available in the statistical literature, ranging from the information criteria based on the model's fit to the indexes based on prediction accuracy, input selection should be addressed by focusing on statistical testing procedures for variable selection.

In this perspective, this choice can be based on relevance measures (White and Racine, 2001; La Rocca and Perna, 2005b, inter alia) following a strategy which is analogous to that generally employed when selecting a model in the classical regression framework. Unfortunately, this procedure requires specification of a proper set of variables to be tested as irrelevant and this can be a hard task if no *a priori* information is available on the problem at hand. Moreover, practitioners could be tempted to use the following scheme: (i) test a given set of inputs for irrelevance; (ii) discard variables found to be not relevant to the model; (iii) re-estimate the model and test another set of variables for irrelevance; (iv) iterate the procedure until no further variables are found to be relevant. Many authors argue against this backward regression procedure (White and Racine, 2001, inter alia) since it fails to take account of the fact that the level of the tests, in step (iii), should be related to the results of the first test.

In other words, the variable selection problem has the structure of a multiple testing framework and the problem then becomes how to decide which hypotheses to reject, accounting for the multitude of tests and the dependence between them. In such a context, data snooping arises as a serious problem, since a given set of data is used more than once for inference and model selection. In the neural network framework, the problem can be even more dangerous due to the lack of theory supporting the model selection strategy.

In this paper, we focus on the problem of variable selection and propose a strategy for input selection in neural network dynamic regression models, based on relevance measures, with a strong statistical perspective. It is based on multiple testing and, to avoid the data snooping problem (White, 2000; Giacomini and White, 2006), the familywise error rate is controlled by using a recent proposal by Romano and Wolf (2005a, 2005b). The neural network model structure and the complexity of the test procedures distinguish the asymptotic distribution of the test statistics involved from the familiar tabulated distributions. The problem can be overcome by using resampling techniques as a simulation tool to approximate the unknown sampling distributions of the statistical quantities involved in the model selection procedure (Baxt and White, 1995; Giordano et al., 2004; La Rocca and Perna, 2005a, inter alia). Here, to obtain valid asymptotic critical values for the test, we describe the implementation of a subsampling scheme, which is able to deliver consistent results under very weak assumptions.

The paper is organized as follows. In Section 2 neural network modelling is briefly reviewed while in Section 3 the variable selection procedure is presented and discussed. In Section 4 some results on simulated data are reported to show the performance of the proposed approach in finite samples. In Section 5 an application to Euro exchange rate modelling is discussed. Some concluding remarks close the paper.

2 Feedforward Neural Network Models

Let the observed data be the realization of a stationary sequence of random vectors of order $(d+1)$, $\left\{ \mathbf{Z}_t = \left(Y_t, \mathbf{X}_t^T\right)^T \right\}$. The random variables Y_t represent targets (in the neural network jargon). What is usually of interest is the probabilistic relationship with the variables $\mathbf{X}_t$, described by the conditional distribution of the random variable $Y_t | \mathbf{X}_t$. Certain aspects of this probability law play an important role in interpreting what is learned by artificial neural network models. If $\mathbb{E}(Y_t) < \infty$, then $\mathbb{E}(Y_t | \mathbf{X}_t) = g(\mathbf{X}_t)$ and we can write

$$Y_t = g(\mathbf{X}_t) + \varepsilon_t, \tag{1}$$

where $\varepsilon_t \equiv Y_t - g(\mathbf{X}_t)$ and $g : \mathbb{R}^d \to \mathbb{R}$ is a measurable function. Clearly, by construction the error term ε_t is such that $\mathbb{E}(\varepsilon_t | \mathbf{X}_t) = 0$. Model (1) is a dynamic regression model and the vector $\mathbf{X}_t$ can possibly include lagged explanatory variables and lagged dependent variables.

The function g embodies the systematic part of the stochastic relation between Y_t and $\mathbf{X}_t$ and several (parametric) non-linear time series models have recently been studied and proposed. Particularly, in recent years, the discovery of non-linear movements in the financial markets has been greatly emphasized by various researchers and models of type (1) have become increasingly popular among financial analysts (Clements et al., 2004, inter alia). Examples are the bilinear models, the Self-Exciting Threshold AutoRegression (SETAR) and the Smooth Transition AutoRegression (STAR) (Tong, 1990; Tsay, 2002, for a review).

These are model-driven approaches based on a specific type of relation among the variables which require that the non-linear model be specified before the parameters can be estimated. Use of neural network-based models is an alternative option available to researchers for capturing the underlying non-linearity in financial time series. They are data-driven approaches which do not require a pre-specification during the modelling process because they independently learn the relationship inherent in the variables.

There are several features of the artificial neural network-based models that make them attractive as a modelling and forecasting tool. As opposed to the traditional model-based methods, ANN-based models are data-driven and selfadaptive and, as a consequence, the analysis depends on little *a priori* knowledge about the relationship between input and output variables. This is particularly useful in financial engineering applications where often much is assumed and little is actually known about the nature of the processes determining asset prices.

A neural network model may be interpreted as a non-linear regression function characterizing the relationship between the dependent variable Y_t and a d-vector of explanatory variables $\mathbf{X}_t$. The unknown function g in model (1) can be approximated as:

$$f\left(\mathbf{x}, \mathbf{w}\right) = w_{00} + \sum_{j=1}^{r} w_{0j} \psi\left(\tilde{\mathbf{x}}^T \mathbf{w}_{1j}\right), \tag{2}$$

where $\mathbf{w} \equiv \left(w_{00}, w_{01}, \ldots w_{0r}, \mathbf{w}_{11}^T, \ldots, \mathbf{w}_{1r}^T\right)^T$ is a $r(d+2)+1$ vector of network weights, $\mathbf{w} \in \mathbf{W}$ with $\mathbf{W}$ being a compact subset of $\mathbb{R}^{r(d+2)+1}$, and $\tilde{\mathbf{x}} \equiv \left(1, \mathbf{x}^T\right)^T$ is the input vector augmented by a bias component 1. The network (2) has d input neurons, r neurons in the hidden layer and the identity function for the output layer. The (fixed) hidden unit activation function ψ is chosen in such a way that $f\left(\mathbf{x}, \cdot\right) : \mathbf{W} \to \mathbf{R}$ is continuous for each x in the support of the explanatory variables and $f\left(\cdot, \mathbf{w}\right) : \mathbf{R}^d \to \mathbf{R}$ is measurable for each $\mathbf{w}$ in $\mathbf{W}$.

The main difference with a parametric model is that instead of postulating a specific non-linear function, a neural network model is constructed by combining many "simple" non-linear functions via a multi-layer structure. In a feedforward network, the explanatory variables simultaneously activate r hidden units in an intermediate layer through some function ψ, and the resulting hidden-unit activations $\psi\left(\tilde{\mathbf{x}}^T \mathbf{w}_{1j}\right), j = 1, \ldots, r$, then activate output units to produce the network output.

Single hidden layer feedforward neural networks have a very flexible non-linear functional form (Fine, 1999, for a review). The activation functions can be chosen quite arbitrarily and it can be shown that they can arbitrarily closely approximate (in the appropriate corresponding metric) to continuous, or to p-th power integrable, non-linear functions $g(\cdot)$, so long as the activation function ψ is bounded and satisfies the necessary conditions of not being a polynomial (Leshno et al., 1993).

Moreover, with reasonable assumptions on the activation function ψ, a single hidden layer neural network can arbitrarily closely approximate to $g(\cdot)$ as well its derivatives, up to any given order (provided that they exist), as measured by a proper norm (Hornik et al., 1994). Barron (1993) also shows that a feedforward neural network can achieve an approximation rate that grows linearly in r and thus this class of models is relatively more parsimonious than other non-parametric methods in approximating unknown functions. These two properties make feedforward networks an attractive econometric tool in (non-parametric) applications.

Given a training set of n observations, estimation of the network weights (learning) is obtained by solving the optimization problem

$$\min_{\mathbf{w} \in \mathbf{W}} \frac{1}{n} \sum_{t=1}^{n} q\left(Y_t, f\left(\mathbf{X}_t, \mathbf{w}\right)\right), \tag{3}$$

where $q(\cdot)$ is a proper chosen loss function, usually a quadratic function. From an operational point of view, estimates of the parameter vector $\mathbf{w}$ can be obtained by using non-recursive or recursive estimation methods such as the back-propagation (BP) algorithm and Newton's algorithm.

Under general regularity conditions, Kuan and White (1994) show that both the methods deliver a consistent weight vector estimator. That is, a weight vector $\hat{\mathbf{w}}_n$ solving equation (3) exists and converges almost surely to $\mathbf{w}_0$, which solves

$$\min_{\mathbf{w} \in \mathbf{W}} \int q\left(y, f\left(\mathbf{x}, \mathbf{w}\right)\right) d\pi\left(\mathbf{z}\right) \tag{4}$$

provided that the integral exists and the optimization problem has a unique solution vector interior to $\mathbf{W}$.

This is not necessarily true for neural network models in the absence of appropriate restrictions since the parametrization of the network function is not unique and certain simple symmetry operations applied to the weight vector do not change the value of the output. For a sigmoid activation function ψ centered around 0, these symmetry operations correspond to an exchange of hidden units and to multiplying all weights of connections going into and out of a particular hidden unit by -1. The permutability of hidden units generally results in a non-unique $\mathbf{w}_0$ as there are numerous distinct weight vectors yielding identical network outputs. In any case this may not be a main concern for two reasons. Firstly, several authors provide sufficient conditions to ensure uniqueness of $\mathbf{w}_0$ in a suitable parameter space $\mathbf{W}$ for specific network configurations. Particularly, for the case of sigmoidal activation functions with $\psi(-a) = -\psi(a)$, it is possible to restrict attention only to weight vectors with $w_{01} \geqslant w_{02} \geqslant \ldots \geqslant w_{0r}$ (Ossen and Rügen, 1996). Secondly, several global optimization strategies (simulated annealing, genetic algorithms, etc.) are available to avoid being trapped in local minima and they have been successfully employed in neural network modelling.

Moreover, observe that Newton's algorithm is statistically more efficient than the BP algorithm and is asymptotically equivalent to the NLS method. Although recursive estimates are not as efficient as NLS estimates in finite samples, they are useful when online information processing is important. Moreover, recursive methods can facilitate network selection when the estimation process is based on rolling windows of observations. White (1989) also suggests that recursive estimation can be performed up to a certain time point and then an NLS technique can be applied to improve efficiency.

Finally, under general regularity conditions Kuan and White (1994) show that the weight vector estimator is asymptotically normally distributed.

Therefore, given a consistent estimate of the asymptotic variance covariance matrix, one could be tempted to test hypotheses about the connection strengths which would be of great help in defining pruning strategies with a strong inferential basis. This approach in any case could be misleading. The parameters (weights) of the neural network model have no clear interpretation and this makes this class of models completely different from classical nonlinear parametric models. Moreover, the same output of the network can be obtained with very different configurations of the weights. Finally, as a model selection strategy, variable selection and hidden layer size should follow different schemes: variables have a clear interpretation while hidden layer size has no clear meaning and should be considered a smoothing parameter which is fixed to control the trade-off between bias and variability.

As a consequence, we advocate a model selection strategy where an informative set of input variables is selected by looking at its relevance to the model, in a statistical significance testing framework, while the hidden layer size is selected by looking at the fitting or predictive ability of the network.

3 Variable Selection in Neural Network Models

The general idea behind variable relevance analysis is to compute some measures that can be used to quantify the relevance of variables with respect to a given model. Therefore, to select a proper set of input variables, we focus on a selection rule which involves: (i) definition of the variable's relevance to the model; (ii) estimation of the sampling distribution of the relevance measure; (iii) testing the hypothesis that the variable is irrelevant (Baxt and White, 1995; Refenes and Zapranis, 1999; La Rocca and Perna, 2005b).

Following White and Racine (2001), the hypotheses that the independent variable X_j has no effect on Y, in model (1), can be formulated as:

$$\frac{\partial g(\mathbf{x})}{\partial x_j} = 0, \forall x. \tag{5}$$

Of course the function g is unknown but we equivalently investigate the hypotheses

$$f_j(\mathbf{x}; \mathbf{w}_0) = \frac{\partial f(\mathbf{x}; \mathbf{w}_0)}{\partial x_j} = 0, \forall x \tag{6}$$

since f is known and $\mathbf{w}_0$ can be closely approximated. So, if a given variable X_j has no effect on Y we have $\mathbb{E}\left[f_j^2(\mathbf{x}, \mathbf{w}_0)\right] = 0$, where the square function is used to avoid cancellation effects.

In this perspective, the hypothesis that a given set of variables has no effect on Y can be formulated in a multiple testing framework as

$$H_j : \theta_j = 0 \quad vs \quad H_j' : \theta_j > 0, \quad j = 1, 2, \dots, d, \tag{7}$$

where $\theta_j = \mathbb{E}\left[f_j^2\left(\mathbf{x}, \mathbf{w}_0\right)\right]$. Each null H_j in (7) can be tested by using the statistic

$$\hat{T}_{n,j} = n\hat{\theta}_{n,j}, \tag{8}$$

where

$$\hat{\theta}_{n,j} = n^{-1} \sum_{t=1}^{n} f_j^2\left(\mathbf{X}_t, \hat{\mathbf{w}}_n\right), \tag{9}$$

and the vector $\hat{\mathbf{w}}_n$ is a consistent estimator of the unknown parameter vector $\mathbf{w}_0$. Clearly, large values of the test statistics indicate evidence against the hypothesis H_j.

Thus the problem here is how to decide which hypotheses to reject, taking into account the multitude of tests. In such a context, several approaches have been proposed to control the familywise error rate (FWE), defined as the probability of rejecting at least one of the true null hypotheses. The most familiar multiple testing methods for controlling the FWE are the Bonferroni method and the stepwise procedure proposed by Holm (1979). In any case, both the procedures are conservative since they do not take into account the dependence structure of the individual p-values. These drawbacks can be successfully avoided by using a recent proposal by Romano and Wolf (2005a; 2005b), suitable for joint comparison of multiple (possibly misspecified) models.

The algorithm runs as follows. Relabel the hypothesis from H_{r_1} to H_{r_d} in redescending order with respect to the value of the test statistics $\hat{T}_{n,j}$, that is $\hat{T}_{n,r_1} \geq \hat{T}_{n,r_2} \geq \ldots \geq \hat{T}_{n,r_d}$. In the first stage, the stepdown procedure tests the joint null hypothesis that all hypotheses H_j are true. This hypothesis is rejected if $\hat{T}_{n,r1}$ (the maximum over all the d test statistics) is large, otherwise all hypotheses are accepted. In other words, in the first step the procedure constructs a rectangular joint confidence region for the vector $\left(\theta_{r_1}, \ldots, \theta_{r_d}\right)^T$, with nominal joint coverage probability $1 - \alpha$, of the form $\left[\hat{T}_{n,r_1} - c_1, \infty\right) \times \cdots \times \left[\hat{T}_{n,r_d} - c_1, \infty\right)$. The common value c_1 is chosen to ensure the proper joint (asymptotic) coverage probability. If a particular individual confidence interval $\left[\hat{T}_{n,r_j} - c_1, \infty\right)$ does not contain zero, the corresponding null hypothesis H_{r_s} is rejected. Once a hypothesis is rejected, it is removed and the remaining hypotheses are tested by rejecting for large values of the maximum of the remaining test statistics. If the first R_1 relabelled hypotheses are rejected in the first step, then $d - R_1$ hypotheses remain, corresponding to the labels $r_{R_1+1}, \ldots, r_d$. In the second step, a rectangular joint confidence region for the vector $\left(\theta_{R_1+1}, \ldots, \theta_{r_d}\right)^T$ is constructed with, again, nominal joint coverage probability $1 - \alpha$. The new confidence region is of the form $\left[\hat{T}_{n,r_{R_1+1}} - c_2, \infty\right) \times \cdots \times \left[\hat{T}_{n,r_d} - c_2, \infty\right)$, where the common constant c_2 is chosen to ensure the proper joint (asymptotic) coverage probability. Again, if a particular individual confidence interval $\left[\hat{T}_{n,r_j} - c_2, \infty\right)$

does not contain zero, the corresponding null hypothesis H_{r_j} is rejected. The stepwise process is repeated until no further hypotheses are rejected.

Given the probabilistic complexity of the neural network model which makes the use of analytic procedures very difficult, estimation of the quantile of order $1 - \alpha$ is obtained by using subsampling.

The resampling scheme runs as follows. Fix b, the subseries length, and let $\hat{\theta}_{b,j,t} = \hat{\theta}_b (\mathbf{Z}_t, \ldots, \mathbf{Z}_{t+b-1})$ be an estimator of θ_j based on the subseries $(\mathbf{Z}_t, \ldots, \mathbf{Z}_{t+b-1})$. Obtain replicates of $\hat{T}_{b,j,t} = b\hat{\theta}_{b,j,t}$ with $t = 1, \ldots, n - b + 1$.

Then, for $\mathbf{x} \in \mathbb{R}^d$, the true joint cdf of the test statistics evaluated at $\mathbf{x}$ is given by

$$G_n (\mathbf{x}) = \Pr \left\{ \hat{T}_{n,1} \leq x_1, \hat{T}_{n,2} \leq x_2 \ldots, \hat{T}_{n,d} \leq x_d \right\}, \tag{10}$$

and it can be estimated by the subsampling approximation

$$\hat{G}_n (\mathbf{x}) = \frac{1}{n - b + 1} \sum_{t=1}^{n-b+1} \mathbb{I} \left\{ \hat{T}_{b,1,t} \leq x_1, \hat{T}_{b,2,t} \leq x_2, \ldots, \hat{T}_{b,d,t} \leq x_d \right\}, \tag{11}$$

where as usual $\mathbb{I}(\cdot)$ denotes the indicator function.

As a consequence, for $D \subset \{1, \ldots, d\}$, the distribution of the maximum of the test statistics, denoted by $H_{n,D} (x)$, can be estimated by the empirical distribution function $\hat{H}_{n,D} (x)$ of the values $\max \left\{ \hat{T}_{b,t,j}, j \in D \right\}$, that is

$$\hat{H}_{n,D} (x) = \frac{1}{n - b + 1} \sum_{t=1}^{n-b+1} \mathbb{I} \left\{ \max \left\{ \hat{T}_{b,j,t}, j \in D \right\} \leq x \right\}, \tag{12}$$

and the quantile of order $1 - \alpha$ can be estimated as

$$\hat{c}_L (1 - \alpha) = \inf \left\{ x : \hat{H}_{n,D} (x) \geq 1 - \alpha \right\}. \tag{13}$$

The procedure for multiple testing is described in Algorithm 1, where the quantile of order $1 - \alpha$ is estimated by using the subsampling.

The choice of subsampling as a resampling technique is justified for several reasons. Firstly, the method does not require any knowledge of the specific structures of the data and so it is robust against misspecified models, a key property when dealing with artificial neural network models which are intrinsically misspecified. Secondly, the procedure delivers consistent results under very weak assumptions. In our case, by assuming: (i) $b \to \infty$ in such a way that $\frac{b}{n} \to 0$, as $n \to \infty$, (ii) conditions that guarantee asymptotic normality of $\hat{\mathbf{w}}_n$ (Kuan and White, 1994), (iii) smoothness conditions on the test statistics $\hat{T}_{n,j}$ (La Rocca and Perna, 2005b), the subsampling approximation is a consistent estimate of the unknown (multivariate) sampling distribution of the test statistics (Romano and Wolf, 2005a, theorem 8). Finally, subsampling does not change dramatically when moving from *iid* to dependent data. This approach can be used even if the data are *iid*, making the procedure robust

---

Algorithm 1: Multiple testing algorithm.

1: Relabel the hypothesis from H_{r_1} to H_{r_d} in redescending order of the value
    of the test statistics $\hat{T}_{n,j}$, that is $\hat{T}_{n,r_1} \geq \hat{T}_{n,r_2} \geq \ldots \geq \hat{T}_{n,r_d}$.

2: Set $L = 1$ and $R_0 = 0$.

3: **for** $j = R_{L-1} + 1$ to d **do**

4:    **if** $0 \notin \left[T_{n,r_j} - \hat{c}_L\left(1 - \alpha\right), \infty\right)$ **then**

5:       reject H_{r_j}

6:    **end if**

7: **end for**

8: **if** no (further) null hypothesis are rejected **then**

9:    Stop

10: **else**

11:    R_L = number of rejected hypothesis

12:    $L = L + 1$

13:    Go to step 3

14: **end if**

---

against misspecification of this assumption. In fact, for *iid* data the number of subsets of length b which can be formed out of a sample of size n grows very fast with n. Therefore usually, just B random selected subsets are considered for computing the subsampling approximation.

Clearly, the main issue when applying the subsampling procedure lies in choosing the length of the block, a problem which is common to all blockwise resampling techniques. Even if the conditions on b are quite weak, they give no guidelines on its choice and this parameter has to be chosen from the data to hand. Nevertheless, Politis et al. (1999) proposed a number of strategies to select b and theorems that ensure that the asymptotic results are still valid for a broad range of choices for subsample size.

Finally, we remark that as an alternative to subsampling, moving block bootstrap resampling schemes may be employed. However, they require much stronger conditions for consistency (Romano and Wolf, 2005a) and generalizations to more complex data structures, such as nonstationary time series, are not as straightforward as in the case of subsampling (Politis et al., 1999). The derivation of conditions for consistency of moving block schemes in our framework and a comparison with subsampling are still under investigation.

4 Numerical Examples and Monte Carlo Results

To illustrate the performance of the proposed model selection procedure we use simulated data sets generated by models with known structure. The simulated data sets were generated by the following models.

The first DGP (Model M1) is an AR(2) model defined as $Y_t = 0.5Y_{t-1} + 0.5Y_{t-2} + \varepsilon_t$ where ε_t is a Gaussian innovation term with zero mean and variance equal to one.

The second DGP (Model M2) is a neural network with 2 input neurons (lags 1 and 2) and 2 hidden neurons, defined as $Y_t = 1 + \tanh(h_{t1}) + \tanh(h_{t2}) + \varepsilon_t$ where $h_{t1} = 2.7Y_{t-1} + 0.7Y_{t-2} - 1.3$ and $h_{t2} = 0.4Y_{t-1} - 1.9Y_{t-2} - 0.4$.

The third DGP (Model M3) is a neural network with 2 input neurons (lags 2 and 4) and 2 hidden neurons, defined as $Y_t = 1 + \tanh(h_{t1}) + \tanh(h_{t2}) + \varepsilon_t$ where $h_{t1} = 0.7Y_{t-2} - 2.3Y_{t-4} + 0.9$ and $h_{t2} = -2.4Y_{t-2} - 1.9Y_{t-4} + 0.5$.

The last DGP (Model M4) is an LSTAR model of order 1 defined as $Y_t = -0.3 - 0.5Y_{t-1}(1 - \mathcal{G}(Y_{t-1})) + 0.1 + 0.5Y_{t-1}\mathcal{G}(Y_{t-1}) + \varepsilon_t$ where $\mathcal{G}$ is the logistic cumulative distribution function.

To illustrate how the proposed procedure works, we apply it to synthetic time series generated by models M1, M2, M3 and M4. We started the multiple testing procedure for variable selection with tentative neural network models with 6 input neurons associated with the first 6 lags of Y_t. We expect the procedure to be able to select the proper relevant lags for each DGP. For this example, we set $n = 1000$, $b = 250$ and $\alpha = 0.05$. The hidden layer size was determined by using the Schwartz information criterion (SIC) and all the neural networks were estimated by using a square loss function in equation (3). The software procedures were implemented in R. The results are reported in Table 1.

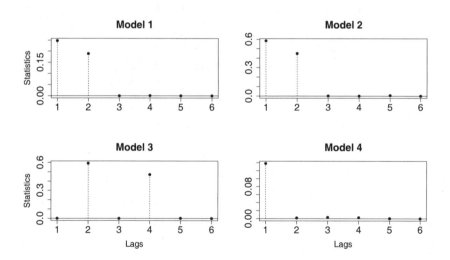

Fig. 1. Values of the statistics T_n at different lags for the models M1(AR of order 2), M2 (ANN with significant lags 1 and 2), M3 (ANN with significant lags 2 and 4), M4 (LSTAR of order 1).

For model M1, after the first step, the procedure rejects the hypothesis that lags 1 and 2 are not relevant and accepts all other hypotheses. At the second step, the remaining lags (3, 4, 5 and 6) are recognized as not relevant and the procedure stops. The results are consistent with the data generating

Table 1. Results of the multiple testing procedure ($n = 1000$, $b = 250$, $\alpha = 0.05$). Figures in bold refer to the rejection of the corresponding hypotheses H_{r_j}. M1(AR of order 2), M2 (ANN with significant lags 1 and 2), M3 (ANN with significant lags 2 and 4), M4 (LSTAR of order 1).

Model	j	$\hat{T}_{n,r_j}$	r_j	$\hat{T}_{n,r_j} - \hat{c}_1$	$\hat{T}_{n,r_j} - \hat{c}_2$
M1	1	**248.189**	1	**189.095**	–
	2	**189.095**	2	**117.146**	–
	3	0.967	4	-70.983	-5.460
	4	0.245	5	-71.704	-6.182
	5	0.187	6	-71.763	-6.241
	6	0.045	3	-71.904	-6.382
M2	1	**583.921**	1	**421.801**	–
	2	**448.842**	2	**286.721**	–
	3	5.786	5	-156.335	-8.582
	4	1.974	3	-160.147	-12.394
	5	0.521	4	-161.599	-13.847
	6	0.452	6	-161.669	-13.916
M3	1	**592.394**	2	**396.316**	–
	2	**471.169**	4	**275.091**	–
	3	1.890	5	−194.188	-6.905
	4	1.789	1	-194.289	-7.006
	5	0.935	3	-195.143	-7.860
	6	0.129	6	-195.949	-8.666
M4	1	**137.973**	1	**50.767**	–
	2	3.478	3	-83.728	-30.167
	3	2.753	4	-84.453	-30.892
	4	1.846	2	-85.360	-31.799
	5	0.462	5	-86.744	-33.183
	6	0.105	6	−87.101	-33.540

process and with the plot reported in Figure 1. Similar conclusions can be drawn for the remaining models.

A moderate Monte Carlo experiment was conducted to evaluate the performance of the proposed procedure in time series of finite length. The design experiment considered four models (M1, M2, M3 and M4), two time series lengths ($n = 1000$ and $n = 2000$), four subseries lengths ($b = 200, 300, 400, 500$ for $n = 1000$ and $b = 250, 500, 750, 1000$ for $n = 2000$). The nominal level of the test procedure was set at 0.05. Due to the nature of the procedure, which is heavily computer intensive, the number of Monte Carlo runs was limited to 200. For the same reason the hidden layer size was not estimated on each series but was fixed for each model. On each series the neural network estimation procedure was repeated 50 times (with random starting values) in order to avoid being trapped in local minima. Moreover, to reduce the computation burden, the neural network models on each subseries were estimated by using as starting values the parameter estimates obtained on the whole series. This approach is analogous to that used by Refenes and Zapranis (1999) for the bootstrap. The results are reported in Tables 2 and 3.

Table 2. Simulations results of the multiple testing procedure ($n = 1000$, $\alpha = 0.05$, Monte Carlo runs=200). Proportions of incorrect identification for lags $1, 2, \ldots, 6$ and observed FWE for models M1 (AR of order 2), M2 (ANN with significant lags 1 and 2), M3 (ANN with significant lags 2 and 4), M4 (LSTAR of order 1). Figures in bold refer to significant lags.

Model	b	1	2	3	4	5	6	FWE
M1	200	**0.000**	**0.000**	0.015	0.025	0.020	0.010	0.070
	300	**0.000**	**0.000**	0.020	0.025	0.015	0.005	0.065
	400	**0.000**	**0.000**	0.020	0.020	0.035	0.010	0.085
	500	**0.000**	**0.000**	0.030	0.030	0.020	0.015	0.095
M2	200	**0.000**	**0.000**	0.000	0.020	0.000	0.000	0.020
	300	**0.000**	**0.000**	0.005	0.025	0.005	0.000	0.035
	400	**0.000**	**0.000**	0.010	0.025	0.005	0.005	0.045
	500	**0.000**	**0.000**	0.010	0.025	0.015	0.005	0.055
M3	200	0.005	**0.000**	0.000	**0.000**	0.000	0.005	0.010
	300	0.020	**0.000**	0.000	**0.000**	0.000	0.005	0.025
	400	0.025	**0.000**	0.000	**0.000**	0.005	0.005	0.035
	500	0.040	**0.000**	0.000	**0.000**	0.015	0.035	0.090
M4	200	**0.075**	0.000	0.000	0.000	0.000	0.000	0.000
	300	**0.050**	0.000	0.000	0.000	0.005	0.000	0.005
	400	**0.090**	0.005	0.005	0.000	0.005	0.000	0.015
	500	**0.075**	0.010	0.005	0.000	0.005	0.010	0.030

Clearly, the procedure seems able to identify the lag structure of the true DGP correctly. For models M1, M2 and M3 the relevant lags were identified in all the Monte Carlo runs. Just in the case of model M4, for $n = 1000$, the percentages are lower than 100%, but in any case they are greater than 91%. The percentages of incorrect lag identification in all the cases considered are very low. The observed FWE is basically close to the nominal level for all models and for all subseries lengths.

Importantly, even if the number of Monte Carlo runs is not very high, there is a slight, monotone, improvement in the FWE when the subseries length increases. This suggests the possible use of the subseries length b as a tuning parameter, used to calibrate the testing scheme in order to attain an observed coverage close to the nominal one (Kokoszka and Wolf, 2004). Since the approach would involve the use of the bootstrap to estimate the observed FWE on the given time series, at the end there is nesting of three very computer-intensive techniques (neural networks, subsampling and bootstrap) which could make the procedure computationally unfeasible.

However, it is worth noting that all the three mentioned schemes have an obvious parallel structure. They can be efficiently implemented on cluster

Table 3. Simulation results of the multiple testing procedure ($n = 2000$, $\alpha = 0.05$, Monte Carlo runs=200). Proportions of incorrect identification for lags $1, 2, \ldots, 6$ and observed FWE for models M1 (AR of order 2), M2 (ANN with significant lags 1 and 2), M3 (ANN with significant lags 2 and 4), M4 (LSTAR of order 1). Figures in bold refer to significant lags.

Model	b	1	2	3	4	5	6	FWE
M1	250	**0.000**	**0.000**	0.015	0.025	0.020	0.010	0.070
	500	**0.000**	**0.000**	0.010	0.035	0.010	0.010	0.065
	750	**0.000**	**0.000**	0.005	0.030	0.010	0.035	0.080
	1000	**0.000**	**0.000**	0.005	0.040	0.020	0.015	0.080
M2	250	**0.000**	**0.000**	0.005	0.000	0.000	0.000	0.005
	500	**0.000**	**0.000**	0.010	0.020	0.005	0.005	0.040
	750	**0.000**	**0.000**	0.015	0.040	0.020	0.010	0.085
	1000	**0.000**	**0.000**	0.020	0.035	0.050	0.010	0.115
M3	250	0.000	**0.000**	0.000	**0.000**	0.000	0.010	0.010
	500	0.005	**0.000**	0.000	**0.000**	0.010	0.010	0.025
	750	0.005	**0.000**	0.010	**0.000**	0.015	0.025	0.055
	1000	0.020	**0.000**	0.005	**0.000**	0.015	0.030	0.070
M4	250	**0.000**	0.000	0.000	0.000	0.000	0.000	0.000
	500	**0.000**	0.015	0.000	0.000	0.010	0.005	0.030
	750	**0.000**	0.020	0.005	0.000	0.015	0.005	0.045
	1000	**0.000**	0.020	0.015	0.005	0.030	0.005	0.075

computers with low parallel overhead since all networks can be independently estimated on the subseries generated by subsampling. It could therefore be possible to exploit the multicore structure of modern computers with a scaling factor expected to be close to one when increasing the number of processing units. These developments are still under investigation.

5 An Application to Euro Exchange Rates

Forecasting and modelling currency exchange rates is an important financial problem that has received much attention especially for its intrinsic difficulty and its practical applications. The statistical distribution of foreign exchange rates and their linear unpredictability have a long history and are recurrent themes in the literature of international finance. The market efficiency hypothesis in its weak form when applied to foreign exchange rates means that there is no information contained in past percentage changes that is relevant for the prediction of future percentage changes. Failure of econometric models and models based on linear time series techniques to deliver superior forecasts to the simple random walk model (Meese and Rogoff, 1983), have forced researchers to use various non-linear techniques. A number of non-linear time series models have been proposed in the recent past for obtaining accurate prediction results in an attempt to build models able to outperform the simple random walk models (Nag and Mitra, 2002; Gencay, 1999; Gradojevic, 2006).

In this exercise, in contrast to approaches that compare out-of-sample predictions of non-linear models to those generated by the random walk model, we again follow White and Racine (2001) and focus on directly testing for unpredictability by conducting tests of significance on models inspired by technical trading rules. Following a standard practice, we construct percentage exchange rate changes ($P_t = \nabla \log S_t$) avoiding potential problems associated with estimation of nonstationary regression functions. Observe that the transformed exchange rate series P_t may not in fact be *iid* series but the subsampling scheme described in Section 3 can successfully cope with this problem.

The non-linear models reported in the literature for exchange rates time series almost always include lagged returns as explanatory variables. Here, we consider lag structures of maximum five lags, since Diebold and Nason (1990) findings were unaffected by using different lag structures. For a given country, we therefore estimate (as a tentative model) the non-linear model

$$P_t = g\left(P_{t-1}, P_{t-2}, P_{t-3}, P_{t-4}, P_{t-5}\right) \tag{14}$$

by using a feedforward neural network with 5 input neurons. The number of hidden units (so-called network complexity) is determined via Schwartz information criterion. It is allowed to range from one to ten hidden units and is computed by using 100 restarts from random initial weights for each network architecture.

The relevance of each lag in equation (14) is tested by using the procedure described in Section 3 where the subseries length is allowed to vary from 250 to 1000 in steps of 250.

The investigation is based on 20 time series of Euro foreign exchange reference rates, as reported by the European Central Bank. The data include both European and non-European countries. The European currencies are the Swiss franc, Cyprus pound, Czech koruna, Danish krone, Pound sterling, Hungarian forint, Icelandic krona, Norwegian krone, Polish zloty, Swedish krona, Slovak koruna while the non-European currencies are the Australian dollar, Canadian dollar, Hong Kong dollar, Japanese yen, South Korean won, New Zealand dollar, Singapore dollar, US dollar and South African rand.

All series span the period from 4 January 1999 to 27 March 2007, totalling 2110 observations. The time series obtained are presented in Table 4 along with their mnemonics.

The summary statistics of the time series considered are reported in Table 5 highlighting slight skewness and high kurtosis. Almost all the series show some degree of non-linearity; the p-value of the Teräsvirta linearity test (Teräsvirta et al., 1993) suggests rejection of the null (at the level equal to 0.05) for all the currencies considered but the US dollar. These results are consistent with those reported in the literature (Brooks, 1996, inter alia) and so the natural question is whether any non-linearity can be exploited to improve

Table 4. Euro exchange rate time series and mnemonics.

Number	Mnemonics	Currency	Number	Mnemonics	Currency
1	CHF	Swiss franc	12	AUD	Australian dollar
2	CYP	Cyprus pound	13	CAD	Canadian dollar
3	CZK	Czech koruna	14	HKD	Hong Kong dollar
4	DKK	Danish krone	15	JPY	Japanese yen
5	GBP	Pound sterling	16	KRW	South Korean won
6	HUF	Hungarian forint	17	NZD	New Zealand dollar
7	ISK	Icelandic krona	18	SGD	Singapore dollar
8	NOK	Norwegian krone	19	USD	US dollar
9	PLN	Polish zloty	20	ZAR	South African rand
10	SEK	Swedish krona			
11	SKK	Slovak koruna			

the model performance. However, parametric non-linear time series models do not seem able to give satisfactory results (Brooks, 1997) while neural networks are favourable tools for successful modelling and forecasting foreign exchange rate time series (Kuan and Liu, 1995, inter alia).

Table 5. Teräsvista linearity test and descriptive statistics for Euro exchange rates.

Currency	Teräsvirta	p-value	min	max	mean	sd	skewness	kurtosis
CHF	188.1758	0.0000	-0.0186	0.0143	0.0000	0.0022	-0.3479	5.5336
CYP	495.7273	0.0000	-0.0107	0.0112	0.000	0.0006	0.0342	119.8337
CZK	142.0412	0.0000	-0.0205	0.0258	-0.0001	0.0034	0.2755	5.1337
DKK	86.4710	0.0010	-0.0011	0.0008	0.0000	0.0002	-0.2241	3.4263
GBP	143.4503	0.0000	-0.0201	0.0287	0.0000	0.0044	0.4197	2.2340
HUF	116.0926	0.0000	-0.0195	0.0478	0.0000	0.0043	1.7630	17.4903
ISK	137.6827	0.0000	-0.0312	0.0439	0.0000	0.0059	0.4180	5.3237
NOK	75.4195	0.0116	-0.0120	0.0222	0.0000	0.0036	0.4459	1.8544
PLN	182.7088	0.0000	-0.0323	0.0414	0.0000	0.0064	0.5964	4.0101
SEK	115.1018	0.0000	-0.0193	0.0187	0.0000	0.0035	0.0032	2.2520
SKK	170.8890	0.0000	-0.0318	0.0185	-0.0001	0.0030	-0.1092	9.8112
AUD	99.4415	0.0000	-0.0284	0.0350	-0.0001	0.0063	0.2327	2.2003
CAD	75.2091	0.0121	-0.0232	0.0448	-0.0001	0.0066	0.4406	2.4853
HKD	69.6243	0.0346	-0.0211	0.0427	0.0001	0.0063	0.2769	1.4650
JPY	104.2077	0.0000	-0.0458	0.0540	0.0001	0.0071	-0.0085	3.9137
KRW	102.9050	0.0000	-0.0266	0.0497	-0.0001	0.0071	0.3804	2.2796
NZD	85.0126	0.0015	-0.0288	0.0336	-0.0001	0.0070	0.2717	1.6084
SGD	82.1992	0.0028	-0.0212	0.0445	0.0000	0.0058	0.3454	2.6225
USD	64.6933	0.0791	-0.0225	0.0420	0.0001	0.0063	0.2727	1.4207
ZAR	286.2377	0.0000	-0.0849	0.0781	0.0002	0.0104	0.5497	7.8242

When using neural networks in foreign exchange rate modelling, Franses and van Griensven (1997) have shown that the choice of explanatory variables appears more important than the choice of the hidden layer size. Moreover, the presence of data snooping can be a serious drawback when modelling this kind of data (Qi and Wu, 2006). Therefore, we believe the proposed multistep procedure could be successfully employed with some advantages over existing techniques.

The results of the procedure are reported in Tables 6–11 in the appendix. There is evidence (at a significance level equal to 0.05) that information contained in past percentage changes is relevant for the prediction of future percentage changes for four out of the twenty countries: Australia, Norway, Canada and Singapore (see Table 8–11). By contrast, for no other currency are significant lags detected by the procedure (see Table 6 and 7). It is interesting to note the result of the Norwegian krone (see Table 11) where lag one was not detected as relevant in the first step and included as relevant in the second step. This stress the advantages of multiple testing procedures over single step ones such as the bootstrap reality check (White, 2000).

The apparent predictability for these countries does not seem to be an artifact of data snooping since it is the result of a testing procedure which is able to keep the familywise error rate under control. Moreover, these conclusions remain stable while changing the subseries length.

Finally, more complex model structures could be easily considered in the described framework, by adding (and testing) as explanatory variables other economic variables (interest rates, etc.) known to have some impact on foreign exchange rates. The relevance of each new explanatory (possibly) lagged variable could be determined along the same lines as those described here. In any case this analysis would be beyond the scope of this paper.

6 Conclusions

In this paper we discussed a multistep testing procedure to select a proper set of inputs in neural network dynamic regression models out of a set (possibly vast) of candidate explanatory variables. The proposed procedure allows data snooping to be undertaken with some degree of confidence that one will not mistake results (sets of relevant variables) that might be generated by pure chance for genuinely good ones. The test is calibrated by using subsampling which gives consistent results under very weak assumptions.

In a simulation study with time series generated by nonlinear autoregressive models, the procedure seems able to identify the lag structure of the true DGP correctly. Moreover, the results of a Monte Carlo experiment show that the observed FWE is close to the nominal level for all models and for all subseries lengths. Thus, subsampling-based inference could be a valuable additional tool to neural network practitioners, allowing the use of inference-based techniques in network modelling where analytical results, if available, are very difficult to derive.

Finally, we applied the novel method to test for predictive power in Euro exchange rates. There is evidence that there does appear to be predictive information in past percentage changes that may be exploited for better point forecasting particularly for Australia, Norway, Canada and Singapore. However, the evolution of foreign exchange market structure implies that the particular forms of these predictive relationships can be expected to have limited life spans.

Acknowledgements

We wish to thank two anonymous referees for insightful and stimulating comments and suggestions. We are also grateful to the Editors who gave us the opportunity to express our intellectual gratitude to Manfred Gilli, to whom we feel indebted for his profound influence in our field.

References

Alvarez-Diaz, M. and Alvarez, A.: 2007, Forecasting exchange rates using an evolutionary neural network, *Applied Financial Economics Letters* **3**, 1744–6554.

Barron, A.R.: 1993, Universal approximation bounds for superposition of a sigmoidal function, *IEEE Transactions on Information Theory* **39**(3), 930–945.

Baxt, W.G. and White, H.: 1995, Bootstrapping confidence intervals for clinical input variable effects in a network trained to identify the presence of acute myocardial infraction, *Neural Computation* **7**, 624–638.

Brooks, C.: 1996, Testing for nonlinearity in daily sterling exchange rates, *Applied Finacial Economics* **6**, 307–317.

Brooks, C.: 1997, Linear and nonlinear (non-)forecastability of high-frequency exchange rates, *Journal of Forecasting* **16**, 125–145.

Clements, M.P., Franses, P.H. and Swanson, M.N.: 2004, Forecasting economic and financial time-series with non-linear models, *International Journal of Forecasting* **20**, 169–183.

Cybenko, G.: 1989, Approximation by superposition of sigmoidal functions, *Math. Contr. Signals Syst.* **2**, 303–314.

Diebold, F.X. and Nason, J.A.: 1990, Nonparametric exchange rate prediction, *J. Int. Economics* **28**, 315–332.

Fine, T.L.: 1999, *Feedforward Neural Network Methodology*, Springer.

Franses, P. H. and van Griensven, K.: 1997, Forecasting foreign exchange rates using neural networks for technical trading rules, *Studies in Nonlinear Dynamics and Econometrics* **2**, 109–114.

Gencay, R.: 1999, Linear, nonlinear and essential foreign exchange rate prediction with simple technical trading rules, *Journal of International Economics* **47**, 91–107.

Giacomini, R. and White, H.: 2006, Tests of conditional predictive ability, *Econometrica* **74**, 1545–1578.

Giordano, F., La Rocca, M. and Perna, C.: 2004, Bootstrap variable selection in neural network regression models, *in* H.-H. Bock, M. Chiodi and A. Mineo (eds), *Advances in Multivariate Data Analysis*, Springer, Heidelberg, pp. 109–120.

Gradojevic, N.: 2006, Nonlinear, hybrid exchange rate modeling and trading profitability in the foreign exchange market, *Journal of Economic Dynamics and Control* **31**, 557–574.

Holm, S.: 1979, A simple sequentially rejective multiple test procedure, *Scandinavian Journal of Statistics* **6**, 65–70.

Hornik, K., Stinchcombe, M., White, H. and Auer, P.: 1994, Degree of approximation results for feedforward networks approximating unknown mappings and their derivatives, *Neural Computation* **6**, 1262–1275.

Kokoszka, P. and Wolf, M.: 2004, Subsampling the mean of heavy-tailed dependent observations, *Journal of Time Series Analysis* **25**(217–234).

Kuan, C.M. and Liu, T.: 1995, Forecasting exchange rates using feedforward networks and recurrent neural networks, *Journal of Applied Econometrics* **10**, 347–364.

Kuan, C.M. and White, H.: 1994, Artificial neural networks: an econometric perspective, *Econometric Reviews* **13**, 1–91.

La Rocca, M. and Perna, C.: 2005a, Neural network modeling by subsampling, *in* J. Cabestany, A. Prieto and F. Sandoval (eds), *Computational Intelligence and Bioinspired Systems*, Vol. 3512 of *Lecture Notes in Computer Science*, Springer, pp. 200–207.

La Rocca, M. and Perna, C.: 2005b, Variable selection in neural network regression models with dependent data: a subsampling approach, *Computational Statistics and Data Analyis* **48**, 415–419.

Lam, M.: 2004, Neural network techniques for financial performance prediction: integrating fundamental and technical analysis, *Decision Support Systems* **37**, 567–581.

Leshno, M., Lin, V., Pinkus, A. and Schocken, S.: 1993, Multilayer feedforward networks with a nonpolynomial activation function can approximate any function, *Neural Networks* **6**, 861–867.

Meese, R. and Rogoff, K.: 1983, Empirical exchange rate models of the seventies. Do they fit out of sample?, *Journal of International Economics* **14**, 3–24.

Nag, A. K. and Mitra, A.: 2002, Forecasting daily foreign exchange rates using genetically optimized neural networks, *Journal of Forecasting* **21**, 501–511.

Ossen, A. and Rügen, S. M.: 1996, An analysis of the metric structure of the weight space of feedforward networks and its application to time series modelling and prediction, *Proceedings of the 4th European Symposium on Artificial Neural Networks (ESANN96)*, Brussels.

Perna, C. and Giordano, F.: 2001, The hidden layer size in feedforward neural networks: a statistical point of view, *Metron* **LIX**, 217–227.

Politis, D.N., Romano, J.P. and Wolf, M.: 1999, *Subsampling*, Springer.

Qi, M.: 1996, Financial applications of artificial neural networks, *in* G.S. Maddala and C.R. Rao (eds), *Handbook of Statistics*, Vol. 14, North Holland, pp. 529–552.

Qi, M. and Maddala, G.S.: 1996, Option pricing using artificial neural networks: The case of S&P 500 index call options, *in* A.P.N. Refenes, Y. Abu-Mostafa, J. Moody and A. Weigend (eds), *Neural Networks in Financial Engineering: Proceedings of the Third International Conference on Neural Networks in the Capital Markets*, World Scientific, New York, pp. 78–91.

Qi, M. and Wu, Y.: 2006, Technical trading-rule profitability, data snooping, and reality check: evidence from the foreign exchange market, *Journal of Money, Credit and Banking* **38**, 2135–2158.

Refenes, A., Burgess, A. and Bentz, Y.: 1997, Neural networks in financial engineering: a study in methodology, *IEEE Transactions on Neural Networks* **8**, 1222–1267.

Refenes, A.P.N. and Zapranis, A.D.: 1999, Neural model identification, variable selection and model adequacy, *Journal of Forecasting* **18**, 299–332.

Romano, J.P. and Wolf, M.: 2005a, Exact and approximate stepdown methods for multiple hypothesis testing, *JASA* **100**, 94–108.

Romano, J.P. and Wolf, M.: 2005b, Stepwise multiple testing as stepwise multiple testing as formalized data snooping, *Econometrica* **73**, 1237–1282.

Teräsvirta, T., Lin, C.F. and Granger, C.W.J.: 1993, Power of the neural network linearity test, *Journal of Time Series Analysis* **14**, 209–220.

Thawornwong, S. and Enke, D.: 2004, The adaptive selection of financial and economic variables for use with artificial neural networks, *Neurocomputing* **56**, 205–232.

Tong, H.: 1990, *Non-linear time series: a dynamical system approach*, Oxford University Press.

Tsay, R.: 2002, *Analysis of financial time series*, Wiley, NY.

West, D., Dellana, S. and Qian, J.: 2005, Neural network ensemble strategies for financial decision applications, *Computers & Operations Research* **32**, 2543–2559.

White, H.: 1989, Learning in artificial neural networks: a statistical prespective, *Neural Computation* **1**, 425–464.

White, H.: 2000, A reality check for data snooping, *Econometrica* **68**, 1097–1127.

White, H. and Racine, J.: 2001, Statistical inference, the bootstrap, and neural network modeling with application to foreign exchange rates, *IEEE Transactions on Neural Networks* **12**, 657–673.

Appendix

Table 6. Currencies of European countries. Multiple testing results for the model $P_t = g\left(P_{t-1}, P_{t-2}, P_{t-3}, P_{t-4}, P_{t-5}\right)$ estimated by using an ANN with 5 input neurons and hidden layer size selected via SIC.

Currency	Size	r_j	T_n	$T_n - \hat{c}_1$ (250)	$T_n - \hat{c}_1$ (500)	$T_n - \hat{c}_1$ (750)	$T_n - \hat{c}_1$ (1000)
CHF	1	3	6.266	-4.175	-0.473	-0.102	-3.112
		4	0.511	-9.930	-6.861	-5.038	-9.484
		2	0.157	-10.284	-7.304	-5.135	-9.899
		1	0.140	-10.301	-7.309	-5.616	-9.912
		5	0.125	-10.317	-7.328	-6.250	-9.934
CYP	2	1	25.535	-29.796	-28.010	-29.496	-16.326
		3	22.220	-33.112	-31.337	-32.819	-19.648
		4	18.671	-36.660	-34.892	-36.371	-23.195
		2	15.677	-39.654	-37.872	-39.357	-26.178
		5	0.760	-54.571	-52.797	-54.274	-41.099
CZK	1	5	8.248	-0.156	-1.652	-2.292	-2.350
		2	6.198	-2.206	-3.733	-4.384	-4.657
		4	2.643	-5.761	-7.279	-8.101	-8.278
		3	1.813	-6.591	-8.169	-9.001	-9.187
		1	0.001	-8.403	-10.373	-10.906	-11.277
DKK	1	1	10.132	-4.151	-7.104	-6.215	-1.632
		4	1.233	-13.050	-15.923	-15.099	-10.513
		5	0.426	-13.857	-16.768	-15.924	-11.318
		2	0.123	-14.160	-17.075	-16.227	-11.629
		3	0.073	-14.210	-17.119	-16.277	-11.679
GBP	1	2	0.984	-9.500	-10.410	-10.627	-9.254
		5	0.289	-10.195	-11.110	-11.327	-9.954
		1	0.275	-10.210	-11.126	-11.344	-9.970
		3	0.066	-10.419	-11.334	-11.551	-10.176
		4	0.050	-10.435	-11.352	-11.569	-10.195
HUF	1	2	8.221	-5.773	-7.595	-2.628	-3.969
		5	3.392	-10.603	-12.893	-7.897	-9.217
		1	1.095	-12.899	-14.887	-9.981	-11.173
		4	0.189	-13.805	-15.883	-10.964	-12.174
		3	0.127	-13.867	-15.911	-10.986	-12.210
ISK	2	4	17.952	-22.789	-16.635	-15.331	-23.779
		3	17.928	-22.813	-16.643	-15.352	-23.847
		1	16.512	-24.229	-18.030	-16.718	-25.217
		5	12.001	-28.740	-22.512	-21.281	-29.793

continued on next page

continued from previous page

Currency	Size	r_j	T_n	$T_n - \hat{c}_1$ (250)	$T_n - \hat{c}_1$ (500)	$T_n - \hat{c}_1$ (750)	$T_n - \hat{c}_1$ (1000)
		2	4.805	-35.936	-29.725	-28.446	-36.951
PLN	1	5	8.980	-3.641	-8.406	-1.171	-0.495
		4	2.443	-10.178	-14.692	-7.828	-7.349
		1	2.376	-10.245	-14.900	-7.947	-7.357
		2	0.927	-11.694	-16.249	-9.452	-8.893
		3	0.628	-11.993	-16.545	-9.744	-9.216
SEK	1	3	8.980	-3.641	-11.062	-8.126	-7.813
		2	2.443	-10.178	-11.593	-8.650	-8.317
		1	2.376	-10.245	-12.408	-9.468	-9.14
		4	0.927	-11.694	-13.022	-10.083	-9.759
		5	0.628	-11.993	-14.465	-11.526	-11.20
SKK	1	3	11.100	-1.266	-6.237	-8.649	-11.359
		4	5.969	-6.396	-6.660	-13.783	-16.497
		1	0.432	-11.933	-11.267	-19.336	-22.030
		2	0.067	-12.299	-12.141	-19.694	-22.396
		5	0.005	-12.360	-12.676	-19.753	-22.455

Table 7. Currencies of non European countries. Multiple testing results for the model $P_t = g\left(P_{t-1}, P_{t-2}, P_{t-3}, P_{t-4}, P_{t-5}\right)$ estimated by using an ANN with 5 input neurons and hidden layer size selected via SIC.

Currency	Size	r_j	T_n	$T_n - \hat{c}_1$ (250)	$T_n - \hat{c}_1$ (500)	$T_n - \hat{c}_1$ (750)	$T_n - \hat{c}_1$ (1000)
HKD	2	4	12.931	-14.108	-20.005	-8.927	-10.158
		5	5.488	-21.551	-27.406	-16.457	-17.597
		2	5.288	-21.751	-27.669	-16.643	-17.836
		3	1.755	-25.285	-31.113	-20.115	-21.298
		1	0.292	-26.747	-32.620	-21.633	-22.81
JPY	1	2	1.894	-5.413	-2.725	-2.113	-0.982
		5	1.154	-6.153	-3.486	-2.882	-1.724
		4	0.025	-7.282	-4.610	-4.024	-2.866
		3	0.019	-7.288	-4.616	-4.030	-2.871
		1	0.002	-7.305	-4.632	-4.046	-2.887
KRW	1	3	6.587	-9.180	-16.074	-7.734	-1.245
		4	4.072	-11.695	-18.558	-10.368	-3.767
		1	3.086	-12.680	-19.617	-11.312	-4.786
		2	1.250	-14.516	-21.464	-13.129	-6.612
		5	0.043	-15.723	-22.630	-14.351	-7.813
NZD	1	2	5.302	-6.695	-1.405	-2.349	-1.604
		1	1.055	-10.942	-6.555	-6.986	-6.882
		5	0.668	-11.329	-9.211	-9.684	-9.604
		3	0.326	-11.671	-10.438	-11.30	-10.870
		4	0.002	-11.994	-10.879	-11.954	-11.416
USD	1	4	5.235	-1.748	-2.717	-2.282	-0.125
		2	2.710	-4.273	-5.287	-5.216	-2.622
		5	1.183	-5.800	-6.841	-5.537	-4.236
		3	0.484	-6.499	-7.523	-5.796	-4.875
		1	0.117	-6.867	-7.875	-5.829	-5.226
ZAR	3	1	21.496	-21.861	-8.165	-10.130	-15.112
		4	19.385	-23.972	-10.275	-12.059	-17.247
		5	17.355	-26.002	-12.319	-13.744	-19.247
		2	12.638	-30.719	-17.036	-18.678	-23.961
		3	5.333	-38.025	-24.335	-26.234	-31.277

Table 8. Singapore dollar (SGD). Multiple testing results for the model $P_t = g(P_{t-1}, P_{t-2}, P_{t-3}, P_{t-4}, P_{t-5})$ estimated by using an ANN with 5 input neurons and one hidden neuron selected via SIC. Figures in bold refer to relevant lags.

b	$\hat{T}_{n,r_j}$	r_j	$\hat{T}_{n,r_j} - \hat{c}_1$	$\hat{T}_{n,r_j} - \hat{c}_2$
250	**9.641**	**2**	**2.775**	–
	7.025	**1**	**0.158**	–
	0.866	3	-6.000	-5.761
	0.234	5	-6.632	-6.393
	0.156	4	-6.711	-6.471
500	**9.642**	**2**	**4.562**	–
	7.025	**1**	**1.945**	–
	0.866	3	-4.214	-3.303
	0.234	5	-4.846	-3.935
	0.156	4	-4.925	-4.013
750	**9.641**	**2**	**3.273**	–
	7.025	**1**	**0.657**	–
	0.866	3	-5.502	-2.318
	0.234	5	-6.134	-2.950
	0.156	4	-6.212	-3.029
1000	**9.641**	**2**	**2.769**	–
	7.025	**1**	**0.153**	–
	0.866	3	-6.006	-1.930
	0.234	5	-6.638	-2.562
	0.156	4	-6.716	-2.640

Table 9. Australian dollar (AUD). Multiple testing results for the model $P_t = g\left(P_{t-1}, P_{t-2}, P_{t-3}, P_{t-4}, P_{t-5}\right)$ estimated by using an ANN with 5 input neurons and one hidden neuron selected via SIC. Figures in bold refer to relevant lags.

b	$\hat{T}_{n,r_j}$	r_j	$\hat{T}_{n,r_j} - \hat{c}_1$	$\hat{T}_{n,r_j} - \hat{c}_2$
250	**24.195**	**2**	**13.825**	–
	19.391	**4**	**9.021**	–
	4.996	3	-5.374	-3.855
	0.489	1	-9.881	-8.362
	0.126	5	-10.244	-8.726
500	**23.853**	**2**	**12.483**	–
	19.089	**4**	**7.720**	–
	4.952	3	-6.417	-5.838
	0.463	1	-10.906	-10.326
	0.116	5	-11.253	-10.674
750	**24.066**	**2**	**9.053**	–
	19.196	**4**	**4.184**	–
	5.018	3	-9.995	-9.940
	0.436	1	-14.577	-14.522
	0.111	5	-14.902	-14.847
1000	**23.855**	**2**	**5.744**	–
	19.100	**4**	**0.989**	–
	4.994	3	-13.116	-13.070
	0.448	1	-17.662	-17.616
	0.115	5	-17.995	-17.949

Table 10. Canadian dollar (CAD). Multiple testing results for the model $P_t = g(P_{t-1}, P_{t-2}, P_{t-3}, P_{t-4}, P_{t-5})$ estimated by using an ANN with 5 input neurons and one hidden neuron selected via SIC. Figures in bold refer to relevant lags.

b	$\hat{T}_{n,r_j}$	r_j	$\hat{T}_{n,r_j} - \hat{c}_1$	$\hat{T}_{n,r_j} - \hat{c}_2$
250	**10.838**	**1**	**3.033**	–
	1.668	2	- 6.137	-5.551
	0.602	5	-7.203	-6.617
	0.525	3	-7.280	-6.694
	0.405	4	-7.400	-6.814
500	**10.843**	**1**	**1.285**	–
	1.666	2	-7.892	-5.157
	0.604	5	-8.954	-6.219
	0.530	3	-9.029	-6.293
	0.401	4	-9.157	-6.422
750	**10.845**	**1**	**2.382**	–
	1.674	2	-6.789	-5.842
	0.599	5	-7.864	-6.916
	0.535	3	-7.928	-6.981
	0.406	4	-8.057	-7.110
1000	**10.830**	**1**	**3.131**	–
	1.663	2	-6.036	-1.585
	0.595	5	-7.104	-2.653
	0.522	3	-7.177	-2.726
	0.407	4	-7.292	-2.841

Table 11. Norwegian krone (NOK). Multiple testing results for the model $P_t = g(P_{t-1}, P_{t-2}, P_{t-3}, P_{t-4}, P_{t-5})$ estimated by using an ANN with 5 input neurons and one hidden neuron selected via SIC. Figures in bold refer to relevant lags.

b	$\hat{T}_{n,r_j}$	r_j	$\hat{T}_{n,r_j} - \hat{c}_1$	$\hat{T}_{n,r_j} - \hat{c}_2$	$\hat{T}_{n,r_j} - \hat{c}_3$
250	**16.224**	**4**	**7.536**	–	–
	13.137	**3**	**4.449**	–	–
	10.331	**2**	**1.643**	–	–
	6.609	**1**	-2.079	**1.752**	–
	0.023	5	-8.665	-4.834	-4.334
500	**16.207**	**4**	**8.448**	–	–
	13.102	**3**	**5.343**	–	–
	10.321	**2**	**2.562**	–	–
	6.594	**1**	-1.165	**1.251**	–
	0.022	5	-7.737	-5.321	-4.869
750	**16.221**	**4**	**8.972**	–	–
	13.072	**3**	**5.823**	–	–
	10.360	**2**	**3.111**	–	–
	6.613	**1**	-0.636	**1.896**	–
	0.022	5	-7.227	-4.695	-3.483
1000	**16.227**	**4**	**8.177**	–	–
	13.172	**3**	**5.122**	–	–
	10.354	**2**	**2.304**	–	–
	6.607	**1**	-1.443	**1.184**	–
	0.023	5	-8.028	-5.400	-3.617

Testing Uncovered Interest Rate Parity and Term Structure Using Multivariate Threshold Cointegration

Jaya Krishnakumar and David Neto

Department of Econometrics, University of Geneva, 40, Bd. du Pont d'Arve, CH-1211, Geneva 4, Switzerland
jaya.krishnakumar|david.neto@metri.unige.ch

Summary. A multivariate threshold vector error correction model (TVECM) is formulated to examine the expectation hypothesis of the term structure (EHTS) of interest rates and uncovered interest rate parity (UIRP) for U.S. and Swiss rates. Tests for no cointegration, for the number of cointegrating relations and for the presence of threshold effects are discussed within the framework of this TVECM with more than one cointegrating relationship, allowing for the possibility of a fewer number of cointegrating relations in one regime compared to the other. The results conclude that all the three possible cointegrating relations are accepted. This is consistent with both the UIRP and the EHTS hypothesis. A strong evidence for a threshold effect is also found.

Key words: Threshold cointegration, uncovered interest rate parity, term structure.

Preamble

We are indeed very happy to contribute to this book in honour of Professor Manfred Gilli. Manfred is not only a distinguished colleague but also a close friend of ours. Jaya and Manfred have been together in the same Department for quite a long period; have had numerous useful discussions on various econometric and computational issues; have been in the same Masters and Ph.D. dissertation committees; have helped in the organisation of each other's conferences and the most interesting part of all this is no doubt the exchange of ideas on some fundamental philosophical questions about scientific research. Jaya and David would like to express their great affection and esteem for Manfred through their contribution in this dedicated volume.

1 Introduction

Studies using cointegration tests for an empirical verification of economic "laws" involving nonstationary variables have grown rapidly in recent years. The area of international finance is particularly rich in such relationships that have important implications for policy design and evaluation. Purchasing Power Parity (PPP), Uncovered Interest Rate Parity (UIRP) or Expectation Hypothesis of Term Structure of interest rates (EHTS) are examples of "laws" that a policy-maker would be interested in knowing whether they can be assumed to be true or not in practice. In simple terms, the first one states that the exchange rate between the currencies of two countries should reflect the relative price levels. Uncovered Interest Rate Parity equates the interest rate differential between two countries to the forward premium or the expected change in spot exchange rates (under rational expectations). Finally, the term structure of interest rates deals with the relationship between long-term and short-term interest rates and predicts a stationary spread.

The above relationships have either been examined individually or in combination by various researchers in the field, though studies that follow the latter approach are limited compared to the former. Given the vast amount of literature dealing with one relationship, we will not even attempt to review it here. The reader is referred to Campbell and Shiller (1987,1991), Hall et al. (1992), and Engel (1996) among others for surveys on empirical evidences about UIRP. In this paper, our main focus is on the simultaneous verification of several relationships using a cointegrated system. Accordingly, in what follows, we will first look at studies that explore a combination involving two or more of the above relationships and summarize their findings. We do not claim it to be exhaustive as we are only concerned with works that follow the cointegration approach which forms the basis of our extension.

Johansen and Juselius (1992) represents one of the earliest investigations testing PPP and UIRP between U.K. and its trading partners using a five-element vector consisting of two prices, two interest rates and the exchange rate. The authors confirm the presence of two cointegrating relations by means of a likelihood procedure within a vector error correction model. They accept interest rate differential but not the PPP. They also find that there is a stationary relationship which includes PPP with some combination of the interest rate differentials. Bekaert et al. (2002) test UIRP and EHTS and find evidence against the latter but inconclusive results for the former. UIRP is seen to be currency dependent but not horizon dependent. Brüggemann and Lütkepohl (2005) find evidence for UIRP and EHTS for U.S. and the Euro zone.

Most of the studies found in the literature, including the ones cited above, have tested linear relationships in their empirical model. In a linear formulation, the above mentioned economic "laws" have more often been rejected than accepted, especially while testing a single relationship at a time. A possible explanation for this is that the cointegration relation may not be operational at all times but only comes into play when the equilibrium error becomes "large". In other words when the error is small, divergence from the law can occur within limits. This implies non-linearity in the relationship by which the equilibrium error can be nonstationary as long as it does not cross a threshold level beyond which it starts behaving like a stationary process. Indeed, Ballie and Kiliç (2006) find strong evidence of such a phenomenon in the case of UIRP: whenever the risk premium is not high enough, deviation from UIRP is possible and there is no adjustment towards the equilibrium. They put forth many possible explanations for such a divergence - transaction costs, Central Bank intervention and limits to speculation.

The possibility of nonlinear or threshold behaviour in adjusting to equilibrium is found to be more and more appropriate in many empirical studies in finance. From a theoretical point of view, threshold cointegration was first introduced in the economic literature by Balke and Fomby (1997). The authors cover a wide range of configurations in their paper including multiple regimes and the possibility of having a unit root in one of the regimes. Enders and Siklos (2001) use both self-exciting TAR (SETAR) and Momentum TAR (MTAR) processes to describe the disequilibrium and propose a methodology to test for cointegration but without a real asymptotic test theory. Recall that the SETAR model specifies the lagged dependent variable as the transition variable while the MTAR model uses its lagged variation. Lo and Zivot (2001) apply and test threshold adjustment in international relative prices.

Hansen and Seo (2002) and Gonzalo and Pitarakis (2006) provide statistics and asymptotic theory for testing the existence of a threshold effect in the error correction model. Gonzalo and Pitarakis (2006) also investigate the stochastic properties of the error correction model with threshold effect and in particular the stability conditions. In Seo (2006) a test for linear cointegration is developed for the equilibrium error specified as a SETAR process and the law of one price hypothesis is examined when threshold dynamics is introduced. Kapetanios and Shin (2006) investigate a test similar to that of Seo (2006) for a three-regime SETAR model where the corridor regime follows a random walk. Note that two pitfalls are combined when one wishes to test for a unit root in a threshold model where the transition variable is the lagged dependent variable itself. In addition to the well known issue of non-identification of the threshold parameter under the null hypothesis inherent in TAR models (Hansen, 1996, 1997), the transition variable is also nonstationary under the null. Thus, although the model of Caner and Hansen (2001) and Gonzalo and Pitarakis (2006) covers a large set of processes, it does not include this particular case considered in Seo (2006) and in Kapetanios and Shin (2006). However, as discussed in Caner and Hansen (2001), one could always take the lagged variation of the threshold variable in order to ensure its stationarity. Enders and Siklos (2001) note that this choice could often be suitable in economic models.

Except for Gonzalo and Pitarakis (2006), all the above studies assumed a single cointegrating relationship. This underlying assumption simplifies both the estimation procedure and inference. Although determining the number of cointegrating relationships for a set of integrated variables and estimating them has been an important area of research in standard cointegration theory (one can cite Johansen (1988, 1991), Phillips (1991), and Phillips and Durlauf (1986) among others), there are few developments in the threshold cointegrated context. Gonzalo and Pitarakis (2006) propose a methodology to test for cointegrating rank in a threshold vector error correction model (TVECM) by directly estimating the unknown ranks of the coefficient matrices using a model selection approach introduced by Gonzalo and Pitarakis (1998, 2002).

In a recent paper, Krishnakumar and Neto (2007) consider estimation and inference procedures for a TVECM with more than one cointegrating relation and develop a test for the cointegrating rank. In this paper, we apply this test for analysing and testing EHTS along with UIRP between U.S. and Switzerland. If both were true, we would have three cointegrating relations namely EHTS for U.S., the same for CH and UIRP between the two. Our empirical results validate the economic laws tested and the threshold behaviour.

Our paper is organised as follows. In Section 2 we give a formal specification of the two economic laws to be tested. These laws give rise to three possible relationships with four interest rates. In section 3, we then present the general formulation of the threshold vector error correction model. We successively consider the test of no cointegration and the number of cointegrating relations (reduced rank tests). It is interesting to point out here that, in the threshold case, we also allow for the possibility of a unit root (absence of cointegration) in one of the regimes and hence fewer number of cointegrating relations in one regime compared to the other. Section 4 presents the empirical results. The paper ends with some concluding remarks in Section 5.

2 The Economic Relations

2.1 Expectation Hypothesis of the Term Structure of Interest Rates

Let $R_t^{(n)}$ be the continuously compounded n-period interest rate and let $R_t^{(n_0)}$ be a shorter-term interest rate, say n_0-period interest rate. The expectations hypothesis of the uncovered interest rate parity leads to the following relation:

$$R_t^{(n)} = \frac{n_0}{n} \sum_{i=0}^{n/n_0 - 1} E_t R_{t+n_0 i}^{(n_0)} + a^{(n_0, n)}, \tag{1}$$

where n_0/n is an integer, E_t denotes the expectation conditional on the information available at time t, and where $a^{(n_0,n)}$ is a time constant premium which could vary with the maturities n and n_0. This premium may incorporate risk considerations or investors' liquidity preferences. Campbell and Shiller (1987) show that if the short-term interest rate is $I(1)$, then EHTS implies that the long-term rate will also be $I(1)$ and the spread will be stationary, i.e. if $R_t^{(n)} \sim I(1)$ and $R_t^{(n_0)} \sim I(1)$, then $S_t^{(n_0,n)} = R_t^{(n)} - R_t^{(n_0)} \sim I(0)$. So, there is a cointegration relationship between these two rates with the cointegration vector $(1, -1)$. Hall et al. (1992) extend their investigations on expectations theory of the term structure to the multivariate case, considering N maturities $R_t^{(n_j)}$, with $n_0 \leq n_1 \leq \ldots \leq n_j \leq \ldots \leq n_{N-1}$.[1] They state that if the short-term rate $R_t^{(n_0)}$ is $I(1)$, then all the others will also be $I(1)$ and that the spreads with the short-term rate will all be $I(0)$. In other words, there will be $(N-1)$ cointegration relationships. The cointegration space is generated by the columns of the $(N \times N - 1)$ matrix below :

$$A_0 = \begin{bmatrix} 1 & 1 & \cdots & 1 \\ -1 & & & \\ & -1 & & \\ & & \ddots & \\ & & & -1 \end{bmatrix},$$

where blank elements are zeros.

2.2 Uncovered Interest Rate Parity

The uncovered interest rate parity hypothesis (UIRP) at the n_0-period horizon states that

$$\frac{1}{n_0}\left(E_t e_{12,t+n_0} - e_{12,t}\right) = R_{1,t}^{(n_0)} - R_{2,t}^{(n_0)} + b^{(n_0)}, \tag{2}$$

where $R_{1t}^{(n_0)}$ and $R_{2t}^{(n_0)}$ are time-t continuously compounded n_0-period interest rates of country 1 and 2, respectively, e_{12} is the logarithm of the spot exchange rate denominated in the currency of country 1, and $b^{(n_0)}$ a constant risk premium.

Following Bekaert et al. (2002), the UIRP at a short horizon n_0 and the EHTS at a long horizon n imply UIRP at the long horizon n. Let us write down the EHTS at the long horizon n for the both countries:

$$R_{j,t}^{(n)} = \frac{n_0}{n}\sum_{i=0}^{n/n_0-1} E_t R_{j,t+n_0 i}^{(n_0)} + a_j^{(n_0,n)}, \qquad \text{for } j = 1, 2, \tag{3}$$

combining the above two equations with the short-term UIRP (2), we get the following relation:

$$R_{1,t}^{(n)} - R_{2,t}^{(n)} = \frac{n_0}{n}\sum_{i=0}^{n/n_0-1} E_t\left(R_{1,t+n_0 i}^{(n_0)} - R_{2,t+n_0 i}^{(n_0)}\right) + a_1^{(n_0,n)} - a_2^{(n_0,n)} \tag{4}$$

$$= \frac{1}{n}\sum_{i=1}^{n} E_t \Delta e_{12,t+n_0 i} + a_1^{(n_0,n)} - a_2^{(n_0,n)} + b^{(n_0)}$$

Few works investigate this tight connection between EHTS and UIRP hypothesis. Only Bekaert et al. (2002) and Wolters (2002) have recently pointed

---

[1] Note that in this general case, there are $N-1$ premia $a^{(n_j,n)}$ for $j = 0, ..., N-1$.

it out. The implication of (4) is that if the exchange rate changes are $I\,(0)$, then $R_{1,t}^{(n)} - R_{2,t}^{(n)}$ will be $I\,(0)$. Thus, for a system with N maturities per country (1 and 2), there should be at most $2N - 1$ cointegration relations according to UIRP and EHTS. In this paper we will test the above relations for 2 countries and 2 maturities (short and long term).

3 The Econometric Framework

3.1 The Model

The following general notation is used. We denote the projection matrix associated with a $(p \times r)$ matrix A as P_A, the orthogonal complement of A as $A_\perp$ which is a $(p \times p - r)$ matrix of full column rank such that $A'A_\perp = 0$. We will use $I\,(1)$ and $I\,(0)$ to represent time series that are integrated of order 1 and 0, respectively. Throughout the paper, integrals are taken over the unit interval and we use "$\Longrightarrow$" to indicate convergence in distribution as sample size T tends to infinity.

Let us consider the p-dimensional finite-order VAR model for an $I\,(1)$ vector X_t:

$$A\,(L)\,X_t = \mu + \varepsilon_t, \quad t = 1, 2, ..., T,$$

where $A\,(L) = I_p - A_1 L - ... - A_K L^K$, the first K data points X_{1-K}, X_{1-K+1}, ..., X_0 are fixed, and where ε_t are iid $N\,(0, \Omega)$. The error correction form of the previous VAR is given by:

$$\Delta X_t = \mu + \Pi X_{t-1} + \Lambda_1 \Delta X_{t-1} + ... + \Lambda_{K-1} \Delta X_{t-K+1} + \varepsilon_t, \quad t = 1, 2, ..., T.$$

The hypothesis of cointegration implies a matrix Π of reduced rank such that $\Pi = \Gamma A'$ where Γ and A are $(p \times r)$ matrices with $r < p$. Normalizing matrix A as $A' = [I_r, -\beta]$ (triangular representation), where β is $(r \times p)$, we have the following particular form of the loading matrix Γ, $\Gamma' = [-I_r, 0]$. Now consider the r equilibrium errors $A'X_t = z_t$ and let them follow a stationary threshold autoregressive (TAR) model of order 1:

$$(\Phi_1\,(L)\,z_t)\,\mathrm{II}_{\{s_{t-d} \leq \theta\}} + (\Phi_2\,(L)\,z_t)\,\mathrm{II}_{\{s_{t-d} > \theta\}} = \varepsilon_{rt},$$

where $\Phi_j\,(L) = 1 - \phi_j L$ for $j = 1, 2$, ϕ_j are diagonal matrices, ε_{rt} is a $(r \times 1)$ vector, s_{t-d} (with $d > 0$) is a stationary and ergodic transition variable and θ a threshold such that $P(s_{t-d} < \theta) = F\,(\theta) \in [\tau, 1 - \tau]$. We write $\varepsilon_{(p-r)t} = B'\Delta X_t$ with $B' = [0, I_{p-r}]$, and $K\,(L)\,\eta_t = \varepsilon_{(p-r)t}$ with $K\,(L) = I_{p-r} - K_1 L - ... - K_{q-1} L^{q-1}$. It is shown in Krishnakumar and Neto (2007) that the vector error correction model (VECM) corresponding to the TAR specification is given by:

$$\Delta X_t = \mu + \Pi^{(1)} X_{t-1} \mathrm{II}_{\{s_{t-d} \leq \theta\}} + \Pi^{(2)} X_{t-1} \mathrm{II}_{\{s_{t-d} > \theta\}} + \Lambda\,(L)\,\Delta X_{t-1} + \varepsilon_t \quad (5)$$

where $\Pi^{(j)} = \Gamma^{(j)}A'$, $\Gamma^{(j)} = \begin{pmatrix} -\Phi_j(1) \\ 0 \end{pmatrix}$, for $j = 1,2$ and $\Lambda(L) = \begin{pmatrix} 0 & \beta\sum_{i=1}^{q-1}K_iL^{i-1} \\ 0 & \sum_{i=1}^{q-1}K_iL^{i-1} \end{pmatrix}$.

Note that the threshold only affects the matrix of loadings in this setting by giving rise to two different $\Gamma^{(j)}$'s in the two regimes of the above threshold vector error correction model (TVECM) (5). However, if z_t followed a TAR whose order is greater than one, then the threshold would also affect the coefficients of the lags ΔX_{t-1-i} (see Krishnakumar and Neto (2007)).

There is an additional consideration to be noted here in view of our practical application in which we will be dealing with interest rates. It is generally agreed that interest rates do not contain time trends but one cannot ignore the possibility of a drift in the spreads due to the presence of a premium. Thus we decide to remove the intercept from our empirical error correction model and only include it in the cointegrating equations.

3.2 Estimation

Exploiting the triangular normalisation of the cointegrated system, Krishnakumar and Neto (2007) propose an iterative procedure based on the work of Ahn and Reinsel (1990). They provide the estimator of the long-run parameters β as well as the loadings $\Gamma^{(j)}$, and their asymptotic properties.

Following the reasoning of the previous section, our TVECM is written as:

$$\Delta X_t = \mu^{(1)}\mathbb{1}_{\{s_{t-d}\leq\theta\}} + \Pi^{(1)}X_{t-1}\mathbb{1}_{\{s_{t-d}\leq\theta\}} +$$
$$\mu^{(2)}\mathbb{1}_{\{s_{t-d}>\theta\}} + \Pi^{(2)}X_{t-1}\mathbb{1}_{\{s_{t-d}>\theta\}} + \Lambda(L)\Delta X_{t-1} + \varepsilon_t,$$

where $\mu^{(j)} = \Gamma^{(j)}\beta_0$ and β_0 is the $(r \times 1)$ vector representing the intercept terms of the cointegration relations. Since the threshold does not affect the coefficients of the lagged variations in our TVECM, we concentrate out the short-term parameters in a first step by estimating the following two models:

$$\Delta X_t = \Lambda_{11}\Delta X_{t-1} + ... + \Lambda_{1q}\Delta X_{t-q-1} + \epsilon_{1t} \tag{6}$$

and

$$X_{t-1} = \Lambda_{21}\Delta X_{t-1} + ... + \Lambda_{2q}\Delta X_{t-q-1} + \epsilon_{2t-1}. \tag{7}$$

Denoting by ΔX_t^* and X_{t-1}^*, the residuals of the previous regressions and writing $X_{t-1}^{*(j)} = X_{t-1}^*\mathbb{1}_{\{\cdot\}}$ and $\iota_p^{(j)} = \iota_p\mathbb{1}_{\{\cdot\}}$, where ι_p is a p−vector of ones, the error correction part of TVECM is given by:

$$\Delta X_t^* = \left(\mathbf{X}_{t-1}^{*\prime} \otimes \mathbf{\Gamma}\right) vec(\mathbf{B}) + \varepsilon_t^* \tag{8}$$

where $\mathbf{X}_{t-1}^* = \left(\iota_p^{(1)\prime} \; X_{t-1}^{*(1)\prime} \; \iota_p^{(2)\prime} \; X_{t-1}^{*(2)\prime}\right)'$, $\mathbf{B} = \begin{pmatrix} \mathbf{A}' & 0 \\ 0 & \mathbf{A}' \end{pmatrix}$, $\mathbf{\Gamma} = \left(\Gamma^{(1)} \; \Gamma^{(2)}\right)$, $\mathbf{A}' = (\beta_0 \; A')$, and $Var(\varepsilon_t^*) = \Omega$.

The iterative estimation procedure described in Krishnakumar and Neto (2007) can be outlined as follows:

For a given threshold.

Step 1. $\Gamma^{(1)}$ and $\Gamma^{(2)}$ can be consistently estimated from a regression of ΔX_t^* on $\left(X_{t-1}^{*(1)\prime}, X_{t-1}^{*(2)\prime} \right)'$, from $\Delta X_t^* = \Pi^{(1)} X_{t-1}^{*(1)} + \Pi^{(2)} X_{t-1}^{*(2)} + \varepsilon_t^*$. The triangular normalisation implies that the first r columns of $\Pi^{(1)}$ and $\Pi^{(2)}$ are equal to $\Gamma^{(1)}$ and $\Gamma^{(2)}$ respectively. Thus, one can use these columns of $\hat{\Pi}^{(1)}$ and $\hat{\Pi}^{(2)}$ as estimators of the loading factors $\Gamma^{(1)}$ and $\Gamma^{(2)}$.

Step 2. Using the consistent estimator of Ω from the estimated variance-covariance matrix of the residuals from step 1: $\hat{\Omega}(\theta) = T^{-1}\sum_t^T \hat{\varepsilon}_t^* \hat{\varepsilon}_t^{*\prime}$, we obtain a feasible estimator of $\hat{\beta}(\Gamma, \Omega, \theta)$ by concentrated maximum likelihood. See Krishnakumar and Neto (2007) for a closed form of this estimator, its variance and its limiting distribution.

Step 3. Using $\hat{\beta}$, we can re-estimate $\Gamma^{(j)}$ by running regression (8) : $\Delta X_t^* = \Gamma \mathbf{B} \mathbf{X}_{t-1}^* + \varepsilon_t^*$.

Steps 1 to 3 could be iterated in order to achieve numerical convergence of the estimates of long-run parameters and loading factors (each iteration step yields an improvement of the likelihood function).

For an unknown threshold.

When the threshold is unknown, it has to be searched over a grid on $[\tau, 1 - \tau]$. In practice, this grid is given by the values taken by the selected transition variable. However, the choice of τ is somewhat arbitrary; empirical investigations take τ to be 5, 10 or 15 percent. In general, it is chosen in order to ensure that there are enough observations in each regime and that the limits for the test statistics used in inference are nondegenerate (see Andrews (1993), Hansen (1996)). In practice, various values must be tried in empirical applications to check the robustness of the results to the selected value of τ.

The concentrated likelihood function is $L_T(\theta) = -\frac{T}{2}\log\left|\hat{\Omega}(\theta)\right| - \frac{Tp}{2}$, and so an estimator of θ is given by $\hat{\theta} = \arg\min \log\left|\hat{\Omega}(\theta)\right|$ (see Tsay (1989), Enders and Siklos (2001), Hansen and Seo (2002)).

3.3 Testing No-cointegration

Stacking the observations and vectorizing (8), we get:

$$vec(\Delta X^*) = Z^{(1)}\pi^{(1)} + Z^{(2)}\pi^{(2)} + vec(\varepsilon^*) \tag{9}$$

where $\pi^{(j)} = vec\left(\Pi^{(j)\prime}\right)$, $Z^{(j)} = \left(I_p \otimes X_{-1}^{*(j)}\right)$ and where $X_{-1}^{*(j)}$ stacks $\left(\iota_p^{(j)\prime} \ X_{t-1}^{*(j)\prime}\right)$ on $t = 1...T$. The estimator of $\pi^{(j)}$ is given by

$$\hat{\pi}^{(j)} = \left(I_p \otimes \left(X_{-1}^{*(j)\prime} X_{-1}^{*(j)}\right)^{-1} X_{-1}^{*(j)\prime}\right) vec\left(\Delta X^*\right).$$

Therefore, the null of no-cointegration is merely given by $H_0 : \mathbf{R}\pi = 0$ where $\pi = \left(\pi^{(1)\prime} \ \pi^{(2)\prime}\right)'$ is a $2p^2$-vector and $\mathbf{R} = \left(I_{p^2} \ -I_{p^2}\right)$ is the selection matrix. Gonzalo and Pitarakis (2006) use the following supremum Wald statistic to test H_0 :

$$\sup \mathcal{W}_0 = \sup_{F(\theta) \in [\tau \ , \ 1-\tau]} \mathcal{W}_{0,T} \left(F\left(\theta\right)\right), \tag{10}$$

where

$$\mathcal{W}_{0,T}\left(F\left(\theta\right)\right) = \left(\hat{\pi}^{(1)} - \hat{\pi}^{(2)}\right)' \left(\mathbf{RSR}'\right)^{-1} \left(\hat{\pi}^{(1)} - \hat{\pi}^{(2)}\right),$$

with $\mathbf{S} = diag\left(\hat{\Omega} \otimes X_{-1}^{*(1)\prime} X_{-1}^{*(1)} \ , \ \hat{\Omega} \otimes X_{-1}^{*(2)\prime} X_{-1}^{*(2)}\right)$. Under some conditions on the transition process s_{t-d}, Gonzalo and Pitarakis (2006) provide the limiting distribution of this statistic under the null $H_0 : \Pi^{(1)} = \Pi^{(2)} = 0$:

$$\sup \mathcal{W}_0 \implies \sup_{F(\theta) \in [\tau, 1-\tau]} \frac{1}{F(\theta)(1-F(\theta))}$$
$$\times tr\left\{\left(\int \bar{B}\left(u\right) dK\left(u, F\left(\theta\right)\right)'\right)' \left(\int \bar{B}\left(u\right) \bar{B}\left(u\right)'\right)^{-1} \left(\int \bar{B}\left(u\right) dK\left(u, F\left(\theta\right)\right)'\right)\right\},$$

where $K\left(u, F\left(\theta\right)\right)$ is a Kiefer process given by $K\left(u, F\left(\theta\right)\right) = B\left(u, F\left(\theta\right)\right) - F\left(\theta\right) B\left(u, 1\right)$, $\bar{B}\left(u\right) = B\left(u\right) - \int B\left(u\right) dr$ denotes a p-dimensional standard demeaned Brownian motion and $B\left(u, F\left(\theta\right)\right)$ a p-dimensional two-parameter standard Brownian motion (See Gonzalo and Pitarakis (2006) and Caner and Hansen (2001), for a proof).

Gonzalo and Pitarakis (2006) only give the critical values of this statistic for the bidimensional case. We reproduce their values and complete their tables by providing critical values for various dimensions p of the model ($p = 1, \ldots, 6$). The asymptotic distribution is calculated by Monte Carlo simulations. The stochastic integrals are evaluated at 10,000 points over the argument r and 100 steps over the argument $F\left(\theta\right)$. The critical values are computed as the corresponding empirical quantiles from 10,000 replications and reported in Table A.1 of the Appendix. These values are also reported for various ranges $[\tau \ , 1 - \tau]$. Caner and Hansen (2001) discuss the inconsistency of the tests when one chooses $\tau = 0$ in threshold models. Indeed, because the critical value of the statistics increases as τ decreases, the rejection of the null requires a larger value of the statistic as τ tends to 0. It follows that τ should be set in the interior of $(0, 1)$.

3.4 Testing Reduced Rank

Once the presence of cointegration is verified, the cointegration ranks of the matrices $\Pi^{(1)}$ and $\Pi^{(2)}$ should be tested. The simplest configuration occurs when there are the same number of cointegrating relationships active in both regimes. Formally it means $rk\left(\Pi^{(1)}\right) = rk\left(\Pi^{(2)}\right) \leq p-1$ (where $rk\left(\cdot\right)$ stands for the rank function), with $\Pi^{(1)} \neq \Pi^{(2)}$. According to the TVECM (5), we have $rk\left(\Pi^{(1)}\right) = rk\left(\Pi^{(2)}\right) = r$. This is a relatively straightforward generalization of the asymmetric adjustment for more than one cointegrating relationship. The problem becomes more complicated if we consider a possible unit root in one of the regimes of the TAR process. This possibility was initially considered in univariate TAR models by Balke and Fomby (1997) and González and Gonzalo (1998) and more recently studied by Caner and Hansen (2001) and Kapetanios and Shin (2006). For instance in a TAR(1) model, this local nonstationary feature occurs when $(\phi_j)_k = 1$ and $(\phi_{j'})_k < 1$ for $j \neq j'$, where $(\phi_j)_k$ is the $k-th$ element of the diagonal of the matrix ϕ_j, for $k \in \{1, ..., r\}$, such that the $k-th$ univariate TAR behaves like a unit root process in one regime. Caner and Hansen (2001) called this process a partial unit root process. In an unpublished paper, González and Gonzalo (1998) investigate threshold unit root (TUR) processes and provide various useful theoretical results. In particular, they have shown that under additional assumptions, even when one of the two regimes has a unit root, a TUR(1) process could still be covariance stationary (see also Gonzalo and Pitarakis (2006)). Following the terminology of Caner and Hansen (2001), when the disequilibrium is a TUR process, this situation can be called one of partial cointegration. Presence of unit roots in the threshold model for the disequilibrium induces some consequences on ranks of $\Pi^{(j)}$ matrices in the TVECM.

Let us assume that $rk\left(\Pi^{(1)}\right) = p - 1 > rk\left(\Pi^{(2)}\right)$. This means that at least one column of $\Pi^{(2)}$ is equal to zero. In other words, some cointegration relationships are inactive in the second regime, *i.e.* one component of z_t behaves like a unit root process in the second regime. In the general case of S regimes, if the TAR(ℓ) has S unit roots (at most one unit root per regime), then S columns of the loading factor matrices $\Gamma^{(j)}$ (and hence S columns of $\Pi^{(j)}$) will be null. This statement comes from the assumption that ϕ_j are diagonal matrices:

$$\Gamma^{(j)} = \begin{pmatrix} -\Phi_j\left(1\right) \\ 0 \end{pmatrix}, \text{ for } j = 1, 2.$$

As an example, let us consider a 4-dimensional integrated system with three cointegration relationships :

$$A'X_t = z_t,$$

$$\Delta X_{4t} = z_{4t},$$

where $X_t = (X_{1t}\ X_{2t}\ X_{3t}\ X_{4t})'$, $z_t = (z_{1t}\ z_{2t}\ z_{3t})'$, $A' = (I_3\ -\beta)$, with β a 3-dimensional vector of parameters, , z_{4t} a (finite autocorrelated) stationary process such that $K\left(L\right)z_{4t} = \varepsilon_{4t}$, and where z_t is described by a trivariate TAR(1) process as follows:

$$\Delta z_{1t} = \rho_1^{(1)} z_{1t-1}^{(1)} + \rho_1^{(2)} z_{1t-1}^{(2)} + \varepsilon_{1t},$$
$$\Delta z_{2t} = \rho_2^{(1)} z_{2t-1}^{(1)} + \rho_2^{(2)} z_{2t-1}^{(2)} + \varepsilon_{2t},$$
$$\Delta z_{3t} = \rho_3^{(1)} z_{3t-1}^{(1)} + \rho_3^{(2)} z_{3t-1}^{(2)} + \varepsilon_{3t},$$

We have $\rho_k^{(j)} = \left(\left(\phi_j\right)_k - 1\right)$ with the stationarity conditions $\rho_k^{(1)} < 0$, $\rho_k^{(2)} < 0$ and $\left(\rho_k^{(1)} + 1\right)\left(\rho_k^{(2)} + 1\right) < 1$ for $k = 1, 2, 3$. Now, if we assume $\rho_2^{(2)} = 0$ (i.e. $\left(\phi_2\right)_2 = 1$), then $\Pi^{(2)} = \Gamma^{(2)} A'$ will be given by:

$$\Pi^{(2)} = \begin{pmatrix} \gamma_{11}^{(2)} & 0 & 0 \\ 0 & 0 & 0 \\ 0 & 0 & \gamma_{33}^{(2)} \\ 0 & 0 & 0 \end{pmatrix} \begin{pmatrix} 1 & 0 & 0 & \beta_{11} \\ 0 & 1 & 0 & \beta_{21} \\ 0 & 0 & 1 & \beta_{31} \end{pmatrix} = \begin{pmatrix} \gamma_{11}^{(2)} & 0 & 0 & \gamma_{11}^{(2)}\beta_{11} \\ 0 & 0 & 0 & 0 \\ 0 & 0 & \gamma_{33}^{(2)} & \gamma_{33}^{(2)}\beta_{31} \\ 0 & 0 & 0 & 0 \end{pmatrix},$$

such that $rk\left(\Pi^{(2)}\right) = 2$ whereas $rk\left(\Pi^{(1)}\right) = 3$.

Thus, for a given A, if $rk\left(A\right) > rk\left(\Pi^{(j)}\right)$ for some regime $j = 1, 2$, it means that the TAR process on the equilibrium errors z_t has a unit root.

Krishnakumar and Neto (2007) propose a methodology inspired from this developed in Lütkepohl and Saikkonen (2000) to test $H_0 : r = r_0$ against the alternative $H_1 : r = r_0 + 1$ (for $r \geq 1$). Through Monte-Carlo experiments, the authors show that while in absence of partial unit root (i.e. $rk\left(\Pi^{(1)}\right) = rk\left(\Pi^{(2)}\right)$) their rank test works as well as the Likelihood Ratio (LR) test, it does a better job when there are unit roots (i.e. $rk\left(\Pi^{(1)}\right) \neq rk\left(\Pi^{(2)}\right)$). The methodology is briefly reviewed below.

Noting that $P_A + P_{A_\perp} = I_p$, inserting it in the TVECM equation and splitting the two regimes, we can write for a given matrix A (and $A_\perp$):

$$\Delta X_t^{*(1)} = \mu^{(1)} + \kappa_1 z_{t-1}^{(1)} + \lambda_1 \nu_{t-1}^{(1)} + \varepsilon_t^{*(1)}, \tag{11}$$
$$\Delta X_t^{*(2)} = \mu^{(2)} + \kappa_2 z_{t-1}^{(2)} + \lambda_2 \nu_{t-1}^{(2)} + \varepsilon_t^{*(2)},$$

where $\kappa_j = \Pi^{(j)} A\left(A'A\right)^{-1}$, $\lambda_j = \Pi^{(j)} A_\perp \left(A'_\perp A_\perp\right)^{-1}$, and $\nu_t^{(j)} = A'_\perp X_t^{*(j)}$. Then, under $H_0 : r = r_0 = rk\left(\Pi^{(j)}\right)$, we have $\lambda_j = 0$, and $\kappa_j = \Gamma^{(j)}$, whereas, under the alternative, some columns of $A_\perp$ will be cointegrating vectors so that $\lambda_j \neq 0$. Thus, one can base the rank test on testing $H_0 : \lambda_j = 0$. For an easier implementation of the test, following Lütkepohl and Saikkonen (2000), we multiply each equation by $\Gamma_\perp^{(j)'}$ to give:

$$\Delta \tilde{X}_t^{*(j)} = \tilde{\mu}^{(j)} + \tilde{\kappa}_j z_{t-1}^{(j)} + \tilde{\lambda}_j \nu_{t-1}^{(j)} + \tilde{\varepsilon}_t^{*(j)}, \tag{12}$$

where for any variable x, we denote $\tilde{x} = \Gamma_{\perp}^{(j)\prime}x$ and for any parameter a, we denote $\tilde{a} = \Gamma_{\perp}^{(j)\prime}a$. Thus, the null becomes $H_0 : \tilde{\lambda}_j = 0$. Piling up the observations, we have the following regression model for both regimes

$$\Delta \tilde{X}^{*(j)} = \left(\tilde{\mu}^{(j)\prime} \otimes \iota_T\right) + z^{(j)}\tilde{\kappa}_j^{\prime} + \nu^{(j)}\tilde{\lambda}_j^{\prime} + \tilde{\varepsilon}^{*(j)}$$

where $\Delta \tilde{X}^{*(j)}$ and $\tilde{\varepsilon}^{*(j)}$ are $(T \times 4 - r_0)$ matrices which stack the observations on $\Delta \tilde{X}_t^{*(j)\prime}$ and $\tilde{\varepsilon}_t^{(j)\prime}$, respectively. The $(T \times r_0)$ and $(T \times 4 - r_0)$ matrices $z^{(j)}$ and $\nu^{(j)}$ stack $z_{t-1}^{(j)}$ and $\nu_{t-1}^{(j)}$, respectively. Given A (and $A_{\perp}$), partitioned regression results yield:

$$vec\left(\hat{\underline{\lambda}}_j\right) - vec\left(\underline{\lambda}_j\right) = \left(I_{4-r_0} \otimes (\nu^{(j)\prime}M\nu^{(j)})^{-1}\nu^{(j)\prime}M\right)vec\left(\tilde{\varepsilon}^{*(j)}\right),$$

for each regime, where $\lambda_j = \tilde{\lambda}_j^{\prime}$, $M = I_T - P_{Z^{(j)}}$, where $Z^{(j)} = \left(\iota_T \; z^{(j)}\right)$. Once again a supremum statistic is used because the threshold appears as a nuisance parameter. The supremum Wald statistic for each regime is:

$$\sup \mathcal{W}^{(j)}\left(r_0\right) = \sup_{F(\theta)\in[\tau\,,\,1-\tau]} \mathcal{W}_T^{(j)}\left(F\left(\theta\right),r_0\right)$$

with

$$\mathcal{W}_T^{(j)}\left(F\left(\theta\right),r_0\right) = vec\left(\hat{\underline{\lambda}}_j\right)^{\prime}Var\left(vec\,\hat{\underline{\lambda}}_j\right)^{-1}vec\left(\hat{\underline{\lambda}}_j\right).$$

In order to produce a feasible version of the statistic, we require estimators of $\mathbf{A}$ and $\Gamma^{(j)}$. One can use the maximum likelihood procedure described in Section 3.2. The limiting distributions of $\sup \mathcal{W}^{(1)}$ and $\sup \mathcal{W}^{(2)}$ are derived and tabulated in Krishnakumar and Neto (2007). The critical values are reported in the Table A.2 in the Appendix.

4 Empirical Analysis

4.1 The Data

In order to carry out a joint test of the EHTS and UIRP hypothesis for U.S. and Switzerland, we use monthly interest rate series over the period January 1986 to February 2007, representing 254 observations. The short-term interest rates for both countries are the three-month rates (denoted by R_{us} and R_{ch} respectively) and the long-term interest are the one-year rates (denoted by r_{us} and r_{ch} respectively). Data are taken from the Datastream database. The series are depicted in Figure 1.

It is generally agreed that interest-rates are $I(1)$ without linear trend. For this reason the unit root test results, which confirm the difference-stationary nature of each variable, are not reported here to save space. So our X vector $X_t = (R_{ch,t}, r_{ch,t}, R_{us,t}, r_{us,t})^{\prime}$ is of order 4 ($p = 4$). According to the

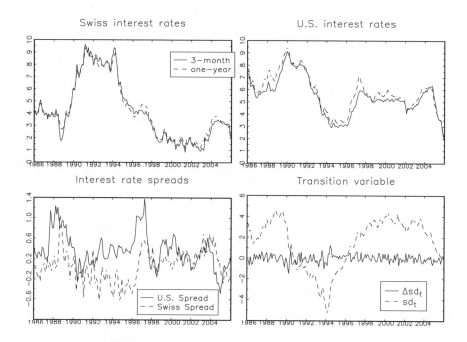

Fig. 1. Interest rates, interest rate spreads, and transition variables.

theory outlined in Section 2, we expect to find at most three cointegrating relationships in each regime of the TVECM. The optimal lag of TVECM is selected using the Akaike information criterion (AIC). Our results indicate the presence of only one lag on ΔX_t.

4.2 Estimation and Testing

Our model does not allow several transition variables (for example one for each cointegration relation) but only one for the full system. While the estimation of a TVECM with more than one transition variable can be envisaged without any major difficulty (van Tol and Wolff, 2005), inference based on such a model requires fundamental modifications. Our inference is based on Tsay's (1989) formulation of the multivariate TAR model, and on the TVECM used in Gonzalo and Pitarakis (2006) which assume stationarity and ergodicity of the transition variable, crucial for the derivation of the asymptotic properties. Hence we keep the same transition variable for all our cointegrating relations.

However, the choice of this (unique) transition variable is somewhat tricky. One could try several variables depending on the economic context and select the "best" one based on an information criterion such as the Bayesian information criterion (BIC) or the AIC. We explored the one period, two period, three period lags of the variation of the interest rate parity, as well as the one period lag of the level of the interest rate parity as different possibilities for the transition variable and report the corresponding estimation results in Table A.3 of the Appendix. The choice of the interest rate parity is justified by the fact that it incorporates money market information from both countries and especially the U.S. market's influence on Swiss interest rates. Hence we believe that it is an economically relevant variable in our context.[2] Surprisingly, all the above variants of the transition variable are more or less equivalent with respect to the BIC information criterion.

We therefore continue with the results for the lagged variation of the interest rate parity $\Delta (r_{us} - r_{ch})_{t-1} = s_{t-1}$ as the transition variable. Note that, we take the change in parity rather than the parity itself in order to ensure stationarity of our transition variable even under the null of no cointegration (see the fourth graph in Figure 1). This variant is named a Momentum-TAR (M-TAR) model used by Enders and Granger (1998) and Enders and Siklos (2001). Furthermore, the limiting distribution of the test statistics would change in case the transition variable is not stationary. Seo (2006) provides a no-cointegration test for a TVECM with only one cointegrating relation when the transition variable is non-stationary under the null. The above mentioned studies report better results in M-TAR compared to TAR in an empirical application of these models to the term structure interest rates.

The value of the supremum Wald test of no-cointegration ($r = r_0 = 0$), obtained over 202 grid points on the empirical distribution of our threshold variable, i.e. $F_T (s_{t-1})$, is $\sup \mathcal{W}_0 = 173.25$ which indicates a clear rejection of the null hypothesis of no-cointegration (See Table A.1). Note that our grid set corresponds to 70% of the estimated values of the transition variable which are taken as possible values of the threshold. The discarded values are the largest and smallest 15% of s_{t-1}, i.e. $\tau = 0.15$. Note that the conclusions of all the tests presented do not change with $\tau = 5\%$ and 10%. Thus, we present the results only for $\tau = 15\%$, which seems a reasonable choice regarding our sample size, without loss of generality.

Next, we apply the rank testing procedure described in the previous section on our data. For this purpose, the TVECM model is estimated under the triangular normalisation, over the above-mentioned grid points on the threshold parameter. For each estimated model, we compute the statistics $\sup \mathcal{W}_T^{(1)}$ and $\sup \mathcal{W}_T^{(2)}$. Since the interest rates are highly heteroskedastic, standard deviations of parameter estimates are computed using the Newey-West method (Newey and West, 1987) and heteroskedasticity-consistent statistic values are obtained.[3] The results are summarized in Table 1.

---

[2] Results do not change if we use the lagged variations of U.S. or Swiss spreads as the transition variable. To save space, we did not report all this results.

[3] Indeed, the tests on the residuals of our TVECM (not reproduced here) conclude that some of them are affected by ARCH effects, and one of them still presents autocorrelation.

Table 1. Statistic values for rank cointegration tests.

$H_0 : r_0$	$\sup \mathcal{W}^{(1)}$	$\sup \mathcal{W}^{(2)}$
$r_0 = 1$	65.090	72.941
$r_0 = 2$	51.766	29.655
$r_0 = 3$	7.193	6.637

Note: $\tau = 0.15$.

From the above results, we can conclude (at the 5% level) that there are three cointegration relationships in each regime (see Table A.2 for the critical values). Given a cointegrating rank of 3 in each regime, we then test whether there is threshold non-linearity or not. For this purpose, we once again use the supremum Wald statistic (10). We find that symmetry is rejected in our data with $\sup \mathcal{W}_T = 40.01$ (see Table A.1). The estimates of the loadings $\hat{\Gamma}^{(j)} = \begin{pmatrix} -\hat{\Phi}_j(1) \\ \mathbf{0}_{(1 \times 3)} \end{pmatrix}$ are given bellow:

$$
\hat{\Gamma}^{(1)} = \begin{pmatrix} -0.078 & 0 & 0 \\ {\scriptstyle(0.014)} & & \\ 0 & -0.068 & 0 \\ & {\scriptstyle(0.014)} & \\ 0 & 0 & -0.180 \\ & & {\scriptstyle(0.037)} \\ 0 & 0 & 0 \end{pmatrix}, \quad \hat{\Gamma}^{(2)} = \begin{pmatrix} -0.021 & 0 & 0 \\ {\scriptstyle(0.019)} & & \\ 0 & -0.033 & 0 \\ & {\scriptstyle(0.018)} & \\ 0 & 0 & -0.142 \\ & & {\scriptstyle(0.053)} \\ 0 & 0 & 0 \end{pmatrix},
$$

with $F(\hat{\theta}) = 0.643$ ($\hat{\theta} = -0.445$).

The above estimates strongly suggests that the speed of adjustment is greater when the variation in the difference in long-term rates gap change is slightly negative, whereas the second regime appears to be a more persistent regime.

Based on our results, we can say that our data are consistent with UIRP hypothesis and EHTS for these two countries. Indeed, the long-run matrix estimate is

$$
\hat{A}' = \begin{pmatrix} 1 & 0 & 0 & \hat{\beta}_1 \\ 0 & 1 & 0 & \hat{\beta}_2 \\ 0 & 0 & 1 & \hat{\beta}_3 \end{pmatrix} \begin{pmatrix} R_{ch,t} \\ r_{ch,t} \\ R_{us,t} \\ r_{us,t} \end{pmatrix} = \begin{pmatrix} 1 & 0 & 0 & -0.887 \\ & & & {\scriptstyle(0.197)} \\ 0 & 1 & 0 & -0.939 \\ & & & {\scriptstyle(0.186)} \\ 0 & 0 & 1 & -1.009 \\ & & & {\scriptstyle(0.018)} \end{pmatrix} \begin{pmatrix} R_{ch,t} \\ r_{ch,t} \\ R_{us,t} \\ r_{us,t} \end{pmatrix},
$$

and noting that the three cointegrating relations specified by our triangular TVECM are given by $r_{ch,t} + \beta_2 r_{us,t}$, $R_{us,t} + \beta_3 r_{us,t}$ and $R_{ch,t} + \beta_4 r_{ch,t}$ where $\beta_4 = -\frac{\beta_1}{\beta_2}$, for parity, U.S. spread, and Swiss spread, respectively, we get $\hat{\beta}_4 = -0.944$ with a standard deviation $\hat{\sigma}_{\beta_4} = 0.119$. This latter value is obtained using the δ-method.[4] Thus these estimates are very close to one that

---

[4] Let $\hat{\beta}$ be an estimator of β whose covariance matrix $V\left(\hat{\beta}\right)$ and $g(\cdot)$ a differentiable function, we have $V\left(g\left(\hat{\beta}\right)\right) = \frac{\partial g(\beta)}{\partial \beta'} V\left(\hat{\beta}\right) \frac{\partial g'(\beta)}{\partial \beta}$.

argues in favor of the both hypotheses UIRP and EHTS. To confirm this, we perform a joint test of the UIRP hypothesis and EHTS from our TVECM with a cointegration rank $r = 3$. We need to test $H_0 : \beta_1 = \beta_2 = \beta_4 = -1$. Using $\beta_4 = -\frac{\beta_1}{\beta_2}$, and combining it with H_0, i.e. $\beta_2 = -1$, we get $\beta_1 = -1$ as well. Then, the test can be carried out directly from the TVECM for

$$H_0 : \begin{pmatrix} 1 & 0 & 0 & \beta_1 \\ 0 & 1 & 0 & \beta_2 \\ 0 & 0 & 1 & \beta_3 \end{pmatrix} = \begin{pmatrix} 1 & 0 & 0 & -1 \\ 0 & 1 & 0 & -1 \\ 0 & 0 & 1 & -1 \end{pmatrix},$$

using a Likelihood Ratio (LR) statistic. We obtain $LR = -32.20$ with a p-value $= 0.99$ (which comes from a $\chi^2_{(3)}$ distribution) and hence the null hypothesis cannot be rejected. This conclusion is consistent with different values of $\tau : 5\%, 10\%$ and 15%.

5 Conclusion

This paper examines the expectation hypothesis of the term structure of interest rates and uncovered interest rate parity for U.S. and Switzerland in a threshold vector error correction framework. For this purpose, we begin by looking at estimation and inference for a multivariate TVECM allowing for more than one cointegrating relationship. In particular, we discuss tests for no cointegration and for the number of cointegrating relations. We allow for the possibility of a unit root in one of the regimes implying fewer number of cointegrating relations in one regime compared to the other. Using the tests proposed in Gonzalo and Pitarakis (2006) and Krishnakumar and Neto (2007), our results reveal that the data are consistent with both the UIRP and EHTS hypotheses. Moreover, we find evidence of asymmetry in the loading factors.

As a topic of further research, one may consider several thresholds in the TAR specification as well as several transition variables. This latter issue requires an extension of the asymptotic theory used in this paper.

Acknowledgements

We are grateful to two anonymous referees for many valuable comments. This work was supported by grant No. 100012-112038 from the Swiss National Fund for Scientific Research.

References

Ahn, S.K. and Reinsel, G.C.: 1990, Estimation for partially nonstationary multivariate autoregressive models, *Journal of the American Statistical Association* **85**(411), 813–823.

Andrews, D.W.K.: 1993, Tests for parameter instability and structural change with unknown change point, *Econometrica* **61**, 821–856.

Balke, N.S. and Fomby, T.B.: 1997, Threshold cointegration, *International Economic Review* **38**, 627–645.

Ballie, R.T. and Kiliç, R.: 2006, Do asymmetric and nonlinear adjustments explain the forward premium anomaly?, *Journal of International Money and Finance* **25**(1), 22–47.

Bekaert, G., Wei, M. and Xing, Y.: 2002, Uncovered interest rate parity and the term structure, *Working paper 8795*, NBER.

Brüggemann, R. and Lütkepohl, H.: 2005, Uncovered interest rate parity and the expectations hypothesis of the term structure: Empirical results for the U.S. and Europe, *Applied Economics Quarterly* **51**(2), 143–154.

Campbell, J.Y. and Shiller, R.J.: 1987, Cointegration and tests of present value models, *Journal of Political Economy* **95**(5), 1062–1088.

Campbell, J.Y. and Shiller, R.J.: 1991, Yield spreads and interest rate movements: A bird's eye view, *Review of Economic Studies* **58**, 495–514.

Caner, M. and Hansen, B.E.: 2001, Threshold autoregression with a unit root, *Econometrica* **69**(6), 1555–1596.

Enders, W. and Granger, C.W.J.: 1998, Unit-root test and asymmetric adjustment with an example using the term structure of interest rates, *Journal of Business & Economic Statistics* **16**(3), 304–311.

Enders, W. and Siklos, P.L.: 2001, Cointegration and threshold adjustment, *Journal of Business & Economic Statistics* **19**(2), 166–176.

Engel, C.: 1996, The forward discount anomaly and the risk premium: A survey of recent evidence, *Journal of Empirical Finance* **3**, 123–192.

González, M. and Gonzalo, J.: 1998, Threshold unit root processes, *Working paper*, U. Carlos III de Madrid, Madrid.

Gonzalo, J. and Pitarakis, P.Y.: 1998, Specification via model selection in vector error correction models, *Economics Letters* **60**, 321–328.

Gonzalo, J. and Pitarakis, P.Y.: 2002, Estimation and model selection based inference in single and multiple threshold models, *Journal of Econometrics* **110**, 319–352.

Gonzalo, J. and Pitarakis, P.Y.: 2006, Threshold effects in multivariate error correction models, *in* T.C. Mills and K. Patterson (eds), *Handbook of Econometrics, Volume 1: Econometric Theory*, Palgrave, Basingstoke, pp. 578–609.

Hall, A.D., Anderson, H.M. and Granger, C.W.J.: 1992, A cointegration analysis of the treasury bill yields, *Review of Economics and Statistics* **74**, 116–126.

Hansen, B.E.: 1996, Inference when a nuisance parameter is not identified under the null hypothesis, *Econometrica* **64**(2), 413–430.

Hansen, B.E.: 1997, Inference in TAR models, *Studies in Nonlinear Dynamics and Econometrics* **2**(1), 1–14.

Hansen, B.E. and Seo, B.: 2002, Testing for two-regime threshold cointegration in vector error-correction models, *Journal of Econometrics* **110**, 293–318.

Johansen, S.: 1988, Statistical analysis of cointegration vectors, *Journal of Economic Dynamics and Control* **12**, 231–254.

Johansen, S.: 1991, Estimation and hypothesis testing of cointegration vectors in Gaussian vector autoregressive models, *Econometrica* **59**(6), 1551–1580.

Johansen, S. and Juselius, K.: 1992, Testing structural hypotheses in a multivariate cointegration analysis of the PPP and the UIP for UK, *Journal of Econometrics* **53**, 211–244.

Kapetanios, G. and Shin, Y.: 2006, Unit root tests in three-regime SETAR models, *Econometrics Journal* **9**(2), 252–278.

Krishnakumar, J. and Neto, D.: 2007, Estimation and testing in threshold cointegrated systems using reduced rank regression, *Working paper*, Department of Econometrics, University of Geneva, Geneva.

Lo, M.C. and Zivot, E.: 2001, Threshold cointegration and nonlinear adjustment to the law of one price, *Macroeconomic Dynamics* **5**, 533–576.

Lütkepohl, H. and Saikkonen, P.: 2000, Testing for the cointegration rank of a VAR process, *Journal of Econometrics* **95**, 177–198.

Newey, W.K. and West, K.D.: 1987, A simple, positive semi-definite, heteroskedasticity and autocorrelation consistent covariance matrix, *Econometrica* **55**(3), 703–708.

Phillips, P.C.B.: 1991, Optimal inference in cointegrated systems, *Econometrica* **59**(3), 283–306.

Phillips, P.C.B. and Durlauf, S.N.: 1986, Multiple time series regression with integrated processes, *Review of Economic Studies* **53**, 473–496.

Seo, M.: 2006, Bootstrap testing for the null of no cointegration in a threshold vector error correction model, *Journal of Econometrics* **134**, 129–150.

Tsay, R.S.: 1989, Testing and modeling threshold autoregressive processes, *Journal of the American Statistical Association* **84**(405), 231–240.

van Tol, M.R. and Wolff, C.C.: 2005, Forecasting the spot exchange rate with the term structure of forward premia: Multivariate threshold cointegration, *Discussion Paper 4958*, CEPR.

Wolters, J.: 2002, Uncovered interest rate parity and the expectation hypothesis of the term structure: Empirical results for the US and Europe, *in* I. Klein and S. Mittnik (eds), *Contribution to Modern Economics: From Data Analysis to Economic Policy*, Kluwer, Boston, pp. 271–282.

Appendix

Table A.1. Critical values for the sup-Wald statistic $\sup\limits_{F(\theta)\in[\tau,1-\tau]} \mathcal{W}_{0,T}\left(F\left(\theta\right)\right)$ to test H_0 : $\pi^{(1)} = \pi^{(2)} = 0$ in the model (9) $vec\left(\Delta X^*\right) = Z^{(1)}\pi^{(1)} + Z^{(2)}\pi^{(2)} + vec\left(\varepsilon^*\right)$, where ΔX^* is p-dimensional and where $Z^{(j)}$ includes an intercept.

	90%	95%	99%	90%	95%	99%	90%	95%	99%
p		1			2			3	
$[\tau, 1-\tau]$									
$[0.15, 0.85]$	6.618	8.179	11.329	13.502	15.544	19.935	22.223	24.552	28.998
$[0.10, 0.90]$	7.053	8.529	11.666	13.967	15.991	20.380	22.805	25.204	29.669
$[0.05, 0.95]$	7.503	8.940	12.016	14.637	16.549	21.088	23.485	25.862	30.238
p		4			5			6	
$[\tau, 1-\tau]$									
$[0.15, 0.85]$	33.011	35.718	41.553	45.467	48.720	55.155	59.491	62.941	70.528
$[0.10, 0.90]$	33.665	36.318	42.108	46.334	49.424	55.601	60.411	64.199	71.284
$[0.05, 0.95]$	34.421	36.936	43.122	47.275	50.183	56.477	61.542	65.277	71.849

Note: Calculated from 10,000 simulations.

Table A.2. Critical values for threshold Wald statistic $\sup\limits_{F(\theta)\in[\tau,1-\tau]} \mathcal{W}_T^{(j)}\left(F\left(\theta\right), r_0\right)$ to test $H_0 : \tilde{\lambda}_j = 0$ i.e. $r = r_0$ with r the number of of cointegration relationships in the model (12) $\Delta \tilde{X}_t^{*(j)} = \tilde{\mu}^{(j)} + \tilde{\kappa}_j^{(1)} z_{t-1}^{(j)} + \tilde{\lambda}_j \nu_{t-1}^{(j)} + \tilde{\varepsilon}_t^{*(j)}$, for $j = 1, 2$, where $\Delta \tilde{X}_t^{*(j)}$ is p-dimensional.

	90%	95%	99%	90%	95%	99%	90%	95%	99%
$p - r$		1			2			3	
$[\tau, 1-\tau]$									
$[0.15, 0.85]$	8.107	9.609	12.941	17.091	19.193	23.660	29.481	32.082	37.730
$[0.10, 0.90]$	8.421	9.974	13.414	17.805	20.002	24.439	30.857	33.750	39.370
$[0.05, 0.95]$	8.790	10.288	13.788	18.661	20.782	25.702	32.124	35.240	40.954
$p - r$		4			5			6	
$[\tau, 1-\tau]$									
$[0.15, 0.85]$	44.962	48.156	54.768	64.349	68.371	75.403	86.866	91.392	100.238
$[0.10, 0.90]$	46.990	50.419	57.077	67.464	71.465	78.779	91.105	95.881	104.086
$[0.05, 0.95]$	49.196	52.507	59.959	70.793	74.709	82.514	95.720	100.344	109.310

Source: Krishnakumar and Neto (2007), table 2.

Table A.3. Estimates of the TVECM for various transition variables.

s_t	$\Delta\left(r_{us}-r_{ch}\right)_{t-1}$	$\Delta\left(r_{us}-r_{ch}\right)_{t-2}$	$\Delta\left(r_{us}-r_{ch}\right)_{t-3}$	$\left(r_{us}-r_{ch}\right)_{t-1}$
$\sup \mathcal{W}_0$	173.246	151.936	164.199	158.080
		$H_0 : r_0$		
$r_0 = 1$	$65.090^{(1)}$	69.799	75.639	69.549
	$72.941^{(2)}$	71.005	49.146	64.083
$r_0 = 2$	51.766	37.767	38.473	43.922
	29.655	35.413	21.570	41.161
$r_0 = 3$	7.193	4.363	6.618	5.940
	6.637	7.105	6.902	10.667
		ESTIMATES for $(r_0 = 3, \tau = 0.15)$		
β_1	-0.887	-1.148	-1.440	-1.270
	(0.197)	(0.199)	(0.223)	(0.216)
β_2	-0.939	-1.000	-1.302	-1.131
	(0.186)	(0.152)	(0.181)	(0.177)
β_3	-1.009	-1.011	-1.014	-1.009
	(0.018)	(0.017)	(0.017)	(0.018)
β_4	0.944	1.147	1.106	1.122
	(0.119)	(0.104)	(0.092)	(0.099)
LR	-32.196^a	-30.982^a	-12.769^a	-30.337^a
θ	-0.445	0.252	0.592	0.724
BIC^*	-11.700	-11.711	-11.715	-11.708
$\sup \mathcal{W}_T$	40.007	26.404	34.567	25.978

[1] Sup-Wald Statistic for regime 1: $\sup \mathcal{W}^{(1)}$.

[2] Sup-Wald Statistic for regime 2: $\sup \mathcal{W}^{(2)}$.

* $BIC = \log\left|\hat{\Omega}\right| + \frac{(q+r)d^2}{T}\log(T)$, where q is the order of the VAR and p the dimension of the model.

[a] The null hypothesis $H_0 : \beta_1 = \beta_2 = \beta_3 = -1$ could not be rejected at 1%, 5% and 10%.

Classification Using Optimization: Application to Credit Ratings of Bonds

Vladimir Bugera, Stan Uryasev, and Grigory Zrazhevsky

University of Florida, ISE, Risk Management and Financial Engineering Lab,
uryasev@ufl.edu

Summary. The classification approach, previously considered in credit card scoring, is extended to multi class classification in application to credit rating of bonds. The classification problem is formulated as minimization of a penalty constructed with quadratic separating functions. The optimization is reduced to a linear programming problem for finding optimal coefficients of the separating functions. Various model constraints are considered to adjust model flexibility and to avoid data overfitting. The classification procedure includes two phases. In phase one, the classification rules are developed based on "in-sample" dataset. In phase two, the classification rules are validated with "out-of-sample" dataset. The considered methodology has several advantages including simplicity in implementation and classification robustness. The algorithm can be applied to small and large datasets. Although the approach was validated with a finance application, it is quite general and can be applied in other engineering areas.

Key words: Bond ratings, credit risk, classification.

1 Introduction

We consider a general approach for classifying objects into several classes and apply it to a bond-rating problem. Classification problems become increasingly important in the decision science. In finance, they are used for grouping financial instruments according to their risk, or profitability characteristics. In the bonds rating problem, for example, the debt instruments are arranged according to the likelihood of debt issuer to default on the defined obligation.

Mangasarian et al. (1995) used a *utility function* for the failure discriminant analysis (applications to breast cancer diagnosis). The utility function was considered to be linear in control parameters and indicator variables and it was found by minimizing the error of misclassification. Zopounidis and Doumpos (1997), Zopounidis et al. (1998) and Pardalos et al. (1997) used linear utility functions for trichotomous classifications of credit card applications. Konno and Kobayashi (2000) and Konno et al. (2000) considered utility functions, quadratic in indicator parameters and linear in decision parameters. The approach was tested with the classification of enterprises and breast cancer diagnosis. Konno and Kobayashi (2000) and Konno et al. (2000) imposed convexity constraints on utility functions in order to avoid discontinuity of discriminant regions. Similar to Konno and Kobayashi (2000) and Konno et al. (2000), Bugera et al. (2003) applied a quadratic utility function to trichotomous classification, but instead of convexity constraints, monotonicity constraints reflecting experts' opinions were used. The approach by Bugera et al. (2003) is closely related to ideas by Zopounidis and Doumpos (1997), Zopounidis et al. (1998) and Pardalos et al. (1997); it considers a multi-class classification with several levelsets of the utility function, where every levelset corresponds to a separate class.

We extend the classification approach considered by Bugera et al. (2003). Several innovations improving the efficiency of the algorithm are suggested:

- A *set* of utility functions, called *separating functions*, is used for classification. The separating functions are quadratic in indicator variables and linear in decision variables. A set of optimal separating functions is found by minimizing the misclassification error. The problem is formulated as a linear programming problem w.r.t. decision variables.
- To control flexibility of the model and avoid overfitting we impose various *new constraints* on the separating functions.

Controlling flexibility of the model with constraints is crucially important for the suggested approach. Quadratic separating functions (depending upon the problem dimension) may have a very large number of free parameters. Therefore, a tremendously large dataset may be needed to "saturate" the model with data. Constraints reduce the number of degrees of freedom of the model and adjust "flexibility" of the model to the size of the dataset.

This paper is focused on a numerical validation of the proposed algorithm. We rated a set of international bonds using the proposed algorithm. The dataset for the case study was provided by the research group of the RiskSolutions branch of Standard and Poor's, Inc. We investigated the impact of model flexibility on classification characteristics of the algorithm and compared performance of several models with different types of constraints. Experiments showed the importance of constraints adjusting the flexibility of the model. We studied "in-sample" and "out-of-sample" characteristics of the suggested algorithm. At the first stage of the algorithm, we minimized the empirical risk, that is, the error of misclassification on a training set (in-sample error). However, the real objective of the algorithm is to classify objects outside the training set with a minimal error (out-of-sample error). The in-sample error is always not greater than the out-of-sample error. Similar issues were studied in the Statistical Learning Theory (Vapnik, 1998). For validation, we used the leave-one-out cross-validation scheme. This technique provides the highest confidence level while effectively improving the predicting power of the model.

The advantage of the considered methodology is its simplicity and consistency if compared to the proprietary models which are based on the combination of various techniques, such as expert models, neural networks, classification trees etc.

The paper is organized as follows. Section 2 provides a general description of the approach. Section 3 considers constraints applied to the studied optimization problem. Section 4 discusses techniques for choosing model flexibility. Section 5 explains how the errors estimation was done for the models. Section 6 discusses the bond-rating problem used for testing the methodology. Section 7 describes the datasets used in the study. Section 8 presents the results of computational experiments and analyses obtained results. We finalize the paper with concluding remarks in Section 9.

2 Description of Methodology

The *object space* is a set of elements (objects) to be classified. Each element of the space has n quantitative characteristics describing properties of the considered objects.

We represent objects by n-dimensional vectors and the object space by an n-dimensional set $\Psi \subset \mathrm{R}^n$. The classification problem assigns elements of the object space Ψ to several classes so that each class consists of elements with similar properties. The classification is based on available prior information. In our approach, the prior information is provided by a set of objects $S = \{\mathbf{x}_1, .., \mathbf{x}_m\}$ with known classification (*in-sample dataset*). The purpose of the methodology is to develop a classification algorithm that assigns a class to a new object based on the in-sample information.

Let us consider a classification problem with objects having n characteristics and J classes. Since each object is represented by a vector in a multi-dimensional space R^n, the classification can be defined by an integer-valued function $f_0(\mathbf{x})$, $\mathbf{x} \in \mathrm{R}^n$. The value of the function defines the class of an object. We call $f_0(\mathbf{x})$ a *classification function*. This function splits the object space into J non-intersecting areas:

$$\mathrm{R}^n = \bigcup_{i=1}^{J} F_i, \; F_i \cap F_j = \emptyset, \; F_i \neq \emptyset, \; F_j \neq \emptyset, \; i \neq j \quad , \tag{1}$$

where each area F_i consists of elements belonging to the corresponding class i:

$$F_i = \{ \mathbf{x} \in \mathrm{R}^n | \; f_0(\mathbf{x}) = i \} \quad . \tag{2}$$

We can approximate the classification function $f_0(\mathbf{x})$ using optimization methods. Let $F(\mathbf{x})$ be a cumulative distribution function of objects in the object space R^n, and Λ be a parameterized set of discrete-value approximating functions. Then, the function $f_0(\mathbf{x})$ can be approximated by solving the following minimization problem:

$$\min_{f \in \Lambda} \int_{R^n} Q\left(f\left(\mathbf{x}\right) - f_0\left(\mathbf{x}\right)\right) dF\left(\mathbf{x}\right), \tag{3}$$

where $Q\left(f\left(\mathbf{x}\right) - f_0\left(\mathbf{x}\right)\right)$ is a penalty function defining the value of misclassification for a single object $\mathbf{x}$. The optimal solution $f\left(\mathbf{x}\right)$ of optimization problem (3) gives an approximation of the function $f_0\left(\mathbf{x}\right)$. The main difficulty in solving problem (3) is the discontinuity of functions $f\left(\mathbf{x}\right)$ and $f_0\left(\mathbf{x}\right)$ leading to non-convex optimization. To circumvent this difficulty, we reformulate problem (3) in a convex optimization setting.

Let us consider a classification function $\bar{f}(\mathbf{x})$ defining a classification on the object space R^n. Suppose we have a set of continuous functions $U_0(\mathbf{x}), ..., U_J\left(\mathbf{x}\right)$. We call them *separating functions for the classification function* $\bar{f}(\mathbf{x})$ if for every object $\mathbf{x}^*$ from class $i = \bar{f}(\mathbf{x}^*)$ values of functions with numbers lower than i are positive:

$$U_0\left(\mathbf{x}^*\right), ..., U_{i-1}\left(\mathbf{x}^*\right) > 0; \tag{4}$$

values of functions with numbers higher or equal to i are negative or zeros:

$$U_i\left(\mathbf{x}^*\right), ..., U_J\left(\mathbf{x}^*\right) \leq 0. \tag{5}$$

If we know the functions $U_0(\mathbf{x}), ..., U_J\left(\mathbf{x}\right)$ we can classify objects according to the following rule:

$$\left\{ \begin{array}{ll} \forall p = 0, .., i-1 : U_p(\mathbf{x}^*) > 0 \\ \forall q = i, .., J : \quad U_q\left(\mathbf{x}^*\right) \leq 0 \end{array} \right\} \Leftrightarrow \left\{ \bar{f}\left(\mathbf{x}^*\right) = i \right\}. \tag{6}$$

The class number of an object $\mathbf{x}^*$ can be determined by the number of positive values of the functions: an object in i class has exactly i positive separating functions. Figure 1 illustrates this property.

Suppose we can represent any classification function $f\left(\mathbf{x}\right)$ from a parameterized set of functions Λ by a set of separating functions. By constructing a penalty for the classification $f_0\left(\mathbf{x}\right)$ using separating functions, we can formulate optimization problem (3) with respect to the parameters of the separating functions.

Suppose the classification function $\bar{f}(\mathbf{x})$ is defined by a set of separating functions $U_0(\mathbf{x}), ..., U_J\left(\mathbf{x}\right)$. We denote by $D_{f_0}\left(U_0, .., U_J\right)$ an integral function measuring deviation of the classification (implied by the separating functions $U_0(\mathbf{x}), ..., U_J\left(\mathbf{x}\right)$) from the true classification defined by $f_0(\mathbf{x})$:

$$D_{f_0}\left(U_0, .., U_J\right) = \int_{R^n} Q\left(U_0\left(\mathbf{x}\right), .., U_J\left(\mathbf{x}\right), f_0\left(\mathbf{x}\right)\right) dF\left(\mathbf{x}\right) \quad , \tag{7}$$

where $Q\left(U_0(\mathbf{x}), .., U_J(\mathbf{x}), f_0(\mathbf{x})\right)$ is a penalty function.

Further, we consider the following penalty function:

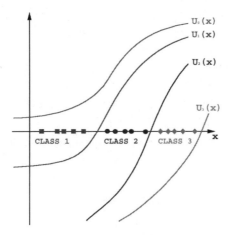

Fig. 1. Classification by separating functions. At each point (object), values of the separating functions with the indices lower than i are positive; the values of the remaining functions are negative or zero. Class of an object is determined by the number of functions with positive values.

$$Q\left(U_0, .., U_J, f_0\left(\mathbf{x}\right)\right) = \sum_{k=0}^{f_0(\mathbf{x})-1} \lambda_k \left(-U_k\left(\mathbf{x}\right)\right)^+ + \sum_{k=f_0(\mathbf{x})}^{J} \lambda_k \left(U_k\left(\mathbf{x}\right)\right)^+, \qquad (8)$$

where $(y)^+ = \max\left(0, y\right)$, and λ_k, $k = 0, .., J$ are positive parameters. The penalty function equals zero if $U_k\left(\mathbf{x}\right) \geq 0$ for $k = 0, .., f_0\left(\mathbf{x}\right) - 1$, and $U_k\left(\mathbf{x}\right) \leq 0$ for $k = f_0\left(\mathbf{x}\right), .., J$. If the classification defined by the separating functions $U_0(\mathbf{x}), ..., U_J\left(\mathbf{x}\right)$ according to rule (6) coincides with the true classification $f_0(\mathbf{x})$, then, the value of penalty function (8) equals zero for any object $\mathbf{x}$. If the penalty is positive for an object $\mathbf{x}$ then separating functions misclassify this object. Therefore, the penalty function defined by (8) provides the condition of the correct classification by the separating functions:

$$\{Q\left(U_0, .., U_J, f_0\left(\mathbf{x}\right)\right) = 0\}$$
$$\Updownarrow \qquad (9)$$
$$\left\{ U_k(\mathbf{x}) > 0 > U_l(\mathbf{x}), \ \forall k = 1, .., f_0\left(\mathbf{x}\right) - 1, \forall l = f_0\left(\mathbf{x}\right), .., J \right\} .$$

The choice of penalty function (8) is motivated by the possibility of building an efficient algorithm for the optimization of integral (7) when the separating functions are linear w.r.t. control parameters. In this case, optimization problem (3) can be reduced to linear programming.

After introducing the separating functions we reformulate optimization problem (3) as follows:

$$\min_{U_0, .., U_J} D_{f_0}\left(U_0, .., U_J\right). \qquad (10)$$

This optimization problem finds optimal separating functions. We can approximate the function $D_{f_0}(U_0, .., U_J)$ by sampling objects according to the cumulative distribution function $F(\mathbf{x})$. Assuming that $\mathbf{x}^1, .., \mathbf{x}^m$ are some sample points, the approximation of $D_{f_0}(U_0, .., U_J)$ is given by:

$$\tilde{D}_{f_0}^m(U_0, .., U_J) = \frac{1}{m}\sum_{i=1}^m Q\left(U_0\left(\mathbf{x}^i\right), .., U_J\left(\mathbf{x}^i\right), f_0\left(\mathbf{x}^i\right)\right). \quad (11)$$

Therefore, the approximation of deviation function (7) becomes:

$$\tilde{D}_{f_0}^m(U_0, .., U_J) =$$
$$= \frac{1}{m}\sum_{i=1}^m \left(\sum_{k=0}^{f_0(\mathbf{x})-1} \lambda_k\left(-U_k\left(\mathbf{x}^i\right)\right)^+ + \sum_{k=f_0(\mathbf{x})}^J \lambda_k\left(U_k\left(\mathbf{x}^i\right)\right)^+\right). \quad (12)$$

To avoid possible ambiguity when the value of a separating function equals 0, we introduced a small positive constant δ inside of each term in the penalty function (constant δ has to be chosen small enough in order not to have a significant impact on the final classification):

$$Q(U_0, .., U_J, f_0(\mathbf{x})) =$$
$$= \sum_{k=0}^{f_0(\mathbf{x})-1} \lambda_k\left(-U_k(\mathbf{x}) + \delta\right)^+ + \sum_{k=f_0(\mathbf{x})}^J \lambda_k\left(U_k(\mathbf{x}) + \delta\right)^+. \quad (13)$$

The penalty function equals zero if $U_k(\mathbf{x}) \geq \delta$ for $k = 0, .., f_0(\mathbf{x}) - 1$, and $U_k(\mathbf{x}) \leq -\delta$ for $k = f_0(\mathbf{x}), .., J$. The approximation of deviation function (7) becomes

$$\tilde{D}_{f_0}^m(U_0, .., U_J) =$$
$$= \frac{1}{m}\sum_{i=1}^m \left(\sum_{k=0}^{f_0(\mathbf{x})-1} \lambda_k\left(-U_k\left(\mathbf{x}^i\right) + \delta\right)^+ + \sum_{k=f_0(\mathbf{x})}^J \lambda_k\left(U_k\left(\mathbf{x}^i\right) + \delta\right)^+\right). \quad (14)$$

Further, we parameterized each separating function by a K-dimensional vector $\boldsymbol{\alpha} \in A \subset \mathrm{R}^K$. Parameter K is defined by the design of the separating functions. Therefore, a set of separating functions $U_{(\boldsymbol{\alpha}^0, .., \boldsymbol{\alpha}^J)}(\mathbf{x}) = \{U_{\boldsymbol{\alpha}^0}, .., U_{\boldsymbol{\alpha}^J}\}$ is determined by a set of vectors $\{\boldsymbol{\alpha}^0, .., \boldsymbol{\alpha}^J\}$. With this parameterization we reformulated the problem (10) as minimization of the convex piece-wise linear functions:

$$\min_{\boldsymbol{\alpha}^0, ..., \boldsymbol{\alpha}^J \in A} \sum_{i=1}^m \left(\sum_{k=0}^{f_0(\mathbf{x}^i)-1} \lambda_k\left(-U_{\boldsymbol{\alpha}^k}\left(\mathbf{x}^i\right) + \delta\right)^+ + \sum_{k=f_0(\mathbf{x}^i)}^J \lambda_k\left(U_{\boldsymbol{\alpha}^k}\left(\mathbf{x}^i\right) + \delta\right)^+\right),$$
$$(15)$$

where $A \subset \mathrm{R}^K$. By introducing new variables σ_i^j, we reduced problem (15) to equivalent mathematical programming problem with the linear objective function:

$$\min_{\alpha,\sigma} \sum_{i=1}^{m} \sum_{j=1}^{J} \lambda_j \sigma_i^j$$
$$\sigma_i^j \geq -U_{\alpha^j}\left(\mathbf{x}^i\right) + \delta, \, j = 0,..,f_0(\mathbf{x}^i) - 1$$
$$\sigma_i^j \geq U_{\alpha^j}\left(\mathbf{x}^i\right) + \delta, \, j = f_0(\mathbf{x}^i),..,J \tag{16}$$
$$\boldsymbol{\alpha}^0,..,\boldsymbol{\alpha}^J \in A$$
$$\sigma_i^1,..,\sigma_i^J \geq 0$$

Further, we consider that the separating functions are linear in control parameters $\alpha_1,..,\alpha_K$. In this case the separating functions can be represented in the following form:

$$U_{\boldsymbol{\alpha}}\left(\mathbf{x}\right) = \sum_{k=1}^{K} \alpha_k g_k\left(\mathbf{x}\right). \tag{17}$$

In this case, optimization problem (16) for finding optimal separating functions can be reduced to the following linear programming problem:

$$\min_{\alpha,\sigma} \sum_{i=1}^{m} \sum_{j=1}^{J} \lambda_j \sigma_i^j$$
$$\sigma_i^j + \sum_{k=1}^{K} \alpha_k^j g_k\left(\mathbf{x}^i\right) \geq \delta, \, j = 0,..,f_0(\mathbf{x}^i) - 1$$
$$\sigma_i^j - \sum_{k=1}^{K} \alpha_k^j g_k\left(\mathbf{x}^i\right) \geq \delta, \, j = f_0(\mathbf{x}^i),..,J \tag{18}$$
$$\boldsymbol{\alpha}^0,..,\boldsymbol{\alpha}^J \in A$$
$$\sigma_i^1,..,\sigma_i^J \geq 0$$

Further, we consider quadratic (in indicator variables) separating functions:

$$U_j\left(\mathbf{x}\right) = \sum_{k=1}^{n} \sum_{l=1}^{n} a_{kl}^j x_i x_j + \sum_{k=1}^{n} b_k^j x_k + c^j, \, j = 0,..,J. \tag{19}$$

Optimization problem (18) with quadratic separating functions is reformulated as follows:

$$\min_{a,b,c,\sigma} \sum_{i=1}^{m} \sum_{j=1}^{J} \lambda_j \sigma_i^j$$
$$\sigma_i^j + \sum_{k=1}^{n} \sum_{l=1}^{n} a_{kl}^j x_k^i x_l^i + \sum_{i=1}^{n} b_k^j x_k^i + c^j \geq \delta, \, j = 0,..,f_0(\mathbf{x}^i) - 1$$
$$\sigma_i^j - \sum_{k=1}^{n} \sum_{l=1}^{n} a_{kl}^j x_k^i x_l^i - \sum_{k=1}^{n} b_k^j x_k^i - c^j \geq \delta, \, j = f_0(\mathbf{x}^i),..,J \tag{20}$$
$$\sigma_i^1,..,\sigma_i^J \geq 0$$

Although there are $J+1$ separating functions in problem (20), only $J-1$ functions are essential for the classification. The functions

$$\sum_{k=1}^{n} \sum_{l=1}^{n} a_{kl}^0 x_k x_l + \sum_{k=1}^{n} b_k^0 x_k + c^0$$

and

$$\sum_{k=1}^{n}\sum_{l=1}^{n} a_{kl}^{J} x_k x_l + \sum_{k=1}^{n} b_k^{J} x_k + c^{J}$$

are boundary functions. For all the classified objects, the value of the first boundary function is positive, and the value of the second boundary function is negative. This can be easily achieved by setting $c^0 = M$ and $c^J = -M$, where M is a sufficiently large number. Thus, these functions can be removed from optimization problem (20). However, these boundary functions can be used for adjusting flexibility of the classification model. In the next section, we will show how to use these functions for imposing the so-called "squeezing" constraints.

For the case $J = 2$ with only two classes, problem (20) finds a quadratic surface $\sum_{k=1}^{n}\sum_{l=1}^{n} a_{kl}^{1} x_k x_l + \sum_{k=1}^{n} b_k^{1} x_k + c^{1} = 0$ dividing the object space R^n into two areas. After solving optimization problem (20) we expect that a majority of objects from the first class will belong to the first area. On these points the function $\sum_{k=1}^{n}\sum_{l=1}^{n} a_{kl}^{1} x_k x_l + \sum_{k=1}^{n} b_k^{1} x_k + c^{1}$ is positive. Similar, for a majority of objects from the second class the function $\sum_{k=1}^{n}\sum_{l=1}^{n} a_{kl}^{1} x_k x_l + \sum_{k=1}^{n} b_k^{1} x_k + c^{1}$ is negative.

For the case with $J > 2$, the geometrical interpretation of optimization problem (20) refers to the partition of the object space R^n into J areas by $J - 1$ non-intersecting quadratic surfaces

$$\sum_{k=1}^{n}\sum_{l=1}^{n} a_{kl}^{j} x_k x_l + \sum_{k=1}^{n} b_k^{j} x_k + c^{j} = 0, \quad j = 1, .., J - 1.$$

Additional feasibility constraints, that assure non-intersection of the surfaces, will be discussed in the following section.

3 Constraints

The considered separating functions, especially the quadratic functions, may be too flexible (have too many degrees of freedom) for datasets with a small number of datapoints. Imposing additional constraints may reduce excessive model flexibility. In this section, we will discuss different types of constraints applied to the model.

Konno and Kobayashi (2000) and Konno et al. (2000) considered convexity constraints on indicator variables of a quadratic utility function. Bugera et al. (2003) imposed monotonicity constraints on the model to incorporate expert preferences.

Constraints play a crucial role in developing the classification model because they reduce excessive flexibility of a model for small training datasets.

Moreover, a classification with multiple separating functions may not be possible for the majority of objects if appropriate constraints are not imposed.

3.1 Feasibility Constraints (F-Constraints)

For classification with multiple separating functions we may potentially come to a possible intersection of separating surfaces. This may lead to inability of the approach to classify some objects. To circumvent this difficulty, we introduce feasibility constraints, that keep the separating functions apart from each other. It makes possible to classify any new point by rule (6). In general, these constraints have the form:

$$U_j(\mathbf{x}) \leq U_{j-1}(\mathbf{x}); \, j = 1, .., J; \, \mathbf{x} \in W \subset R^n \,, \tag{21}$$

where W is a set on which we want to achieve the feasibility. We do not consider $W = R^n$, because it leads to "parallel" separating surfaces, which were studied in the previous work by Bugera et al. (2003). In this section, we consider a set W with a finite number of elements. In particular, we consider the set W being the training set plus the set of objects we want to classify (out-of-sample dataset). In this case, constraints (21) can be rewritten as

$$U_j(\mathbf{x}_i) \leq U_{j-1}(\mathbf{x}_i); \, j = 1, .., J; \, i = 1, .., r \,, \tag{22}$$

where r is a number of objects in the set W. The fact that we use out-of-sample points does not cause any problems because we use the data without knowing their class numbers.

Since classification without feasibility constraints may lead to inability of classifying new objects (especially for small training sets), we will always include the feasibility constraints (22) to classification problem (16). For quadratic separating functions, classification problem (20) with feasibility constraints can be rewritten as

$$\min_{a,b,c,\sigma} \sum_{i=1}^{m} \sum_{j=1}^{J} \lambda_j \sigma_i^j$$
$$\sigma_i^j + \sum_{k=1}^{n} \sum_{l=1}^{n} a_{kl}^j x_k^i x_l^i + \sum_{i=1}^{n} b_k^j x_k^i + c^j \geq \delta, \, j = 0, .., f_0(\mathbf{x}^i) - 1$$
$$\sigma_i^j - \sum_{k=1}^{n} \sum_{l=1}^{n} a_{kl}^j x_k^i x_l^i - \sum_{k=1}^{n} b_i^j x_k^i - c^j \geq \delta, \, j = f_0(\mathbf{x}^i), .., J$$
$$\sum_{k=1}^{n} \sum_{l=1}^{n} a_{kl}^{j-1} x_k^i x_l^i + \sum_{i=1}^{n} b_k^{j-1} x_k^i + c^{j-1} - \sum_{k=1}^{n} \sum_{l=1}^{n} a_{kl}^j x_k^i x_l^i + \sum_{i=1}^{n} b_k^j x_k^i + c^j \geq \delta$$
$$i = 1, .., r, \, j = 1, .., J$$
$$\sigma_i^1, .., \sigma_i^J \geq 0$$
$$\tag{23}$$

3.2 Monotonicity Constraints (M-Constraints)

We use monotonicity constraints to incorporate the preference of greater values of indicator variables. In financial applications, monotonicity with respect to some indicators follows from "engineering" considerations. For instance, in the bond rating problem, considered in this paper, we have indicators (return on capital, operating-cash-flow-to-debt ratio, total assets), greater values of which lead to the higher ratings of bonds. If we enforce the monotonicity of the separating functions with respect to these indicators, objects with greater values of the indicators will have higher ratings. For a smooth function $h(\mathbf{x})$, the monotonicity constraints can be written as the constraint on the non-negativity of the first partial derivative:

$$\frac{\partial h(\mathbf{x})}{\partial x_i} \geq 0, \quad i \in \{1, .., n\}. \tag{24}$$

For the case of linear separating functions,

$$U_\alpha(\mathbf{x}) = \sum_{k=1}^{K} a_k^\alpha x_k + c^\alpha, \tag{25}$$

the monotonicity constraints are

$$a_k \geq 0, k \in K \subset \{1, .., n\}. \tag{26}$$

3.3 Positivity Constraints for Quadratic Separating Functions (P-Constraints)

Unlike the case with the linear functions, imposing exact monotonicity constraints on a quadratic function

$$U_j(\mathbf{x}) = \sum_{k=1}^{n} \sum_{l=1}^{n} a_{kl}^j x_k x_l + \sum_{k=1}^{n} b_k^j x_k + c^j \tag{27}$$

is a more complicated issue (indeed, in general, the quadratic function is not monotonic in the whole space R^n). Instead of imposing exact monotonicity constraints we consider the following constraints (we call them "positivity constraints"):

$$a_{kl}^\gamma \geq 0, b_k^\gamma \geq 0, k, l = 1, .., n. \tag{28}$$

Bugera et al. (2003) demonstrated that the positivity constraints can be easily included into the linear programming formulation of the classification problem. They do not significantly increase the computational time of the classification procedure, but provide robust results for small training datasets and datasets with missing or erroneous information. These constraints impose monotonicity with respect to variables x_i, $i = 1, .., n$ on the positive part $R^+ = \{ \mathbf{x} \in R^n \mid x_k \geq 0, k = 1, .., n \}$ of the object space R^n.

3.4 Gradient Monotonicity Constraints (GM-Constraints)

Another way to enforce monotonicity of quadratic separating functions is to restrict the gradient of separating functions on some set of objects X^* (for example, on a set of objects combining in-sample and out-of-sample points):

$$\left.\frac{\partial U_\alpha(\mathbf{x})}{\partial x^s}\right|_{\mathbf{x}=\mathbf{x}^*} \geq \gamma_s, \ \alpha = 0, .., J, \ s = 1, .., K, \ \mathbf{x}^* \in X^*, \qquad (29)$$

where

$$\left.\frac{\partial U_\alpha(\mathbf{x})}{\partial x^s}\right|_{\mathbf{x}=\mathbf{x}^*} = \left.\frac{\partial\left(\sum\limits_{k=1}^{n}\sum\limits_{l=1}^{n} a_{kl}^\alpha x_k x_l + \sum\limits_{k=1}^{n} b_k^\alpha x_k + c^\alpha\right)}{\partial x^s}\right|_{\mathbf{x}=\mathbf{x}^*} =$$

$$= \sum_{k=1}^{n} (a_{ks}^\alpha + a_{sk}^\alpha) x_k^* + b_s^\alpha. \qquad (30)$$

In constraint (29), $\gamma_s, \ s = 1, .., K$ are nonnegative constants.

3.5 Risk Constraints (R-Constraints)

Another important constraint that we apply to the model is the risk constraint. The risk constraint restricts the average value of the penalty function for misclassified objects. For this purpose we use the concept of Conditional Value-at-Risk (CVaR). The optimization approach for CVaR introduced by Rockafellar and Uryasev (2000), was further developed in Rockafellar and Uryasev (2002) and Rockafellar et al. (2002). Suppose X is a training set and for each object from this set the class number is known. In other words, discrete-value function $f(x)$ assigns the class for each object from set X. Let J be a total number of classes. The function $f(x)$ splits the set X into a set of subsets $\{X_1, .., X_J\}$,

$$X_j = \{\ x \mid x \in X, f(x) = j\ \}. \qquad (31)$$

Let I_j be the number of elements (objects) in the set X_j. We define the CVaR constraints as follows:

$$\left\{\varsigma_j^+ + \frac{1}{\kappa}\frac{1}{I_j} \sum_{\mathbf{x}\in X_j} \left(U_j(\mathbf{x}) - \varsigma_j^+\right)^+\right\} \leq$$
$$\leq \left\{-\varsigma_{j+1}^- - \frac{1}{\kappa}\frac{1}{I_{j+1}} \sum_{\mathbf{x}\in X_{j+1}} \left(-U_j(\mathbf{x}) - \varsigma_{j+1}^-\right)^+\right\} - \delta \qquad j = 1, .., I-1, \quad (32)$$

where ς_j^+ and ς_j^- are free variables; δ and κ are parameters.

For an optimal solution of the optimization problem (16) with the constraints (32), the left-hand part of the inequality is an average of $U_j(\mathbf{x})$ for the largest $\kappa\%$ of objects from the jth class; the right-hand part of the inequality is an average of $U_j(\mathbf{x})$ for the smallest $\kappa\%$ of objects from the $(j+1)^{th}$ class. We will call these values κ-CVaR largest of the jth class and κ-CVaR smallest of the $(j+1)^{th}$ class, correspondingly. The general sense of the constraints is the following: the κ-CVaR largest of the jth class is smaller at least by δ than the κ-CVaR smallest of the $(j+1)^{th}$ class.

3.6 Monotonicity of Separating Function with Respect to Class Number (MSF-Constraints)

By introducing these constraints, we move apart values of the separating functions for the objects belonging to different classes. Moreover, the constraints imply monotonicity of the separating functions with respect to the index denoting the class number. The constraint is set for every pair of the objects belonging to the neighboring classes. For every p and q, so that $f_0(\mathbf{x}^p) + 1 = f_0(\mathbf{x}^q)$, the following constraint is imposed on the optimization model:

$$U_i(\mathbf{x}^p) - U_i(\mathbf{x}^q) \geq \delta_{ipq}, \; i = 0, .., I, \tag{33}$$

where δ_{ipq} are non-negative constants, and I is the number of separating functions. Another way to impose monotonicity on the separating functions with respect to the number of class is to consider δ_{ipq} as variables, and include these variables into the objective function:

$$\sum_{i=1}^{m} \sum_{j=1}^{J} \lambda_j \sigma_i^j - \sum_{i,p,q} \delta_{ipq}. \tag{34}$$

3.7 Model Squeezing Constraints (MSQ-Constraints)

Squeezing constraints (implemented together with feasibility constraints) efficiently adjust the flexibility of the model and control the out-of-sample versus in-sample performance of the algorithm. The constraints have the following form:

$$U_0(\mathbf{x}^p) - U_J(\mathbf{x}^p) \leq S_{f(\mathbf{x}^p)}, \; p = 1, .., m. \tag{35}$$

With these constraints we bound variation of values of different separating functions on each object. Another way to squeeze the spread of the separating functions is to introduce a penalty coefficient for the difference of the functions in (35). In this case the objective function of problem (16) can be rewritten as:

$$\sum_{i=1}^{m} \sum_{j=1}^{J} \lambda_j \sigma_i^j + \sum_{p=1}^{m} \left(m_{f(\mathbf{x}^p)} \left(U_0(\mathbf{x}^p) - U_J(\mathbf{x}^p) \right) \right). \tag{36}$$

The advantage of the squeezing constraints is that it is easy to implement in the linear programming framework.

3.8 Level Squeezing Constraints (LSQ-Constraints)

This type of constraints is very similar to the model squeezing constraints. The difference is that instead of squeezing the boundary functions $U_0(\mathbf{x})$ and $U_J(\mathbf{x})$, we bound the absolute deviation of values of the separating functions from their mean values on each class of objects. The constraints have the following form:

$$\sum_{p:f_0(\mathbf{x}^p)=k} \left| U_j(\mathbf{x}^p) - \frac{1}{I_k} \sum_{q:f_0(\mathbf{x}^q)=k} U_j(\mathbf{x}^q) \right| \leq \gamma_{jk}, \qquad (37)$$

where γ_{jk} are positive constants, and I_k is the number of objects $\mathbf{x}$ in class K. Similar to constraints (35), constants γ_{jk} can be considered as variables and be included into the objective function:

$$\sum_{i=1}^{m} \sum_{j=1}^{J} \lambda_j \sigma_i^j + \sum_{j,k} \gamma_{jk}. \qquad (38)$$

4 Choosing Model Flexibility

This section explains our approach to adjusting flexibility of classification models without strict definitions and mathematical details.

The considered constraints form various models based on the optimization of quadratic separating functions. The models differ in the type of constraints imposed on the separating functions.

The major characteristics of a particular classification model are in-sample and out-of-sample errors. To find a classification model we solve optimization problem (16) constructed for a training dataset. The error of misclassification of a classification model on the training dataset is called the "in-sample" error. The error achieved by the constructed classification model on the objects outside of the training set is called the "out-of-sample" error. The misclassification error can be expressed in various ways. We measure the misclassification (if it is not specified otherwise) by the percentage of the objects on which the computed model gives wrong classification.

Theoretically, classification models demonstrate the following characteristics (see Figure 2). For small training sets, the model fits the data with zero in-sample error, whereas the expected out-of sample error is large. As the size of the training set increases, the expected in-sample error diverges from zero (the model cannot exactly fit the training data). Increasing the size of the training set leads to a larger expected in-sample error and a smaller expected out-of-sample error. For sufficiently large datasets, the in-sample and out-of-sample errors are quite close. In this case, we say that the model is saturated with data.

We say that the class of models A is more flexible than the class of models B if the class A includes the class B. Imposing constraints reduces the flexibility of the model. Figure 3 illustrates theoretical in/out-of-sample characteristics for two classes of models with different flexibilities. For small training sets, the less flexible model gives a smaller expected out-of-sample error (compared to the more flexible model) and, consequently, predicts better than the more flexible model. However, the more flexible models outperform the less flexible models for large training sets (more flexible models have a smaller expected

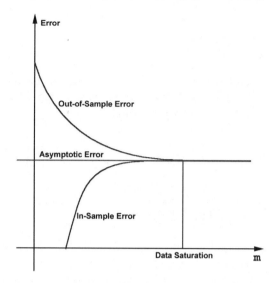

Fig. 2. In-sample and out-of-sample performance vs. size of training dataset.

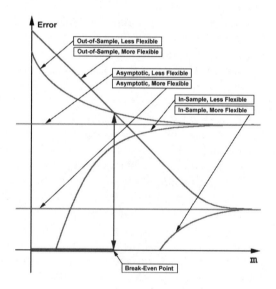

Fig. 3. In-sample and out-of-sample performance for different models.

out-of sample error compared to less flexible models). This is because the less flexible model requires less data for saturation than the more flexible model.

A more flexible model may need a large training dataset to provide a good prediction, while a less flexible model is saturated with data more quickly than a more flexible model. Therefore, for small datasets, less flexible models tend to outperform (out-of sample) more flexible models. However, for large datasets, more flexible models outperform (out-of-sample) less flexible ones.

We demonstrate these considerations with the following example. For illustration purposes we consider four different models (A, B, C and D) with different levels of flexibility. We applied these models to the classification of a dataset containing 278 in-sample datapoints and 128 out-of-sample datapoints. The in-sample dataset is used to construct linear programming problem (20). The solution of this problem is a set of separating quadratic functions (19). The in-sample and out-of-sample classification is done using the obtained separating functions.

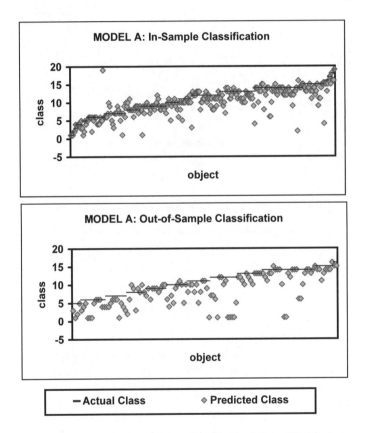

Fig. 4. In-sample and out-of-sample errors for model A. Model A uses parallel separating functions, and is the least flexible model among considered.

Figures 4, 5, 6, and 7 demonstrate the in-sample and out-of-sample performances of the considered models. Although the figures are based on actual computational results related to classification of bonds, they are presented in this paper for illustration purposes only. The horizontal axis of each graph

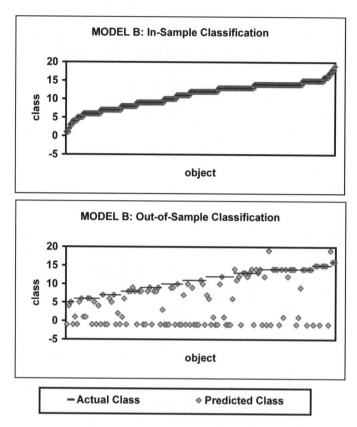

Fig. 5. In-sample and out-of-sample errors for model B. Model B uses separating functions without any constraints, and is the most flexible model. It has a better in-sample performance than Model A, but fails to assign classes for most of out-of-sample objects.

corresponds to the object number, which is ordered according to the object actual class number. The vertical line corresponds to the calculated by the model class number. The solid line on the graph represents the actual class number of an object. The round-point line represents the calculated class. The left graphs show in-sample classification: the class is computed by the separating functions for the objects from the in-sample dataset. The right graphs show out-of-sample classification: the class is computed by the separating functions for the objects from the out-of-sample dataset.

Model A uses "parallel" quadratic separating functions. This case corresponds to the classification model considered by Bugera et al. (2003), where instead of multiple separating functions one utility function was used. The classification with the utility function can be interpreted as a classification with separating functions (15) with the following constraints on the functions

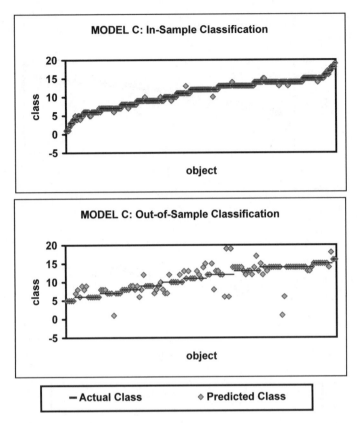

Fig. 6. In-sample and out-of-sample errors for model C. Model C is obtained from Model B by imposing feasibility constraints. This makes the out-of-sample prediction possible for all out-of-sample objects.

$$U_\alpha\left(\mathbf{x}\right) = \sum_{i=1}^{n} \sum_{j=1}^{n} a_{ij}^{\alpha} x_i x_j + \sum_{i=1}^{n} b_i^{\alpha} x_i + c^{\alpha}:$$

$$\forall \alpha_1, \alpha_2 \in \{0, .., J\} : \forall i, j \in \{0, .., n\} : a_{ij}^{\alpha_1} = a_{ij}^{\alpha_2}, b_i^{\alpha_1} = b_i^{\alpha_2} \qquad (39)$$

Constraints (39) make the separating functions parallel in the sense that the difference between the functions remains the same for all the points of the object space.

Figure 4 shows that Model A has large errors for both in-sample and out-of-sample calculations. According to Figure 8, when a model has approximately the same in-sample and out-of-sample errors, the data saturation occurs. Therefore, the out-of-sample performance cannot be improved by reducing flexibility of the model. A more flexible model should be considered.

Model B uses quadratic separating functions without any constraints. This model is more flexible than Model A, because it has more control variables.

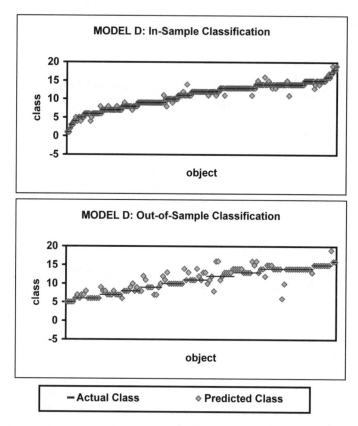

Fig. 7. In-sample and out-of-sample errors for model D. Model D is obtained from Model C by adding CVaR-risk constraints. Model D has the best out-of-sample performance among all the considered models (A, B, C, and D).

According to Figure 5, the in-sample error equals zero for Model B. However, in contrast to the in-sample behavior, the out-of-sample error is larger for Model B than for Model A. Moreover, Model B is not able to assign a class to many of the out-of-sample objects (on the picture the unpredictable result corresponds to the negative number -1 of the class). The separating functions may intersect, and some areas of the object are impossible to classify. For objects from these areas, the condition

$$U_0(\mathbf{x}^*), ..., U_{j-1}(\mathbf{x}^*) > 0 \tag{40}$$

and

$$U_j(\mathbf{x}^*), ..., U_J(\mathbf{x}^*) \leq 0 \tag{41}$$

are not satisfied for any j. It makes the result of the out-of-sample classification uninterpretable.

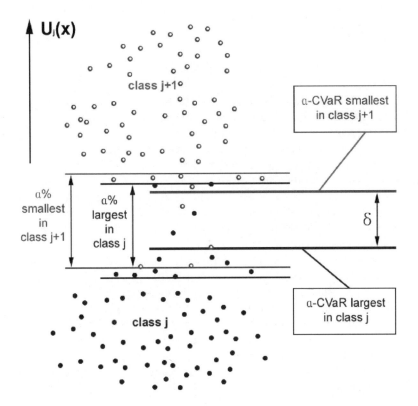

Fig. 8. Nature of risk constraint. For an optimal solution of the optimization problem with the CVaR constraints, the α-CVaR largest of the jth class (the average of $U_j(\mathbf{x})$ for the largest $\alpha\%$ of objects from the jth class) is smaller at least by δ than the α-CVaR smallest of the $(j+1)^{th}$ class (the average of $U_{j+1}(\mathbf{x})$ for the smallest $\alpha\%$ of objects from the $(j+1)^{th}$ class).

Model C is obtained by adding the feasibility constraints

$$U_{j-1}(\mathbf{x}_i) \le U_j(\mathbf{x}_i), \; j \in \{1, .., J\}, \; i = 1, .., m. \tag{42}$$

to Model B. Since Model C is less flexible then Model B, the in-sample error becomes greater than in Model B (see Figure 6). On the other hand, the model has a better out-of-sample performance than Models A and B. Moreover, a feasibility constraint makes the classification possible for any out-of-sample object. Since there is a significant discrepancy between in-sample and out-of-sample performances of the model, it is reasonable to impose more constraints on the model.

Model D is obtained by adding CVaR-risk constraints to Model C:

$$\begin{cases} \varsigma_j^+ + \frac{1}{\alpha}\frac{1}{I_j} \sum_{\mathbf{x}\in X_j} \left(U_j(\mathbf{x}) - \varsigma_j^+\right)^+ \leq \\ -\varsigma_{j+1}^- - \frac{1}{\alpha}\frac{1}{I_{j+1}} \sum_{\mathbf{x}\in X_{j+1}} \left(-U_j(\mathbf{x}) - \varsigma_{j+1}^-\right)^+ \end{cases} - \delta \qquad j = 1, .., I - 1. \qquad (43)$$

Figure 7 shows that this model has higher in-sample error compared to Model B, but the out-of-sample performance is the best among the considered models. Moreover, the risk constraint reduces the number of many-class misclassification jumps. Whereas Model C has 8 out-of-sample objects for which the value of misprediction is more than 5 classes, Model D has only one object with a 5-class misprediction. So, Model D has the best out-of-sample performance among all the considered models. This has been achieved by choosing appropriate flexibility of the model.

A model with low flexibility may fit an in-sample dataset well. For models with a high in-sample error (such as Model A), more flexibility can be added by introducing more control variables to the model. In classification models with separating functions, feasibility constraints play a crucial role because they make classification always possible for out-of-sample objects. An excessive flexibility of the model may lead to "overfitting" and poor prediction characteristics of the model. To remove excessive flexibility of the considered models various types of constraints (such as risk constraints) can be imposed. The choice of the types of constraints, as well as the choice of the class of separating functions, plays a critical role for good out-of-sample performance of the algorithm.

5 Error Estimation

To estimate the performance of the considered models, we use the "leave-one-out" cross validation scheme. For the description of this scheme and other cross validation approaches, see, for instance, Efron and Tibshirani (1994). Let us denote by m a number of objects in the considered dataset. For each model (defined by the classes of the constraints imposed on optimization problem (20)), we performed m experiments. By excluding objects $\mathbf{x}^i$ one by one from the set X, we constructed m training sets,

$$Y_i = X \backslash \left\{\mathbf{x}^i\right\}, i = 1, .., m.$$

For each set Y_i, we solved (20) optimization problems with the appropriate set of constraints and found the optimal parameters of the separating functions $\{U_{\alpha^0}(\mathbf{x}), .., U_{\alpha^J}(\mathbf{x})\}$. Further, we computed the number of misclassified objects M_i from the set Y_i. Let us introduce the variable

$$P_i = \begin{cases} 0, if \ \{U_{\alpha^0}(\mathbf{x}), .., U_{\alpha^J}(\mathbf{x})\} \ correctly \ classifies \ \mathbf{x}^i, \\ 1, otherwise \ . \end{cases} \qquad (44)$$

The in-sample error estimate $E_{in-sample}$ is calculated by the following formula:

$$E_{in-sample} = \frac{1}{m} \sum_{i=1}^{m} \frac{M_i}{(m-1)} = \frac{1}{m(m-1)} \sum_{i=1}^{m} M_i, \tag{45}$$

where M_i is the number of misclassified objects in the set Y_i. In the last formula, the ratio $\frac{M_i}{m-1}$ estimates the probability of an in-sample misclassification in the set Y_i. We calculated the average of these probabilities to estimate the in-sample error.

The out-of-sample error estimate $E_{out-of-samle}$ is defined by the ratio of the total number of the misclassified out-of-sample objects in m experiments to the number of experiments:

$$E_{out-of-sample} = \frac{1}{m} \sum_{i=1}^{m} P_i \quad . \tag{46}$$

The considered leave-one-out cross-validation scheme provides the highest confidence level (by increasing the number of experiments) while effectively improving the predicting power of the model (by increasing the size of the in-sample datasets)

6 Bond Classification Problem

We have tested the approach with a bond classification problem. Bonds represent the most liquid class of the fixed-income securities. A bond is an obligation of the bond issuer to pay cash flow to the bond holder according to the rules specified at the time the bond is issued. A bond pays a specific cash flow (face value) at the time of maturity. In addition, a bond may pay periodic coupon payments.

Although a bond generates a prespecified cash flow stream, it may default, if an issuer gets into financial difficulties or becomes a bankrupt. To characterize this risk, bonds are rated by several rating organizations (*Standard & Poor's* and *Moody's* are major rating companies). Bond rating evaluates the possibility of default of a bond issuer based on the issuer's financial condition and profits potential. The assignment of a rating class is mostly based on the issuer's financial status. A rating organization evaluates the status using expert opinions and formal models based on various factors including financial ratios, such as the ratio of debt to equity, the ratio of current assets to current liabilities, and the ratio of cash flow to outstanding debt.

According to *Standard and Poor's*, bond ratings start at AAA for bonds having the highest investment quality, and end at D for bonds in payment default. The rating may be modified by a plus or minus to show relative standing within the category. Typically, a bond with a lower rating has a lower price than a bond generating the same cash flow, but having a higher rating.

In this study, we have replicated Standard & Poor's ratings of bonds using the suggested classification methodology. The rating replication problem is reduced to a classification problem with 20 classes.

7 Description of Data

For the computational verification of the proposed methodology we used several datasets (A, W, X, Y, Z) provided by the research group of the RiskSolutions branch of Standard and Poor's, Inc. The datasets contain quantitative information about several hundred companies rated by Standard and Poor's, Inc. Each entry in the datasets has fourteen fields that correspond to certain parameters of a specific firm in a specific year. The first two fields are the company name, and the year when the rating was calculated. We used these fields as identifiers of objects for classification. The next eleven fields contain quantitative information about financial performance of the considered company. These fields are used for the decision-making in the classification process. The last field is the credit rating of the company assigned by Standard and Poor's, Inc. Table 1 represents the information used for classification.

Table 1. Data format of an entry of the dataset.

Identifier	Quantitative characteristics	Class
1) Company Name	1) Industry Sector	1) Credit rating
2) Year	2) EBIT interests coverage (times)	
	3) EBITDA interest coverage (times)	
	4) Return on capital (%)	
	5) Operating income/sales (%)	
	6) Free operating cash flow/total debt (%)	
	7) Funds form operations/total debt (%)	
	8) Total debt/capital (%)	
	9) Sales (mil)	
	10) Total equity (mil)	
	11) Total assets (mil)	

We preprocessed the data and rescaled all quantitative characteristics into [-1,1] intervals: all quantitative characteristics are monotonic. Since the total number of rating classes equals 20, we used integer numbers from 1 to 20 to represent the credit rating of an object. The rating is arranged according to credit quality of the objects: the greater value corresponds to the better credit rating.

In the case study, we considered 5 different datasets (A, W, X, Y, Z) to verify the proposed methodology for different sizes of the input data. Each set was split into an in-sample set and an out-of-sample set. The first one was used for developing a model, and the second one was used for the verification of the out-of-sample performance of the developed model. Table 2 contains information about the sizes of the considered datasets.

Table 2. Sizes of the considered datasets.

Dataset	A	W	X	Y	Z
Size of Set	406	205	373	187	108
Size of In-Sample	278	172	315	157	89
Size of Out-of-Sample	128	33	58	30	19

8 Numerical Experiments

For dataset A we performed computational experiments with 16 models generated by all possible combinations of four different types of constraints. For each dataset W, X, Y, and Z we found a model with the best out-of-sample performance.

For dataset A we applied optimization problem (20) with Feasibility Constraints (F). Besides, we applied all possible combinations of the following four types of constraints: Gradient Monotonicity Constraints (GM), Monotonicity of Separation Function w.r.t. Class Constraints (MSF), Risk Constraints (R), and Level Squeezing Constraints (LSQ). Each combination of the constraints corresponds to one of 16 different models, see Table 3. Although we considered other types of constraints (see Section 3.3), we report the results only for the models with F-, GM-, MSF-, R- and LSQ-constraints.

The numerical experiments were conducted with Pentium III , 1.2GHz in C/C++ environment. The linear programming problems were solved by CPLEX 7.0 package. The calculation results for 16 models for dataset A are presented in Table 4 and Figure 9.

Figure 9 represents in-sample and out-of-sample errors for different models. For each model there are three columns representing the computational results. The first column corresponds to the percentage of in-sample classification error, the second column corresponds to the percentage of out-of-sample classification error, and the third column represents the average out-of-sample classification expressed in percents (100% misprediction corresponds to the case when the difference between actual and calculated classes equals 1).

On Figure 9, the models are arranged according to an average of out-of-sample error (see Table 4, column AVR). Models 0000 and 0010 have zero in-sample error. These two models are the most "flexible" – they fit the data with zero objective in (20). The smallest average out-of-sample error is obtained by 0101- model with MSF and LSQ constraints. Table 4 shows that this model has the lowest maximal misprediction. Models 0100 and 0110 have the

Table 3. List of the considered models.

Model Identificator	Types of Constraints Applied			
	GM	MSF	R	LSQ
0000				
0001				YES
0010			YES	
0011			YES	YES
0100		YES		
0101		YES		YES
0110		YES	YES	
0111		YES	YES	YES
1000	YES			
1001	YES			YES
1010	YES		YES	
1011	YES		YES	YES
1100	YES	YES		
1101	YES	YES		YES
1110	YES	YES	YES	
1111	YES	YES	YES	YES

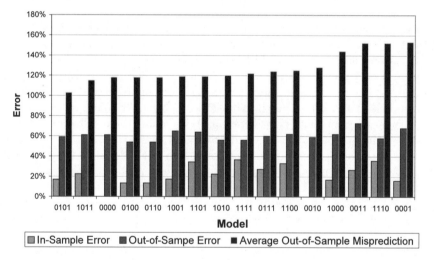

Fig. 9. In-sample and out-of-sample errors for various models.

minimal number of out-of-sample mispredictions (column Error(1)), but the maximal mispredictions (column MAX) for these models are high. It is worth mentioning that imposing constraints on the separating functions increases the computing time. However, for all the considered optimization problems, the classification procedure did not take more than 5 minutes: The CPLEX LP solver, that was used for this study, computed classification model 1111 (all constraints were imposed) in about 5 minutes. Pentium III 1.2Ghz was used for the study. As for all the other models, it took CPLEX LP solver less than 5 minutes.

Table 4. Calculation results for dataset A

Model	Error(In)	AVR	Correct	Error(1)	Error(2)	Error(3)	Error(4)	MAX
0000	0.00%	1.18	37.37%	62.63%	23.23%	9.09%	3.03%	18
0001	16.00%	1.54	30.30%	67.68%	35.35%	22.22%	13.13%	13
0010	0.00%	1.28	39.39%	60.61%	24.24%	10.10%	5.05%	17
0011	26.70%	1.53	25.25%	74.75%	37.37%	18.18%	12.12%	7
0100	13.50%	1.18	44.44%	55.56%	25.25%	12.12%	7.07%	18
0101	17.00%	1.03	39.39%	60.61%	25.25%	12.12%	4.04%	5
0110	13.60%	1.18	44.44%	55.56%	25.25%	12.12%	7.07%	18
0111	27.30%	1.24	38.38%	61.62%	30.30%	13.13%	5.05%	16
1000	16.90%	1.44	36.36%	63.64%	26.26%	11.11%	10.10%	16
1001	17.50%	1.19	33.33%	66.67%	29.29%	11.11%	7.07%	6
1010	22.60%	1.20	42.42%	57.58%	26.26%	12.12%	8.08%	14
1011	22.50%	1.15	37.37%	62.63%	29.29%	11.11%	7.07%	7
1100	33.10%	1.25	36.36%	63.64%	24.24%	14.14%	7.07%	18
1101	34.50%	1.19	34.34%	65.66%	24.24%	14.14%	7.07%	10
1110	35.70%	1.53	40.40%	59.60%	30.30%	19.19%	9.09%	18
1111	36.70%	1.22	42.42%	57.58%	28.28%	18.18%	8.08%	10

Error(In) = percentage of in-sample misclassifications; AVR = an average error (in number of classes); Correct = percentage of correct out-of-sample predictions; Error(1) = percentage of out-of-sample mispredictions; Error(i) = percentage of out-of-sample prediction errors for more than or equal to i classes; MAX = maximal error for out-of-sample classification.

For the datasets W, X, Y, and Z we imposed various constraints, and selected the models with best performance. For each dataset we choose the best performing models and compared performances of the models with reference models used by *RiskSolutions group* at *Standard and Poor's*, see Table 5. For each data set the table contains the following performance information about the best chosen and the reference model: the average error (expressed in number of classes); the standard deviation; the percentage of correct out-of-sample predictions, the distribution of out-of-sample prediction (expressed in percentage of out-of-sample predictions with a deviation less than or equal to i classes for $i=1,2,3$); and the description of the constraints included in the best model.

The results of this analysis show that the proposed methodology of classification with the separating functions is competitive with the reference model available from *RiskSolutions group* at *Standard and Poor's*. Although the proposed models do not provide better results compared to the reference model in all the cases, the proposed algorithms have a better or at least comparable performance, for small datasets Z and W.

Table 5. Comparison of best found model with reference model for sets W, X, Y and Z.

Model	W		X		Y		Z	
In-Sample	172		315		157		89	
Out-of-Sample	33		58		30		19	
Model	Best	Refer.	Best	Refer.	Best	Refer.	Best	Refer.
AVR	1.82	1.18	1.67	1.02	1.93	1.13	2.21	2.32
STDV	1.72	0.92	2.45	1.07	2.60	1.50	2.20	2.50
Correct	33.3%	21.2%	24.1%	37.9%	30.0%	50.0%	15.8%	21.1%
1-Class area	42.4%	72.7%	53.4%	70.7%	60.0%	70.0%	42.1%	63.2%
2-Class area	69.7%	87.9%	86.2%	94.8%	80.0%	83.3%	78.9%	68.4%
3-Class area	84.8%	100.0%	94.8%	96.6%	86.7%	86.7%	84.2%	68.4%
4-Class area	93.9%	100.0%	96.6%	98.3%	86.7%	96.7%	84.2%	78.9%
Constraints	Feasibility		Feasibility		Feasibility		Feasibility	
	MSF		MSF		MSF		CVaR	MSF

For each dataset the table has two columns. The first column corresponds to the best found model; the second column corresponds to the reference model from RiskSolutions. Each column contains information about out-of-sample performance of the corresponding model.

AVR = an average error (in number of classes); STDV = a standard deviation; Correct = percentage of correct out-of-sample predictions; i-Class area = percentage of out-of-sample predictions with a deviation less than or equal to i classes; Constraints = description of the constraints included in the best model

9 Concluding Remarks

We extended the methodology discussed in previous work (Bugera et al., 2003) to the case of multi-class classification. The extended approach is based on finding "optimal" separating functions belonging to a prespecified class of functions. We have considered the models with quadratic separating functions (in indicator parameters) with different types of constraints. As before, selecting a proper class of constraints plays a crucial role in the success of the suggested methodology: by imposing constraints, we adjust the flexibility of a model to the size of the training dataset. The advantage of the considered methodology is that the optimal classification model can be found by linear programming techniques, which can be efficiently used for large datasets.

We have applied the developed methodology to a bond rating problem. We have found that for the considered dataset, the best out-of-sample characteristics (the smallest out-of sample error) are delivered by the model with MSF and LSQ constraints on the control variables (coefficients of the separating function).

The study was focused on the evaluation of computational characteristics of the suggested approach. As before, despite the fact that the obtained results are data specific, we conclude that the developed methodology is a robust classification technique suitable for a variety of engineering applications. The advantage of the proposed classification is its simplicity and consistency if compared to the proprietary models which are based on the combination of various techniques, such as expert models, neural networks, classification trees etc.

References

Bugera, V., Konno, H. and Uryasev, S.: 2003, Credit cards scoring with quadratic utility function, *Journal of Multi-Criteria Decision Analysis* **11**(4-5), 197–211.

Efron, B. and Tibshirani, R.J.: 1994, *An Introduction to the Bootstrap*, Chapman & Hall, New York.

Konno, H., Gotoh, J. and Uho, T.: 2000, A cutting plane algorithm for semidefinite programming problems with applications to failure discrimination and cancer diagnosis, *Working Paper 00-5*, Tokyo Institute of Technology, Center for research in Advanced Financial Technology, Tokyo.

Konno, H. and Kobayashi, H.: 2000, Failure discrimination and rating of enterprises by semi-definite programming, *Asia-Pacific Financial Markets* **7**, 261–273.

Mangasarian, O., Street, W. and Wolberg, W.: 1995, Breast cancer diagnosis and prognosis via linear programming, *Operations Research* **43**, 570–577.

Pardalos, P.M., Michalopoulos, M. and Zopounidis, C.: 1997, On the use of multi-criteria methods for the evaluation of insurance companies in Greece, *in* C. Zopounidis (ed.), *New Operational Approaches for Financial Modeling*, Physica, Berlin-Heidelberg, pp. 271–283.

Rockafellar, R.T. and Uryasev, S.: 2000, Optimization of conditional value-at-risk, *The Journal of Risk* **2**(3), 21–41.

Rockafellar, R.T. and Uryasev, S.: 2002, Conditional value-at-risk for general loss distributions, *Journal of Banking and Finance* **26**(7), 1443–1471.

Rockafellar, R.T., Uryasev, S. and Zabarankin, M.: 2002, Generalized deviations in risk analysis, *Finance and Stochastics* **10**, 51–74.

Vapnik, V.: 1998, *Statistical Learning Theory*, John Wiley & Sons, New York.

Zopounidis, C. and Doumpos, M.: 1997, Preference desegregation methodology in segmentation problems: The case of financial distress, *in* Zopounidis C. (ed.), *New Operational Approaches for Financial Modeling*, Physica, Berlin-Heidelberg, pp. 417–439.

Zopounidis, C., Pardalos, P., Doumpos, M. and Mavridou, T.: 1998, Multicriteria decision aid in credit cards assessment, *in* C. Zopounidis and P. Pardalos (eds), *Managing in Uncertainty: Theory and Practice*, Kluwer Academic Publishers, New York, pp. 163–178.

Evolving Decision Rules to Discover Patterns in Financial Data Sets

Alma Lilia García-Almanza, Edward P.K. Tsang, and Edgar Galván-López

Department of Computer Science, University of Essex, Wivenhoe Park, Colchester
CO4 3SQ, U.K. algarc@essex.ac.uk, edward@essex.ac.uk,
egalva@essex.ac.uk

Summary. A novel approached, called *Evolving Comprehensible Rules* (ECR), is
presented to discover patterns in financial data sets to detect investment opportunities. ECR is designed to classify in extreme imbalanced environments. This is
particularly useful in financial forecasting given that very often the number of profitable chances is scarce. The proposed approach offers a range of solutions to suit the
investor's risk guidelines and so, the user could choose the best trade-off between
miss-classification and false alarm costs according to the investor's requirements.
The Receiver Operating Characteristics (ROC) curve and the Area Under the ROC
(AUC) have been used to measure the performance of ECR. Following from this
analysis, the results obtained by our approach have been compared with those one
found by standard Genetic Programming (GP), EDDIE-ARB and C.5, which show
that our approach can be effectively used in data sets with rare positive instances.

Key words: Evolving comprehensible rules, machine learning, evolutionary algorithms.

1 Introduction

In this work, we propose to evolve decision rules by using Genetic Program-
ming(GP) (Koza, 1992). The result of this is a reliable classifier which is able
to detect positive cases in imbalanced environments. This work is aided by
Machine Learning (ML), an Artificial Intelligence field that has been successfully applied to financial problems (Chen, 2002; Chen and Wang, 2004; Tsang
and Martinez-Jaramillo, 2004). ML embraces techniques to extract rules and
patterns from data. These, like other forecasting techniques, extend past experiences into the future. However, in rare event detection, the imbalance
between positive and negative cases poses a serious challenge to ML (Japkowicz, 2000; Weiss, 2004; Batista et al., 2004; McCarthy et al., 2005). This is due
to negative classifications are favoured given that these have a high chance of
being correct, as we shall explain later.

This work is organised as follows. Section 2 provides an introduction of the Machine Learning and Evolutionary Computation systems. Section 3 describes some metrics to measure the performance of a learning classifier system and provides a description of the Receiver Operating Characteristic (ROC) curve. Section 4 describes in detail our approach. Finally, Section 5 describes the results found by our approach and we draw some conclusions in Section 6.

2 Previous Work

ML, a field of Artificial Intelligence (Mitchell, 1997), is a multidisciplinary area which embraces probability and statistics, complexity theory, Evolutionary Computation (EC) and other disciplines. ML is often used to create computer programs that form new knowledge (which is acquired by information and data analysis) for a specific application. Our approach is based on supervised learning, a branch of ML in which the system is fed by a training data set which comprises examples with inputs and the corresponding desired outputs.

In this paper, we focus on using GP to evolve rules. Other methods have also been used to evolve decision rules; for instance, Corcoran and Sen (1994) used a Genetic Algorithm (GA) to evolve a set of rules. For this purpose, they treated it as an optimisation problem. That is, the goal of the GA is to maximise the number of correct classifications of a training data set. The contribution of their approach was to evolve rules with continue variables rather than using a binary representation. Bobbin and Yao (1999) were interested in evolving rules for nonlinear controllers. In addition, their approach was able to offer rules that can be interpreted by humans.

Jong and Spears (1991) proposed an application called GA batch-incremental concept learner (GABIL). The idea is to evolve a set of rules using GAs. That is, GAs evolve fixed-length rules for attributes whose values are nominal. Each member in the population is composed by a variable number of rules which means that the individual's size is variable too.

Following the same idea, Janikow (1993) proposed Genetic-based Inductive Learning (GIL). Janikow proposed three types of operations: (a) rule set level, (b) rule level and (c) condition level. Each of them contains specific operations. For the former, the operations involve rules exchange, rules copy, new event, rule generalisation (generalise two rules picked at random) and rule specialisation. The operations in the rule level are in charge of introduce or drop conditions. Finally, there are 3 operators for the condition level: reference change, reference extension and reference restriction.

Using Evolutionary Algorithms, Kwedlo and Kretowski (1999) proposed an approach which novelty was to use multivariate discretisation. The proposed approach has the ability to search simultaneously for threshold values for all the attributes that hold continuous values.

Fidelis et al. (2000) proposed an approach based on GAs in the hope that it can discover comprehensible IF-THEN rules. The novelty of their approach was that it allowed to have a fixed length at genotype level but the number of rules conditions (which is mapped at the phenotype level) is variable. The key ingredient that allows this feature in the genotype-phenotype mapping was to allow an element at the genotype level that the authors called "weight". As the authors explained in their paper, the results regarding to the accuracy were promising in one of the data sets but more importantly, they showed how a GA was able to find concise rules that are more understandable in terms of complexity.

Bojarczuk et al. (1999) were interested in discovering rules to classify 12 different diseases. Moreover, the authors were also interested in producing rules that were comprehensible for the final user. The latter requirement, as the authors expressed in their work, is that the resulting rule can be "readable" and so, it can be used as a complement to the user's knowledge. The authors claimed that the results found by their GP system were promising because they reported high accuracy in classifying the diseases.

Using GP, Falco et al. (2001) also evolved classification rules that were comprehensible for the final user. This approach evolves decision trees that classify a single class and so, if there are more than one class to classify the GP is executed as many times as the number of classes. The authors stated that the tree represents a set of rules which is composed by disjunctions and conjunctions.

Niimi and Tazaki (2000) proposed a discovery rule technique by means of generalisation of association rules using an a priori algorithm, then the rules are converted to decision trees which will be used as the initial population in GP. After evolving the potential solutions, the best individual is converted into classification rules.

The works described above suggest that ECs could perform very well on classification tasks. Our approach, as it will be explained in Section 4, has been designed in such a way to produce rules, which detect rare cases in imbalanced environments. The output is a range of classifications that provides to the user the option to choose the best trade-off between miss-classification and false alarm costs according to the user's needs. Moreover, it presents various beneficial features as comprehensible rules.

3 Performance Metrics

The confusion matrix displays the data about actual and predicted classifications done by a classifier Kohavi and Provost (1998). This information is used in supervised learning to determine the performance of classifiers and some learning systems. Given an instance and a classifier, there are four possible results:

- *True positive* (TP): the instance is positive and it is classified as positive.
- *False positive* (FP): the instance is negative and it is classified as positive.

Table 1. Confusion matrix to classify two classes.

	Actual Positive	Actual Negative
Positive Prediction	*True Positive* (TP)	*False Positive* (FP)
Negative Prediction	*False Negative* (FN)	*True Negative* (TN)
	Total Positive	Total Negative

- *False negative* (FN): the instance is positive and it is classified as negative.
- *True negative* (TN): the instance is negative and it is predicted as negative.

Table 1 shows a confusion matrix for two classes. For detailed analysis, we have used other metrics taken the confusion matrix as a basis. These are shown in Table 2.

Table 2. Metrics used for a detail analysis of the results found by our approach.

Accuracy	is the proportion of the total number of predictions that were done correctly. This is determined by the equation: $$Accuracy = \frac{TP+TN}{TP+FP+FN+TN}$$
Recall	also called *sensitivity* or *true positive rate*, is the proportion of positive cases that was correctly identified. This is given by: $$Recall = \frac{TP}{TP+FN}$$
Precision	also called *positive predictive value*, is the proportion of positive cases that were correctly predicted. It is calculated as follows: $$Precision = \frac{TP}{TP+FP}$$
False positive rate	also known as *false alarm rate*, is the proportion of negative cases that were wrongly predicted as positive. It is calculated as follows: $$False\ positive\ rate = \frac{FP}{FP+TN}$$

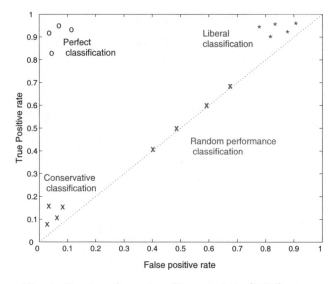

Fig. 1. Receiver Operating Characteristic (ROC) space.

3.1 ROC Space

Receiver Operating Characteristic (ROC) is a technique that plots the performance of a classifier and is able to select the best trade-off between successful cases and false alarms based on benefits and costs (Fawcett, 2004).

The ROC graph is constructed by plotting the *true positive rate* (recall) on the Y-axis and the *false positive rate* (false alarms) on the X-axis (see for instance Hanley and McNeil (1998), Fawcett (2004)). Figure 1 depicts the ROC space. As can be seen, the ROC graph is plotted in the space of (0,0) and (1,1). The performance of the classifier is plotted in (0,0) when it does not find any positive case and it does not report any false alarm. Thus, it gets all the negative cases right but it gets all the positive wrong. The opposite case is at position (1,1), where the totality of the cases are classified as positive.

The performance of a classifier is better than other if it is plotted in the upper left area of the ROC space (Fawcett, 2004). The classifiers whose performance is plotted in the left hand side of the ROC space close to the X-axis, are denominated *conservative*. This is because these make a positive classification just when these have strong indications or evidences; as a consequence, these have few false alarms. On the other hand, classifiers on the upper right hand side of the ROC space are called *liberal* because these make positive classifications with unsubstantial evidence. Finally, the diagonal line that goes from position (0,0) to (1,1) describes classifiers whose performance is no different from those made by random predictions.

3.2 Area Under the ROC Curve (AUC)

One of the most used ROC metrics is the Area Under the Curve (AUC) which indicates the quality of the classifier (Vanagas, 2004). Huang and Ling (2005) showed theoretically and empirically that AUC is a better measure than accuracy.

When $AUC = 1$, then classification is perfect. In general, the closer AUC is to 1 the better performance of the classifier is. When a classifier's AUC is close to 0.5, it represents a random classifier performance.

3.3 Choosing the Best Operating Point

ROC can be used to estimate the best threshold of the classifier by calculating the best balance between the cost of misclassifying positive and negative cases. To calculate the best trade-off, let us define the following variables:

μ The cost of false positive or false alarm
β The cost of false negative
ρ the percentage of positive cases

Thus, the slope is calculated by the following formula:

$$slope = \mu \cdot (1 - \rho)/(\beta \cdot \rho).$$

The point where the line is tangent to the curve indicates the threshold of the best trade-off between misclassifications and false alarms costs.

We have selected ROC as a performance measure for the following reasons. Firstly, ROC is suitable for measuring the performance of classifiers for imbalanced data sets. Secondly, it is able to measure the performance of a classifier that is based on thresholds. Thirdly, ROC offers the possibility to the user (investor) to tune the parameters in a way to choose the best trade-off between miss-classification and false alarms.

4 Evolving Comprehensible Rules

To detect important movements in financial stock prices, our approach is inspired by two previous works: EDDIE (Tsang et al., 1998; Li, 2001; Tsang et al., 2004, 2005) and Repository Method (Garcia-Almanza and Tsang, 2006b,a). EDDIE is a financial forecasting tool that trains a GP using a set of examples. Every instance in the data set is composed by a set of attributes or independent variables and a *signal* or *desired output*. The independent variables are indicators derived from financial technical analysis. These indicators have been used to identify patterns that can suggest future activity (Sharpe et al., 1995). The signal is calculated looking ahead in a future horizon of n units of time, trying to detect an increase or decrease of at least $r\%$. However, when the value of r is very high, which implies an important movement in the stock price, the number of positive cases is extremely small and it becomes very difficult to detect these events. To deal with these special cases, we have proposed a method to discover patterns in financial data sets. This method was designed to detect cases in extreme imbalanced environments.

Our approach, that we have called *Evolving Comprehensible Rules* (ECR), evolves a set of decision rules by using GP and receives feedback from a key element that we called repository. The idea behind using GP in our approach is to being able to represent rules as tree-like structures and so, the GP will create comprehensible rules that the final user could analyse. The resulting rules can be used to create a range of classifications that allows the user to choose the best trade-off between the misclassifications and the false alarms cost. So, our approach has the novelty of offering a range of classifications that best suits the user risk guidelines.

Our approach is composed by the following steps:

1. *Initialisation of population.* Initialise the population using the grow method (Koza, 1992, pages 91-94) (i.e., individuals could have any shape). The objective of this procedure is to create a collection of candidate solutions. We propose to create a population of decision trees using the Discriminator Grammar (DG) (Garcia-Almanza and Tsang, 2006b) (Figure 3 depicts the Grammar used in our approach). This grammar[1] produces decision trees that classify or not a single class. By using DG, it is simple to get the rules that are embedded in the individuals. Moreover, it is in charge of producing valid individuals. In other words, assuring a valid structure. Figure 2 shows a typical individual created with DG.

2. *Extraction of rules.* Once the initial population has been created, before the algorithm starts evaluating, the system decomposes each individual in its corresponding rules and these are evaluated. Given that the system will deal with many rules, we need to define a precision threshold to keep only those rules that have good performance. Let us define a rule $R_k \in T$ as the minimal set of conditions which intersection satisfies the decision tree T. In other words, the tree could contain one or more rules. A decision tree is satisfied when at least one of its rules is satisfied. A rule is satisfied when all its conditions are satisfied. Figure 2 shows a decision tree that holds three rules and as can be seen each of them is able to satisfy the decision tree. To recognise every rule it is necessary to identify the minimal sets of conditions that satisfy the tree T.

3. *Rule simplification.* The aim of rule simplification is to remove noisy conditions (which refers to rules that regardless its values are always satisfied) and redundant conditions. Redundant conditions are those which are repeated or report the same event, e.g., $R_1 = \{Var_1 > 0.5 \text{ and } Var_1 > 0.7\}$ where the first condition is redundant. None of them (noisy and redun-

---

[1] The term grammar refers to a representation concerned with the syntactic components and the regulation that specify it (Chomsky, 1965).

dant conditions) affect the decision of the rule. The simplification of rules is an important process because it allows to recognise the real variables and interactions involved in that rule. Furthermore, it allows to identify the duplication of rules in the repository. This is one of the important elements of our approach because it assures to collect different rules.

4. *Adding new rules in the repository.* The process detects new rules by comparing the new ones with those rules that are stored in the repository. If the rule is totally new, then it is added in the repository. If there is a similar rule in the repository but the new one offers better performance then the old rule is replaced by the new rule. In any other case the rule will be discarded.

5. Once the evolutionary search starts, we create the next population as follows:

   • *Evolving population.* The resulting rules that are stored in the repository are taken as parents and then we apply mutation and use hill-climbing[2] to create the new offsprings.

   • In case there are no enough rules in the repository to create the population, then the remaining individuals are created at random (this is determined by the parameter called "repository random"). This also allows to prevent losing variety. The process is repeated from step 2 until the algorithm has reached the maximum number of generations.

6. Once the evolutionary process has finished, *ECR* is tested by using the corresponding testing data set. It is evaluated by using the rules that are stored in the repository. Those rules are grouped according to their precision in order to classify the cases (i.e., $Precision = \{1, .95, \cdots, .05, 0\}$).

5 Results and Discussion

To conduct our experiments we have used 3 examples based on 2 data sets. These are explained as follows:

Barclays complete - 1,718 cases. The data sets to train and test our approach came from the London stock market from Barclays' stock (from March, 1998 to January, 2005). The attributes of each record are composed by indicators derived from financial technical analysis. Technical analysis is used in financial markets to analyze the stock price behaviour. This is mainly based on historical prices and volume trends (Sharpe et al., 1995). The indicators were calculated on the basis of the daily *closing price*,[3] volume and some

---

[2] Hill-climbing is a stochastic technique to find the global optimum where the search is given by trying one step in each direction and then choosing the steepest one (Langdon and Poli, 2002).

[3] The settled price at which a traded instrument is last traded at on a particular trading day.

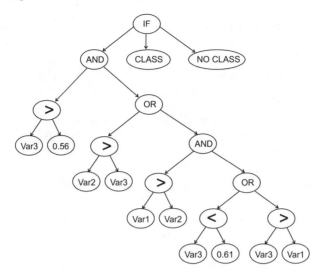

Fig. 2. A typical individual created with Discriminator Grammar. The rules contained within the tree are: $R_1 = \{Var_3 > 0.56 \wedge Var_2 > Var_3\}$, $R_2 = \{Var_3 > 0.56 \wedge Var_1 > Var_2 \wedge Var_3 < 0.61\}$, $R_3 = \{Var_3 > 0.56 \wedge Var_1 > Var_2 \wedge Var_3 > Var_1\}$.

Fig. 3. Discriminator Grammar.

G	$\rightarrow$	<Root>
<Root>	$\rightarrow$	"If-then-else",<Conjunction> | <Condition>,"Class","No Class"
<Condition>	$\rightarrow$	<Operation>, <Variable>, <Threshold> | <Variable>
<Conjunction>	$\rightarrow$	"AND"|"OR",<Conjunction>|<Conditional>,
		<Conjunction>|<Conditional>
<Operation>	$\rightarrow$	"<" | ">"
<Variable>	$\rightarrow$	var_1 | var_2 | ... var_n
<Threshold>	$\rightarrow$	Real number

financial indices as the FTSE.[4] We looked for an increase of 15% in the stock price in an horizon of 10 days. The training data set contains 887 cases (where 39 cases are positive which represents 4.3% of this data set) and the testing data set contains 831 instances (where 24 cases are positive which represents 2.8% of this data set).

Arbitrage - 1,641 cases. This data set consists of 1,000 cases (where 242 cases are positive which represents 24.2% of this data set) for the training data set and the testing data set consists of 641 cases (where 159 cases are positive which represents 24.80% of this data set).

---

[4] An index of 100 large capitalization companies stock on the London Stock Exchange, also known as "Footsie".

Table 3. Parameters used for the data sets called Barclays complete and variation of Barclays.

Parameters	Values
Population size	1000
Maximum number of generations	25
Hill-Climbing Probability	.14
Maximum number of rules in the repository	2500
Precision threshold	.08
Repository random	.80

Table 4. Parameters used for the data set called Arbitrage.

Parameters	Values
Population size	1000
Maximum number of generations	30
Hill-Climbing Probability	.05
Maximum number of rules in the repository	2000
Precision threshold	.10
Repository random	.80

Variation of Barclays' data set - 400 cases. This data set is a variation of the first data set. It is a sample of 400 cases for each data set (training and testing). In this case, we have 15 positive cases for the training data set (which correspond to 3.7%) and there are 13 positive cases for the testing data set (which correspond to 3.2%).

5.1 Performance Analysis

Let us start analysing the results found by our approach, ECR, in the first example (i.e., using the complete Barclays data set). In Figure 4, we show the plot produce by the results found by ECR and in Table 5 we show the parameters used by our approach. Notice how the results produced by our approach are fine balance between the conservative and liberal space. The area under the curve (AUC) plotted when using the results found by our approach is equal to 0.80. Now, let us discuss the points that form the curve. For instance, when the precision threshold is equal 0.50, the method has a high recall given that the system was able to detect 79% of the positive examples. This result has been found without sacrifing a good accuracy (i.e., 77%).

When the investor's risk-guidelines are conservative, the system is able to offer him/her a suitable classification. For instance, take a look when the precision threshold is 0.70. The system is able to classify up to a quarter of positive cases with a very high accuracy (92%). For values where the precision threshold is greater or equal to 0.40, then the classifier's performance tends to decrease because the number of new positive cases that are detected are based on a serious decrease of accuracy.

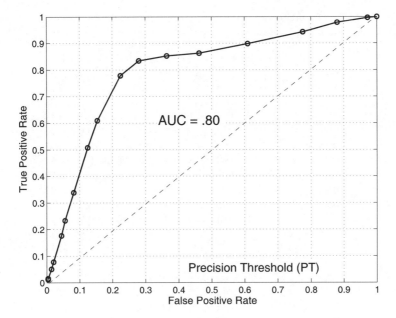

Fig. 4. ROC curve for the Barclays complete data set.

Moreover, we tested the same data set using a standard GP. To obtain meaningful results, we performed a series of 20 runs. Table 6 displays the parameters used by the standard GP. The results found by the GP are: TP=3, FP=68, FN=21, TN=739. Thus, the false positive rate = 0.084 and true positive rate =0.12 which is under the ROC curve produced by the results found by our classifier (see Figure 4). As it can be observed the standard GP classification is conservative despite the fact that its fitness function is the $\sqrt{Recall \cdot Precision}$ which has been used in ML to deal with imbalanced data sets (Kubat et al., 1998) and it is able to encourage the classification of positive cases. On the other hand, ECR not just has better performance but it also offers a range of classification to suit the user's risk guidelines.

Now, let us focus our attention in the second example (using the Arbitrage data sets - which is an imbalanced data set but less imbalanced than the one used in our previous example). When the precision threshold = 0.95, the system is able to detect 70% of the positive examples with an impressive precision of 92% and an accuracy of 91%. As can be seen from Table 5, almost the totality (96%) of the positive cases is achieved using a precision threshold of 0.60 with an accuracy of 65%. At this point, the performance of the classifier becomes liberal.

However, it is fair to say that the approach proposed by Tsang et al. (2005), called EDDIE-ARB, achieved a precision of 100% and a recall of 42%. Nevertheless, our approach (ECR) achieved a precision of 96% with the same recall reported by Tsang and co-workers. The advantage of our approach is

Table 5. Results found by our approach. Precision, recall and accuracy are shown in labels (a), (b) and (c), respectively for the 3 sets of examples.

Precision	Barclays			Arbitrage			Barclays 400		
Threshold	(a)	(b)	(c)	(a)	(b)	(c)	(a)	(b)	(c)
1.00	0.13	0.07	0.96	0.97	0.62	0.90	0.08	0.55	0.78
0.95	0.13	0.07	0.96	0.92	0.70	0.91	0.08	0.55	0.78
0.90	0.13	0.08	0.96	0.84	0.76	0.90	0.08	0.56	0.78
0.85	0.12	0.11	0.95	0.75	0.77	0.88	0.08	0.63	0.76
0.80	0.12	0.13	0.95	0.67	0.82	0.85	0.07	0.71	0.67
0.75	0.11	0.20	0.93	0.58	0.87	0.81	0.06	0.81	0.58
0.70	0.12	0.25	0.92	0.53	0.89	0.77	0.06	0.84	0.54
0.65	0.11	0.35	0.90	0.46	0.95	0.71	0.05	0.89	0.47
0.60	0.11	0.53	0.86	0.42	0.96	0.65	0.04	0.96	0.32
0.55	0.10	0.63	0.83	0.39	0.99	0.61	0.04	0.99	0.22
0.50	0.09	0.79	0.77	0.35	1.00	0.54	0.04	1.00	0.15
0.45	0.08	0.83	0.72	0.32	1.00	0.47	0.04	1.00	0.13
0.40	0.07	0.85	0.65	0.29	1.00	0.39	0.03	1.00	0.10
0.35	0.05	0.86	0.56	0.26	1.00	0.29	0.03	1.00	0.07
0.30	0.04	0.89	0.42	0.25	1.00	0.25	0.03	1.00	0.04
0.25	0.03	0.94	0.25	0.25	1.00	0.25	0.03	1.00	0.03
0.20	0.03	0.98	0.14	0.25	1.00	0.25	0.03	1.00	0.03
0.15	0.03	0.99	0.08	0.25	1.00	0.25	0.03	1.00	0.03
0.10	0.03	0.99	0.06	0.25	1.00	0.25	0.03	1.00	0.03
0.05	0.03	0.99	0.06	0.25	1.00	0.25	0.03	1.00	0.03
0.00	0.03	0.99	0.06	0.25	1.00	0.25	0.03	1.00	0.03

Table 6. GP Summary of Parameters.

Parameter	Value
Population size	1,000
Initialisation method	Growth
Generations	100
Crossover Rate	0.8
Mutation Rate	0.05
Selection	Tournament (size 2)
Control bloat growing	50% of trees whose largest branch exceed 6 nodes are penalized with 20% in its fitness
Fitness Function	$\sqrt{Recall \cdot Precision}$

that is able to produce a range of classifications that allows the user to detect more positive cases according to his requirements. This comes at a cost that the user could deal with more false alarms (i.e., losing precision).

Finally, let us discuss the third example used in this work. For analysis purposes we have used C.5 (Quinlan, 1993) to compare our proposed approach (ECR).

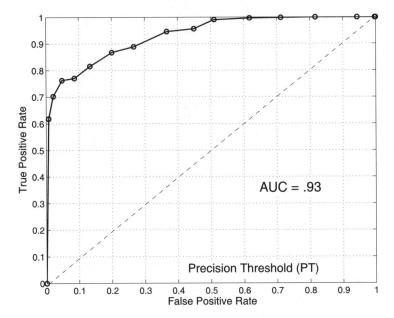

Fig. 5. ROC curve for the Arbitrage data set.

To perform the experiment we used the trial C.5 version. This demonstration version, however, cannot process more than 400 training or testing cases. For such reason, the size of the data sets had to be adjusted to meet this requirement. The new data sets hold 400 records each of them and these are conformed as follows: training data contains 385 negative cases and 15 positive examples and the testing data holds 387 negative cases and 13 positive cases. We performed a series of 20 runs using the new data sets. On the other hand, Quinlan's algorithm was tested using ten-fold cross-validation and standard parameters settings.

The result obtained by C.5 is the following: *True positive* $= 0$, *false positive*$=0$, *false negatives* $= 13$, *true negative* $= 187$. As it can be seen C.5 has an excellent accuracy (96.7%), however, it fails to detect the positive cases. In contrast ECR was able to detect about 63% of positive cases using a precision threshold $= 0.85$ with an accuracy of 76%. The remaining results obtained by ECR are plotted in Figure 6, which shows that AUC $= 0.75$.

6 Conclusions

In this work, we have presented a new approach called *Evolving Comprehensible Rules* (ECR) to classify imbalanced classes. For analysis purposes, we have used 2 data sets and conducted 3 different experiments. Our approach was applied to the data sets in order to discover patterns to classify rare events. The

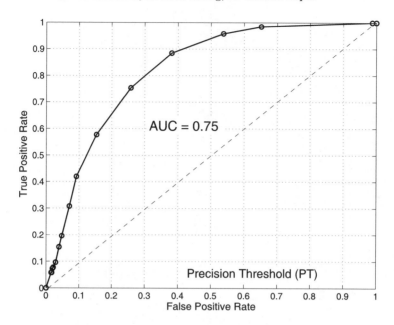

Fig. 6. ROC curve for the Barclays data set - 400 cases.

system's output is a set of decision rules which, by varying a threshold, is able to produce a range of classifications to suit different investor's risk preferences.

The core of our approach is based on GP, which is aided by a repository of rules. The aim of this repository is to collect useful patterns that could be used to produce the following population in the evolutionary process. The main operator of this approach is based on subtree mutation, which produces instances of the collected patterns.

From experimental results, it has been showed that our approach is able to improve the recall in the classification. In other words, it has been able to pick up more positive cases in imbalanced environments. For a detailed analysis, we have plotted the range of classifications in Receiver Operating Characteristic (ROC). This helped us to visualise how the points in the curve are well-distributed. The applicability of such tool can be translated that the investor could choose among different risk strategies.

Following the same idea, we have used Area Under the ROC Curve (AUC) to measure the general performance of our approach (ECR). Finally, to complement our analysis, we have used the standard GP for the first example, a specialized GP proposed by Tsang et al. (2005) for the second example and C.5 for the remaining example to compare with the results found by ECR. For the first example (which is an extremely imbalanced data set), our approach outperformed the traditional GP. For the second example, which is less imbalanced than the former data set, our approach did not outperform Tsang's approach. However, our approach was able to provide a full-range of classifications to meet different users' requirements. Finally, for the last example, we have compared the performance of C.5 with the performance of our approach. ECR was able to classify more positive cases than C.5 in extreme imbalanced data sets. These results support the claim that ECR is effective and practical for classification in imbalanced data sets.

Acknowledgements

The first and the third authors thank to CONACyT for support to pursue graduate studies at University of Essex. The authors would like to thank the anonymous reviewers for their valuable comments.

References

Batista, G.E.A.P.A., Prati, R.C. and Monard, M.C.: 2004, A study of the behavior of several methods for balancing machine learning training data, *SIGKDD Explor. Newsl.* **6**(1), 20–29.

Bobbin, J. and Yao, X.: 1999, Automatic discovery of relational information in comprehensible control rules by evolutionary algorithms, *Proceedings of the Third Australia-Japan Joint Workshop on Intelligent and Evolutionary Systems*, Canberra, Australia, pp. 117–123.

Bojarczuk, C.C., Lopes, H.S. and Freitas, A.A.: 1999, Discovering comprehensible classification rules by using genetic programming: a case study in a medical domain, *in* W. Banzhaf, J. Daida, A.E. Eiben, M.H. Garzon, V. Honavar, M. Jakiela and R.E. Smith (eds), *Proceedings of the Genetic and Evolutionary Computation Conference*, Vol. 2, Morgan Kaufmann, Orlando, Florida, USA, pp. 953–958.

Chen, S.-H.: 2002, *Genetic Algorithms and Genetic Programming in Computational Finance*, Kluwer Academic.

Chen, S.-H. and Wang, P.P.: 2004, *Computational intelligence in economics and finance*, Springer.

Chomsky, N.: 1965, *Aspects of the theory of syntax*, Cambridge M.I.T. Press.

Corcoran, A.L. and Sen, S.: 1994, Using real-valued genetic algorithms to evolve rule sets for classification, *International Conference on Evolutionary Computation*, pp. 120–124.

Falco, I. De, Cioppa, A. Della and Tarantino, E.: 2001, Discovering interesting classification rules with genetic programming, *Applied Soft Computing* **1**(4), 257–269.

Fawcett, T.: 2004, ROC graphs: Notes and practical considerations for researchers, *Introductory paper*.

Fidelis, M.V., Lopes, H.S. and Freitas, A.A.: 2000, Discovering comprehensible classification rules a genetic algorithm, *Proceedings of the 2000 Congress on*

Evolutionary Computation CEC00, IEEE Press, La Jolla Marriott Hotel La Jolla, California, USA, pp. 805–810.

Garcia-Almanza, A.L. and Tsang, E.P.K.: 2006a, Forecasting stock prices using genetic programming and chance discovery, *12th International Conference on Computing in Economics and Finance*, Limassol.

Garcia-Almanza, A.L. and Tsang, E.P.K.: 2006b, The repository method for chance discovery in financial forecasting, *To appear in KES2006 10th International Conference on Knowledge-Based and Intelligent Information and Engineering Systems*, Springer-Verlag.

Hanley, J.A. and McNeil, B.J.: 1998, The meaning and use of the area under a reciever operating characteristic ROC curve, *Radiology*, Vol. 143, W. Madison, pp. 29–36.

Huang, J. and Ling, C.X.: 2005, Using AUC and accuracy in evaluating learning algorithms, *IEEE Transactions on Knowledge and Data Engineering* **17**(3), 299–310.

Janikow, C.Z.: 1993, A knowledge-intensive genetic algorithm for supervised learning, *Machine Learning* **13**(1-3), 189–228.

Japkowicz, N.: 2000, The class imbalance problem: Significance and strategies, *Proceedings of the 2000 International Conference on Artificial Intelligence (IC-AI'2000)*, Vol. 1, pp. 111–117.

Jong, K.A. De and Spears, W.M.: 1991, Learning Concept Classification Rules using Genetic Algorithms, *Proceedings of the Twelfth International Conference on Artificial Intelligence IJCAI-91*, Vol. 2.

Kohavi, R. and Provost, F.: 1998, Glossary of terms, *Edited for the Special Issue on Applications of Machine Learning and the Knowledge Discovery Process*, Vol. 30.

Koza, J.R.: 1992, *Genetic Programming: On the Programming of Computers by Means of Natural Selection*, The MIT Press, Cambridge, Massachusetts.

Kubat, M., Holte, R.C. and Matwin, S.: 1998, Machine learning for the detection of oil spills in satellite radar images, *Machine Learning*, Vol. 30, pp. 195–215.

Kwedlo, W. and Kretowski, M.: 1999, An evolutionary algorithm using multivariate discretisation for decision rule induction, *Principles of Data Mining and Knowledge Discovery*, pp. 392–397.

Langdon, W.B. and Poli, R.: 2002, *Foundations of Genetic Programming*, Springer Verlag, Berlin, Germany.

Li, J.: 2001, *A genetic programming based tool for financial forecasting*, PhD Thesis, University of Essex, Colchester CO4 3SQ, UK.

McCarthy, K., Zabar, B. and Weiss, G.: 2005, Does cost-sensitive learning beat sampling for classifying rare classes?, *UBDM '05: Proceedings of the 1st international workshop on Utility-based data mining*, ACM Press, New York, NY, USA, pp. 69–77.

Mitchell, T.M.: 1997, *Machine Learning*, McGraw-Hill, Boston, Mass.

Niimi, A. and Tazaki, E.: 2000, Rule discovery technique using genetic programming combined with Apriori algorithm, *in* S. Arikawa and S. Morishita (eds), *Proceedings of the Third International Conference on Discovery Science*, Vol. 1967 of *Lecture Notes in Computer Science*, Springer.

Quinlan, J.R.: 1993, *C.45 Programs for Machine Learning*, Morgan Kaufmann, San Mateo California.

Sharpe, W.F., Alexander, G.J. and Bailey, J.V.: 1995, *Investments*, Prentice-Hall International, Inc, Upper Saddle River, New Jersey 07458.

Tsang, E.P.K., Li, J. and Butler, J.M.: 1998, EDDIE beats the bookies, *International Journal of Software, Practice and Experience*, Vol. 28 of *10*, Wiley, pp. 1033–1043.

Tsang, E.P.K., Markose, S. and Er, H.: 2005, Chance discovery in stock index option and future arbitrage, *New Mathematics and Natural Computation, World Scientific*, Vol. 1 of *3*, pp. 435–447.

Tsang, E.P.K. and Martinez-Jaramillo, S.: 2004, Computational finance, *IEEE Computational Intelligence Society Newsletter*, IEEE, pp. 3–8.

Tsang, E.P.K., Yung, P. and Li, J.: 2004, EDDIE-automation, a decision support tool for financial forecasting, *Journal of Decision Support Systems, Special Issue on Data Mining for Financial Decision Making*, Vol. 37 of *4*.

Vanagas, G.: 2004, Receiver operating characteristic curves and comparison of cardiac surgery risk stratification systems, *Interact CardioVasc Thorac Surg* 3(2), 319–322.

Weiss, G.M.: 2004, Mining with rarity: A unifying framework, *Special issue on learning from imbalanced datasets*, Vol. 6, pp. 7–19.

Part III

Banking, Risk and Macroeconomic Modelling

A Banking Firm Model: The Role of Market, Liquidity and Credit Risks

Brenda González-Hermosillo[1] and Jenny X. Li[2]

[1] Deputy Devision Chief, Global Financial Stability, Monetary and Capital Markets Department, International Monetary Fund, 700 19th Street, NW, Washington, DC 20431. BGONZALEZ@imf.org

[2] Department of Mathematics and Department of Economics, The Pennsylvania State University, University Park, PA 16802, and Yunnan University of Finance and Economics. li@math.psu.edu

Summary. A theoretical and computational investigation on how the expected market, credit and liquidity risks play a role in daily business decisions made by profit maximizing banks is investigated. These risks in effect interact with each other, particularly at times of stress, and they are always considered simultaneously in practice. In all the existing studies in the theory of banking firm, however, not all of these three key risks have been considered simultaneously. A theoretical framework is developed to take a full account for all of these three risks concurrently. The resulting model amounts to a dynamic equilibrium problem which involves a high dimensional nonlinear optimization. This optimization is solved by new numerical methods proposed. The numerical results are calibrated to examine the implications of the model and to shed light on the policy implications.

Key words: Banking-firm, liquidity risk, credit risk.

1 Introduction

In January 2001, the Basel Committee on Banking Supervision issued a proposal for a new Capital Accord which intends to foster strong emphasis on banks' risk management practices as the basis for determining the appropriate level of capital. This accord, also referred to as Basel II, is expected to come into effect by January 1 2009. Once finalized, the revisions will replace the 1988 Capital Accord (or also named Basel I). One of the key differences that these changes will bring about is that the appropriate level of capital for a bank will no longer be determined by a rule such as a fixed share of assets weighted by certain risk bucket, as in Basel I. The revisions under Basel II are profound in that the appropriate level of a bank's capital will depend on the degree of risk that a bank assumes. This paradigm shift means that it is now critical to determine how banks manage risks and to which kind of risk exposures they are subject to.

From a basic perspective, the main sources of financial sector vulnerability are credit risk, market risk and liquidity risk.[3] Operationally, these risks are analyzed by sophisticated banks, and through consulting firms in the case of less sophisticated banks, based on scenario analysis and stress-testing exercises which rely on certain assumptions regarding the probabilities of events. However, the theory behind these probability models has yet to address all three risks simultaneously.

In the academic literature on the theory of the banking firm, the assessment of these three risks has been dealt with largely by focusing on one type of risk at a time, but not all three of them concurrently. For example, credit risk models include Merton (1974) based on pricing theory, and Schwartz and Torous (1992). Models of liquidity risk include Prisman et al. (1986) where the volume of deposits is subject to random shocks; and Diamond and Dybvig (1983), Diamond and Rajan (1999) and Allen and Gale (2002) where the source of risk is given by panic deposit runs and multiple equilibria. Market risk models include Ingersoll (1987) based on capital asset pricing theory, and Pyle (1971) and Hart and Jaffee (1974) based on mean-variance analysis models. Freixas and Rochet (1997) provide a survey of the literature. Some recent analyses dealing with more than one risk at the time is Buchinsky and Yosha (1997) where market risk and default risk are modeled jointly, and Bangia et al. (2002) who model market risk based on the value-of-risk approach and introduce liquidity risk. Peck and Shell (2003) used a sunspot model for the the liquidity risk too.

In this paper, we extend Buchinsky and Yosha's (1997) model and correct some misspecification issues in that model. Specifically, we introduce liquidity risk in the analysis, which allows us to model all the basic risks facing banks. This research is novel in that, for the first time to our knowledge, market risk, credit risk and liquidity risk are examined together in the context of a rigorous theoretical model of the banking firm. As well, the numerical algorithm developed in this paper provides a more general approach for solving high dimensional stochastic dynamic problems.

Beyond the theoretical interest in developing a new and more rigorous model of the banking firm that better mirrors the actual behavior of bank managers, the questions examined in this paper are also of keen interest to bank regulators and policy makers. As recent events have shown in the aftermath of the U.S. subprime mortgage debacle in the summer of 2007, market risks (such as a collapse of the U.S. housing market) can quickly be transmitted to credit risks (as mortgage holders default on their loans) and can lead

---

[3] The Bank for International Settlements identifies more risks, including operational risk which deals with issues such as the failure of computers. However, the main sources of risk facing banks, based on the structure of their balance sheet, are these three.

to liquidity shocks in the system (with banks and other financial institutions becoming unwilling to provide liquidity to other financial firms and to the market more generally). Possibly as a result of asymmetric information.

Moreover, by allowing the representative banking firm in the model to short positions, we also introduce an analytical framework able to examine not only the balance sheets of banks, but also their off-balance sheet positions. Current banking regulatory frameworks have been unable to fully capture the importance of banks' off-balance sheet positions. Another lesson of the recent financial crisis in the Summer of 2007 was that banks moved positions to off-balance sheet special investment vehicles (SIV), which were supposed to be at arms' length to the banks. During the period of stress, however, those positions had to be moved onto the parent bank's balance sheet.

The rest of the paper is arranged as follows: Section 2 gives an introduction of our model, Section 3 describes numerical approaches used to solve the problem. Section 4 gives conclusions of our study with some policy observations.

2 The Model

Core to the operation of banks, and financial institutions in general, is the intermediation of funds and risk management. Bank managers are required to make decisions on a daily basis regarding basic policy tools such as the type and amount of loans offered to customers, in what type of tradable securities to invest and how much of the portfolio to invest on those assets, the amount of liquid assets sufficient to set aside to cover for potential deposit withdrawals, the level of interest rates (above a given risk-free interest rate prevailing in the economy) that should be offered to depositors, and the maximum amount of dividends that can be distributed to shareholders. While bank managers have certain policy tools at their disposal to accomplish these objectives, they also need to manage unforeseen circumstances and conform to certain regulatory requirements which may include, depending on the country, minimum liquidity and capital-asset ratios.

In this model, an infinitely-lived risk-averse representative bank owner derives utility from the consumption of dividend income d_t. In each period, the bank owner invests the money at its disposal, namely its monetary value (or capital) M_t and the newly acquired deposits S_t payable at demand, in three types of investments. The financial intermediary can purchase publicly traded securities a_t which are subject to market risk or the risk that their value may decline with market conditions (such as changes in interest rates, the exchange rate, output growth, etc.). It can also invest funds in loans offered to corporations and/or households l_t which are subject to credit risk or the risk that the loans will not be paid back. For simplicity, it is assumed that the expected recovery rate of the loans is zero and that they are not subject to market risk. In reality, the default risk of borrowers would be expected to be also determined by market conditions (including interest rates and the level of economic activity) and, in that sense, also by market risk. Thus, the distinction between market risk and credit risk made in this study is somewhat artificial despite the fact this distinction is normally made. For example, in Basel II the Bank for International Settlements assumes that loans are only subject to credit risk.

The bank will also invest part of the funds in risk-free liquid assets b_t, such as government bonds. In this model, the amount of liquid assets held by the bank is a function of a regulatory fraction λ with $0 < \lambda < 1$, which is exogenously given to the bank, of total deposits $S(.)$ where the deposits function $S_t(r_t, q_t, Z_{t+1}^s)$, is assumed to depend on the interest rate offered by the bank on deposits r_t (relative to an exogenous risk-free rate r_b), the expected probability of the bank's survival at time t for the next period q_t, and certain random shocks Z_{t+1}^s representing the depositors' liquidity needs which are exogenously determined in the model. Which means the depositors only take the promised interest rate and the bank's survival probability into consideration when he/she makes the deposit decision and regardless the payoff in the event of bankruptcy. We also assume $S_t(r_t, q_t, Z_{t+1}^s)$ is continuously differentiable and strictly increasing in r_t and q_t. We then have:

$$b_t = M_t + S_t(r_t, q_t, Z_{t+1}^s) - a_t - l_t - d_t \geq \lambda S_t(r_t, q_t, Z_{t+1}^s). \tag{1}$$

However, deposit withdrawals can still be larger than the amount of liquid assets required by regulatory authorities. As such, at any point in time, the bank intermediary faces three uncertain outcomes over the next period: the return on traded securities Z_{t+1}^a is an aggregate random variable that affects the entire economy due to market risk, the return on loans Z_{t+1}^l is stochastic due to the credit risk, and Z_{t+1}^s is the stochastic component of deposit withdrawals. Thus, the bank manager's decisions must take into account market, credit and liquidity risks at same time.

The bank's probability of survival at time t for the next period q_t is defined as one minus the probability at time t that the bank will become bankrupt at time $t + 1$. Bankruptcy will occur if at the beginning of period $t + 1$ the bank cannot repay its debt to depositors or, equivalently, if $M_{t+1} < 0$. Actually, $M_{t+1} < 0$ could be viewed as a regulatory bankruptcy, which typically defines a threshold under which bankruptcy occurs. We define this threshold to be zero, but it can be some other finite numbers. Bankruptcy in this sense is distinctly different from insolvency which would require that the present value of the bank's assets fall short of the bank's non-owned liabilities. Thus a bankrupt bank which is unable to pay its obligations at period t, may either be insolvent, illiquid or both.

However, the bank may still be able to operate for at least one more period if it is able to borrow against its tradable securities to pay its depositors, and therefore avert bankruptcy. This model allows for the possibility that a bank can borrow $A(M_t)$ against its tradable securities, by discounting these assets or engaging in repo operations directly in the market or with the central bank.

The bank's ability to borrow against its tradable securities is assumed to be limited by the stock of tradable assets a_t held by the bank at and is assumed to be an increasing convex function of its monetary value. Higher levels of capital will make it easier for the bank to borrow. Without loss of generality, $A(M_t)$ could also take the value of zero if there is no infrastructure for the bank to borrow against its tradable securities either because financial markets are not well developed or because the central bank does not have a discount window facility. The function is assumed to be strictly negative, reflecting the fact that the bank would be short selling publicly traded securities. Thus, more generally, a bank is said to be bankrupt at time $t + 1$ if

$$M_{t+1} - A(M_{t+1}) < 0 \tag{2}$$

depending on the policy choices made by management, but also on the shocks facing the bank making this outcome stochastic. For the simplicity and the bank in our model can not borrow to raise funds, simply because the loans at the window are very short term and often has tighter supervision.

Thus, M_{t+1}, conditioned on survival, is governed by the following law of motion:

$$M_{t+1} = [M_t + S_t(r_t, q_t, Z_{t+1}^s) - a_t - l_t - d_t]r_b + a_t Z_{t+1}^a + l_t Z_{t+1}^l - S_t(r_t, q_t, Z_{t+1}^s)r_t. \tag{3}$$

Again, here Z_{t+1}^a is a random variable representing the one period return on investments made in period t in publicly traded securities, Z_{t+1}^l is a random variable which represents the one period return on bank-specific loans, and Z_{t+1}^s captures the random deposit withdrawals. Let

$$Z_t(t + 1) = (Z_{t+1}^a, Z_{t+1}^l, Z_{t+1}^s)$$

be the risk returns on the bank's risky investments and the risk of deposit withdraws in period t. $Z_t(t + 1)$ are random variables at time t and the joint probability distribution is assumed to be stationary, with a strictly increasing, continuous and differentiable cumulative distribution function, which is known to the bank.

The probability of bankruptcy in period $t + 1$, as perceived in period t, is given by:

$$P\{M_{t+1} - A(M_{t+1}) < 0\} = F(0, M_t, c_t, q_t) \tag{4}$$

where $F(0, M_t, c_t, q_t)$ is induced by the c.d.f of the random shocks Z_{t+1}^a, Z_{t+1}^l, Z_{t+1}^s and the bank policy choices $c_t = (a_t, l_t, d_t, r_t)$. Correspondingly, the probability of survival q_t is given by:

$$1 - P\{M_{t+1} - A(M_{t+1}) < 0\} = q_t. \tag{5}$$

The solution to this equation is consistent with a rational expectations equilibrium in which depositors' beliefs regarding the bank's probability of survival

affects their supply of deposits. It is assumed that the probability of survival
in period $t + 1$ is known in period t and that there is no divergence of opinion
between the bank's management and the depositors regarding q_t, given the
information available.

As discussed below, the bank's behavior is conditioned by their objective
to maximize a utility function which solely depends on dividends d_t provided
that the bank survives or $q_t > 0$. In other words, the degree of soundness of
the bank (the level of q_t) is only important to the bank owners because they
may face deposit withdrawals from investors who assess the probability of the
survival of the bank. Bank managers need to secure the survival of the bank in
order to maximize dividends, but different degrees of soundness or q_t are not
critical to them unless there is market discipline whereby depositors punish
banks with a low q_t. Banks can also offer a higher interest rate on deposits r_t
to compensate depositors for a low q_t, but only to a point. Thus, for example,
if q_t is very low (the probability of survival extremely low) even a high level of
interest rate on deposits r_t will not compensate depositors for the high risk of
loosing the entire nominal value of their savings if deposit guarantees are not
full.[4] The reason for this discontinuity is that the expected return to deposits
from a higher interest rate would not compensate for the loss of the larger
nominal value of the deposits if the bank is going to bankrupt.

We will show that for any given state and policy variables in period t,
there is a unique solution for the survival rate: $q_t = q(M_t, c_t)$.

The following theorem provides the existence and uniqueness of the endogenously determined probability of survival q_t at period $t + 1$.

Theorem 1. *There is a unique q_t between 0 and 1 which satisfies equation
(5) for any given initial capital M_t and policy $c_t = (a_t, l_t, d_t, r_t)$. Furthermore
q_t is differentiable in $(0, 1)$ and continuous on $[0, 1]$.*

Proof. We divide the proof into several steps.
Step 1: $F(\cdot, M_t, c_t, q_t)$ is continuous and differentiable in M_t, c_t and q_t. For
fixed M_t and c_t, since $S(r_t, q_t, Z_{t+1}^s)$ is differentiable in q_t, and by the law of
motion of M_{t+1} in equation (3) M_{t+1} is differentiable in q_t. In addition, $A(\cdot)$
is differentiable in M_{t+1} and $F(\cdot, M_t, c_t, q_t)$ is differentiable in its argument,
thus $F(\cdot, M_t, c_t, q_t)$ is continuous and differentiable in q_t. Following the same
logic, $F(\cdot, M_t, c_t, q_t)$ is continuous and differentiable in M_t and c_t.
Step 2: $F(\cdot, M_t, c_t, q_t)$ is increasing in q_t. Rewrite equation (3) as

$$M_{t+1} = (M_t - d_t)r_b + a_t(z_{t+1}^a - r_b) + l_t(z_{t+1}^l - r_b) - z_{t+1}^s S(r_t, q_t)(r_t - r_b); \quad (6)$$

---

[4] In practice, there may be transaction costs even if deposit guarantees are full.
Alternatively, it may be that the deposit guarantees, even if de jure full, may
be less than fully credible if many banks become bankrupt at the same time.
This is because deposit guarantees are typically in the form of a contingent fund
insufficient to cover for a fully fledged financial crisis. For simplicity, the analysis
here ignores those possibilities.

Since $S(r_t, q_t, Z^s_{t+1})$ is increasing in q_t, and $r_b < r_t$, we know that M_{t+1} is decreasing in q_t. By the assumption that $A(\cdot) \le 0$ and $A'(\cdot) \le 0$, it follows that $M_{t+1} - A(M_{t+1})$ is non-increasing in q_t and hence its c.d.f. is non-decreasing in q_t.

Step 3: Rewrite equation (5) as

$$H(M_t, c_t, q_t) = 1, \tag{7}$$

where $H(M_t, c_t, q_t) \equiv F(0; M_t, c_t, q_t) + q_t$. Then there is a unique $q_t \in [0, 1]$ such that equation (5) holds.

Extend the function $H(M_t, c_t, q_t)$ so that for $q_t \le 0$, $H(M_t, c_t, q_t) \equiv F(0; M_t, c_t, 0) + q_t$, and for $q_t \ge 1$, $H(M_t, c_t, q_t) \equiv F(0; M_t, c_t, 1) + q_t$. Since $F(0; M_t, c_t, q_t)$ is bounded, it follows that $\lim_{q_t \to \infty} H(M_t, c_t, q_t) = \infty$ and $\lim_{q_t \to -\infty} H(M_t, c_t, q_t) = -\infty$. Since $H(M_t, c_t, q_t)$ is continuous and monotone increasing in q_t, there is a unique value of q_t such that equation (7) holds. Also since $F(0; M_t, c_t, q_t) \in [0, 1]$, this value of q_t must also lie in the interval [0,1].

Step 4: Let $q_t \in (0, 1)$. By arguments analogous to those in step 2, it follows that $F(0; M_t, c_t, q_t)$ is strictly increasing in q_t. Since for any $q_t \in (0, 1)$ the function $H(M_t, c_t, q_t)$ is differentiable in (M_t, c_t), by applying the Implicit Function Theorem to (7), we can conclude that $q(M_t, c_t)$ is continuous and differentiable in (M_t, c_t). Also by the continuity of $F(0; M_t, c_t, q_t)$. This completes the proof of the Theorem.

To understand the bank's behavior, we will first define the feasible policy variable which banks will therefore choose to maximize the expected utility function.

A policy variable $c_t = (a_t, l_t, d_t, r_t)$ is feasible with state M_t if it satisfies the following constrains:

$$0 \le d_t \le M_t - A(M_t), \tag{8}$$

$$a_t \ge -A(M_t), \tag{9}$$

$$M_t + S_t(r_t, q_t, Z^s_{t+1}) - a_t - l_t - d_t \ge \lambda S_t(r_t, q_t, Z^s_{t+1}), \tag{10}$$

$$a_t \ge 0, \quad l_t \ge 0, \tag{11}$$

$$r_b < r_t < \bar{r}. \tag{12}$$

The first condition states that the manager cannot distribute dividends in excess of its nominal value plus the amount that it can borrow. The second condition defines a cap on the amount of funds that the bank can borrow against its tradable securities. The third condition refers to a reserve requirement or reserve ratio. The fourth and fifth conditions require that tradable securities and loans are nonnegative. The last condition restricts the range of interest rates on deposits that the bank can offer. The rate offered on deposits is assumed to be bounded on the low side by the risk-free interest rate and $\bar{r}$ on the up side.

The owner of the bank has a utility function $u(.)$ derived from the consumption of dividends. As usual, the utility function is assumed to be continuous, strictly increasing, concave and differentiable. We also assume the utility function has the property $u(0) = 0$. Bankruptcy implies that the utility is zero in all subsequent periods. In each period the owner of the bank chooses a feasible policy $\{c_k = (a_t, l_t, d_t, r_t)\}_{k=t}^{\infty}$ to maximize the expected discounted utility function over subsequent periods subject to the law of motion, and the expectation is taken with respect to the sequence of random variables $\{Z_k\}_{k=t+1}^{\infty}$. In other words, we need to

$$\max_{c_k} E[\sum_{k=t}^{\infty} \beta u(d_k)]. \tag{13}$$

Subject to the law of motion (3), with $\beta \in (0, 1)$ as the discount factor. Solving this optimization problem is equivalent to solving the so-called Bellman functional equation:

$$V(M_t) = \sup_{c_t}\{u(d_t) + \beta E[V(M_{t+1})]\} \tag{14}$$

Thus, we have the following theorem:

Theorem 2. *Assume the supply of deposits is a function which is concave in* r_t *and strictly convex with respect to* q_t*. Then there is a unique function* $V(.)$*, that is strictly increasing, continuous and bounded and that satisfies the above Bellman equation.*

The concavity of $S_t(r_t, q_t, Z_{t+1}^s)$ with respect to r_t implies that, for a given probability of survival, the deposit is marginally diminishing with the increasing interest rate. The convexity of q_t on the other hand means that, for same interest rate the higher probability of survival will bring higher increment of deposit. The proof of the theorem is standard and straight forward.

2.1 A Numerical Approach to the Model

The goal of our analysis that to find the relationship between the probability of survival and capital size, the relationship between the interest rate and capital size and the relationship between all investments and capital size. Unfortunately, the analytic solution for such complicated problem is unattainable, hence a numerical approach is the only choice. The most difficult part of solving the problem is that q_t is endogenously determined. That is, besides finding the solution for the dynamic program, we also have to solve for the equilibrium probability of survival for all feasible bank policies.

2.2 Numerical Algorithm of Solution

A survey of the literature regarding solving Bellman equation can be found in Judd (1998). Also there are other works measuring the financial risk, such as Gilli and Këllezi (2006). However the biggest challenge we face is the curse of dimensionality. Because optimal policy consist of four parameters and if we use the classic discretization method, including the discretization of the capital of the bank, we would need to deal with a problem with five dimensions. Furthermore, there are three i.i.d shocks involved in the model, which will make the computation more complicated. So, the natural value function iteration becomes almost impossible. Our approach is similar to the so called policy iteration. Since the utility function $u(\cdot)$ in our model is concave, monotonically increasing and uniformly bounded, and the constraints for all the components in the optimal policy are convex and bounded, hence there exists an optimal policy to solve this problem. To characterize the optimal policy, we use the envelop condition and first order conditions for each of the components of the optimal policy. This gives the following 4 equations:
Envelop condition:

$$V'(M_t) = u'(d_t) \tag{15}$$

First order condition w.r.t. r_t

$$-\beta E[z_s^t S_t V'(M_{t+1})] - \beta(r_t - r_b)E[z_s^t \frac{\partial S_t}{\partial q_t} \frac{\partial q_t}{\partial r_t} V'(M_t)] \leq 0; \tag{16}$$

First order condition w.r.t. d_t

$$u'(d_t) - \beta r_b E[z_s^t V'(M_{t+1})] - \beta(r_t - r_b)E[z_s^t \frac{\partial S_t}{\partial q_t} \frac{\partial q_t}{\partial d_t} V'(M_{t+1})] \leq 0; \tag{17}$$

First order condition w.r.t. a_t

$$\beta r_b E[V'(M_{t+1})] - \beta r_b E[z_t^a V'(M_{t+1})] - \beta(r_t - r_b)E[z_s^t \frac{\partial S_t}{\partial q_t} \frac{\partial q_t}{\partial a_t} V'(M_{t+1})] \leq 0. \tag{18}$$

And the numerical algorithm can be described as following, where all random numbers are generated by the random generator online:

1. Discretize the interval of capital size. In our model, the capital size will be larger than -0.8. The specific lower bound is chosen due to the form of $A(\cdot)$. If capital size is lower than -0.8, then by the definition of q_t, $q_t = 0$ and $V(M_t) = 0$. Thus a situation in which a bank's capital size is lower than -0.8 is not included in our computation.
2. Get the initial guess of the optimal policy d_t^0, a_t^0, q_t^0 as well as initial guess of q_t for each capital size M_t^i.
3. With the initial guess of $\{d_t^{0,i}\}_{i=1}^N$ and the envelop condition, we can get $(V'(M_t))^{0,i}$ for $i = 1, \ldots, N$.

4. Use linear interpolation to approximate $V'(M_t)^0$, thus we have a continuous function which is the initial approximation of $V'(M_t)$.
5. Use the first order condition of r_t and the initial guess of d_t, a_t, q_t to compute what should be the optimal r_t^i with d_t^0, a_t^0, q_t^0 fixed, then we take this value as $r_t^{i,1}$.
6. Use the first order condition of d_t and the initial guess of $a_t, q_t,$ r_t^1 to compute what should be the optimal d_t^i with $a_t^0, q_t^0,$ r_t^1 fixed and then take this value as $d_t^{i,1}$.
7. Use the first order condition of a_t and initial guess of q_t, r_t^1, d_t^1 to compute what should be the optimal a_t^i with q_t^0, r_t^1, d_t^1 fixed and take this value as $a_t^{i,1}$.
8. With all the components of the optimal policy updated, we can update the survivability of the bank q_t. Then go back to step 2) until the sequence of the optimal policy shows the convergence.

Another approach of course is the quasi-Monte Carlo simulation, which may also provide an efficient numerical algorithm to deal with such a problem and which we are currently investigating.

Function and parameters for the numerical simulation are:

- Let the utility function be: $u(x) = \frac{\log(x+1)}{1+\log(x+1)}$ and $\beta = 0.95$
- Let the deposit function be: $S_t(r_t, q_t, Z_{t+1}^s) = (r_t - r_b)^{0.5} q_t^{1.1} Z_{t+1}^s$
- $\lambda = 0.1$ for reserve requirement
- $A(M_t - d_t) = -[1 + 0.2((M_t - d_t)]$
- Z_{t+1}^a and Z_{t+1}^l both satisfy log-normal distributions.

It is worth pointing out that all the functions here satisfy all previous assumptions. While in Buchinsky and Yosha's paper the deposit function is obviously wrong since it indicates that deposit is positive even when the depositors know the probability of survival of the bank is zero. So their numerical result is questionable. Our deposit function equals zero when q_t equals zero. Figures 1–3 show the relation between the optimal and different size of bank.

3 Conclusion

The primary simulation results indicate that the probability of banks' survival increases with the amount of equity capital. Small intermediaries have a lower probability of survival and therefore must offer a higher interest rate. Also face to a severe negative deposit shock (or withdrawal), the survivability of a bank with relatively high capital size is higher than that of a bank with small capital size. Furthermore banks with relatively large capital endowments can see their probability of survival be reduced if they face a very large deposit withdrawal shock.

The paper demonstrates that it is not satisfactory to model banks as if they only face one or two types of risks. Clearly they face many risks, but the

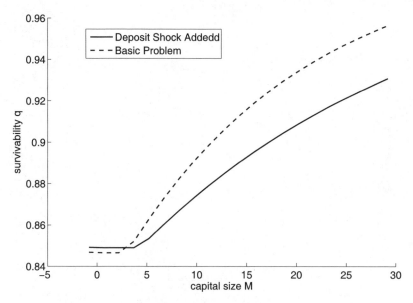

Fig. 1. Probability of survive vs the capital size.

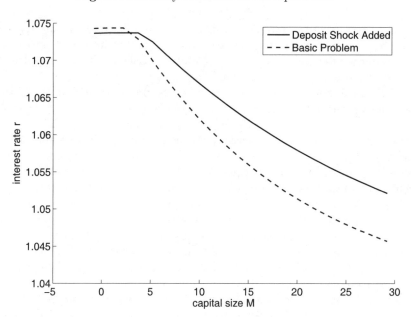

Fig. 2. Interest rate vs the capital size.

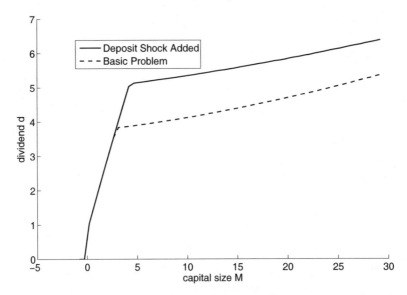

Fig. 3. Dividend vs the capital size.

core risks facing banks represent market, credit and liquidity conditions. Theoretically, it is necessary to gauge these risks together in a more systematic fashion. Only after fully understanding all these risks will bank regulators be able to propose a more coherent and comprehensive regulatory system. The recent financial crisis in mid-2007 illustrates the importance of understanding how banks manage market, credit and liquidity risks. Based on these recent events, it is clear that more research is needed to understand banks' risk management practices and the potential risks facing financial institutions. This is especially the case during times of stress when a given market shock (such as the decline in U.S. real estate prices and the consequent subprime mortgage debacle) can lead to a liquidity shock (as banks see the value of their capital decline or if they face constraints in their ability to sell their assets). This reinforcing mechanism can potentially result in risks of systemic proportions, as financial institutions may need to sell assets that have become illiquid at "fire sale" prices, thus prompting fresh concerns of solvency. Our model still suffers from the oversimplification that market, liquidity and credit risks are assumed to be independent. However, we believe that this is the first time in the literature that all the three key risks facing banks are analyzed simultaneously. Moreover, by allowing banks in our model to short assets, the model is able to capture many of the off-balance sheet positions that banks hold in reality. Future research should relax the assumption of independence among the various risks that banks face, thus providing hints as to the dynamics behind systemic episodes of stress, and consider more than one representative bank since the actions of one bank can affect the prospects of other banks through the interbank market or as a result of asymmetric information.

Acknowledgements

We thank Jia Pan and Pantao Sun for their research assistance.

References

Allen, F. and Gale, D.: 2002, Asset price bubbles and stock market interlinkages, *Working Papers 02-22*, Wharton School Center for Financial Institutions, Unviersity of Pennsylvania.

Bangia, A., Diebold, F.X., Schuermann, T. and Stroughair, J.: 2002, Modeling liquidity risk, with implications for traditional market risk measurement and management, *Risk Management: The State of the Art*, Kluwer, Amsterdam, pp. 1–13.

Buchinsky, M. and Yosha, O.: 1997, Endogenous probability of failure for a financial intermediary: A dynamic model, *Working Papers 1997-23*, Brown University.

Diamond, D.W. and Dybvig, P.H.: 1983, Bank runs, deposit insurance, and liquidity, *The Journal of Political Economy* **91**(3), 401–419.

Diamond, D.W. and Rajan, R.G.: 1999, Liquidity risk, liquidity creation and financial fragility: A theory of banking, *Working Papers 7430*, NBER.

Freixas, X. and Rochet, J.-C.: 1997, *Microeconomics of Banking*, Mit Press, Cambridge, MA.

Gilli, M. and Këllezi, E.: 2006, An application of extreme value theory for measuring financial risk, *Computational Economics* **27**(2-3), 207–228.

Hart, O. and Jaffee, D.: 1974, On the application of portfolio theory to depository financial intermediaries, *Review of Economic Studies* **41**, 129–147.

Ingersoll, J.E.: 1987, *Theory of Financial Decision Making*, Rowman & Littlefield Publishers, Inc., Totowa, NJ.

Judd, K.D.: 1998, *Numerical Methods in Economics*, MIT Press, Cambridge, MA.

Merton, R.C.: 1974, On the pricing of corporate debt: The risk structure of interest rates, *Journal of Finance* **29**, 449–470.

Peck, J. and Shell, K.: 2003, Equilibrium bank runs, *Journal of Political Economy* **111**(1), 103–123.

Prisman, E., Slovin, M.B. and Sushka, M.: 1986, A general model of the banking firm under conditions of monopoly, uncertainty, and recourse, *Journal of Monetary Economics* **17**(2), 293–304.

Pyle, D.: 1971, On the theory of financial intermediation, *Journal of Finance* **26**, 737–748.

Schwartz, E.S. and Torous, W.N.: 1992, Prepayment, default, and the valuation of mortgage pass-through securities, *Journal of Business* **65**(2), 221–239.

Identification of Critical Nodes and Links in Financial Networks with Intermediation and Electronic Transactions

Anna Nagurney and Qiang Qiang

Isenberg School of Management, University of Massachusetts, Amherst, Massachusetts 01003, nagurney@gbfin.umass.edu

Summary. A network performance measure for the evaluation of financial networks with intermediation is proposed. The measure captures risk, transaction cost, price, transaction flow, revenue, and demand information in the context of the decision-makers' behavior in multitiered financial networks that also allow for electronic transactions. The measure is then utilized to define the importance of a financial network component, that is, a node or a link, or a combination of nodes and links. Numerical examples are provided in which the performance measure of the financial network is computed along with the importance ranking of the nodes and links. The results can be used to assess which nodes and links in financial networks are the most vulnerable in the sense that their removal will impact the performance of the network in the most significant way. Hence, the results have relevance to national security as well as implications for the insurance industry.

Key words: Financial networks, financial intermediation, risk management, portfolio optimization, complex networks, supernetworks, critical infrastructure networks, electronic finance, network performance, network vulnerability, network disruptions, network security, network equilibrium, variational inequalities.

1 Introduction

The study of financial networks dates to the 1750s when Quesnay (1758), in his *Tableau Economique*, conceptualized the circular flow of financial funds in an economy as a network. Copeland (1952) subsequently explored the relationships among financial funds as a network and asked the question, *"Does money flow like water or electricity?"* The advances in information technology and globalization have further shaped today's financial world into a complex network, which is characterized by distinct sectors, the proliferation of new financial instruments, and with increasing international diversification of portfolios. Recently, financial networks have been studied using network models with multiple tiers of decision-makers, including intermediaries. For a detailed literature review of financial networks, please refer to the paper by Nagurney (2007) (see also Fei (1960), Charnes and Cooper (1967), Thore (1969), Thore and Kydland (1972), Thore (1980), Christofides et al. (1979), Crum and Nye (1981), Mulvey (1987), Nagurney and Hughes (1992), Nagurney et al. (1992), Nagurney and Siokos (1997), Nagurney and Ke (2001; 2003), Boginski et al. (2003), Geunes and Pardalos (2003), Nagurney and Cruz (2003a; 2003b), Nagurney et al. (2006), and the references therein). Furthermore, for a detailed discussion of optimization, risk modeling, and network equilibrium problems in finance and economics, please refer to the papers in the book edited by Kontoghiorghes et al. (2002).

Since today's financial networks may be highly interconnected and interdependent, any disruptions that occur in one part of the network may produce consequences in other parts of the network, which may not only be in the same region but many thousands of miles away in other countries. As pointed out by Sheffi (2005) in his book, one of the main characteristics of disruptions in networks is "the seemingly unrelated consequences and vulnerabilities stemming from global connectivity." For example, the unforgettable 1987 stock market crash was, in effect, a chain reaction throughout the world; it originated in Hong Kong, then propagated to Europe, and, finally, the United States. It is, therefore, crucial for the decision-makers in financial networks, including managers, to be able to identify a network's vulnerable components in order to protect the functionality of the network. The management at Merrill Lynch well understood the criticality of their operations in World Trade Center and established contingency plans. Directly after the 9/11 terrorist attacks, management was able to switch their operations from the World Trade Center to the backup centers and the redundant trading floors near New York City. Therefore, the company managed to mitigate the losses for both its customers and itself (see Sheffi (2005)).

Notably, the analysis and the identification of the vulnerable components in networks have, recently, emerged as a major research theme, especially in the study of what are commonly referred to as *complex* networks, or, collectively, as *network science* (see the survey by Newman (2003)). However, in order to be able to evaluate the vulnerability and the reliability of a network, a measure that can quantifiably capture the performance of a network must be developed. In a series of papers, Latora and Marchiori (2001; 2003; 2004) discussed the network performance issue by measuring the "global efficiency" in a weighted network as compared to that of the simple non-weighted small-world network. The weight on each link is the geodesic distance between the nodes. This measure has been applied by the above authors to evaluate the importance of network components in a variety of networks, including the (MBTA) Boston subway transportation network and the Internet (cf. Latora and Marchiori (2002; 2004)).

However, the Latora-Marchiori network efficiency measure does not take into consideration the flow on networks, which we believe is a crucial indicator of network performance as well as network vulnerability. Indeed, flows represent the usage of a network and which paths and links have positive flows and the magnitude of these flows are relevant in the case of network disruptions. For example, the removal of a barely used link with very short distance would be considered "important" according to the Latora-Marchiori measure.

Recently, Qiang and Nagurney (2007) proposed a network performance measure that can be used to assess the network performance in the case of either fixed or elastic demands. The measure proposed by Qiang and Nagurney (2007), in contrast to the Latora and Marchiori measure, captures flow information and user/decision-maker behavior, and also allows one to determine the criticality of various nodes (as well as links) through the identification of their importance and ranking. In particular, Nagurney and Qiang (2007a; 2007b; 2007d) were able to demonstrate the applicability of the new measure, in the case of fixed demands, to transportation networks as well as to other critical infrastructure networks, including electric power generation and distribution networks (in the form of supply chains). Interestingly, the above network measure contains, as a special case, the Latora-Marchiori measure, but is general in that, besides costs, it also captures flows and behavior on the network as established in Nagurney and Qiang (2007a; 2007b).

Financial networks, as extremely important infrastructure networks, have a great impact on the global economy, and their study has recently also attracted attention from researchers in the area of complex networks. For example, Onnela et al. (2004) studied a financial network in which the nodes are stocks and the edges are the correlations among the prices of stocks (see also, Kim and Jeong (2005)). Caldarelli et al. (2004) studied different financial networks, namely, board and director networks, and stock ownership networks and discovered that all these networks displayed scale-free properties (see also Boginski et al. (2003)). Several recent studies in finance, in turn, have analyzed the local consequences of catastrophes and the design of risk sharing/management mechanisms since the occurrence of disasters such as 9/11 and Hurricane Katrina (see, for example, Gilli and Këllezi (2006), Loubergé et al. (1999), Doherty (1997), Niehaus (2002), and the references therein).

Nevertheless, there is very little literature that addresses the vulnerability of financial networks. Robinson et al. (1998) discussed, from the policy-making point of view, how to protect the critical infrastructure in the US, including financial networks. Odell and Phillips (2001) conducted an empirical study to analyze the impact of the 1906 San Francisco earthquake on bank loan rates in the financial network within San Francisco. To the best of our knowledge, however, there is no network performance measure to-date that has been applied to financial networks that captures both economic behavior as well as the underlying network/graph structure. The only relevant network study is that by Jackson and Wolinsky (1996), which defines a value function for the network topology and proposes the network efficiency concept based on the value function from the point of view of network formation. In this paper, we propose a novel financial network performance measure, which is motivated by Qiang and Nagurney (2007) and that evaluates the network performance in the context where there is noncooperative competition among source fund agents and among financial intermediaries. Our measure, as we also demonstrate in this paper, can be further applied to identify the importance and the ranking of the financial network components.

The paper is organized as follows. In Section 2, we briefly recall the financial network model with intermediation of Liu and Nagurney (2007). The financial network performance measure is then developed in Section 3, along with the associated definition of the importance of network components. Section 4 presents two financial network examples for which the proposed performance measure are computed and the node and link importance rankings determined. The paper concludes with Section 5.

2 The Financial Network Model with Intermediation and Electronic Transactions

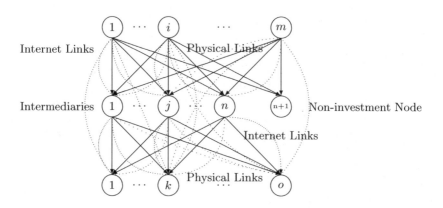

Fig. 1. The structure of the financial network with intermediation and with electronic transactions.

In this section, we recall the financial network model with intermediation and with electronic transactions in the case of known inverse demand functions associated with the financial products at the demand markets (cf. Liu and Nagurney (2007)). The financial network consists of m sources of financial funds, n financial intermediaries, and o demand markets, as depicted in Figure 1. In the financial network model, the financial transactions are denoted by the links with the transactions representing electronic transactions delineated by hatched links. The majority of the notation for this model is given in Table 1.

Table 1. Notation for the financial network model.

Notation	Definition
S	m-dimensional vector of the amounts of funds held by the source agents with component i denoted by S^i
q_i	$(2n + o)$-dimensional vector associated with source agent i; $i = 1, \ldots, m$ with components: $\{q_{ijl}; j = 1, \ldots, n; l = 1, 2;$ $q_{ik}; k = 1, \ldots, o\}$
q_j	$(2m + 2o)$-dimensional vector associated with intermediary j; $j = 1, \ldots, n$ with components: $\{q_{ijl}; i = 1, \ldots, m; l = 1, 2; q_{jkl};$ $k = 1, \ldots, o; l = 1, 2\}$
Q^1	$2mn$-dimensional vector of all the financial transactions/flows for all source agents/intermediaries/modes with component ijl denoted by q_{ijl}
Q^2	mo-dimensional vector of the electronic financial transactions/flows between the sources of funds and the demand markets with component ik denoted by q_{ik}
Q^3	$2no$-dimensional vector of all the financial transactions/flows for all intermediaries/demand markets/modes with component jkl denoted by q_{jkl}
g	n-dimensional vector of the total financial flows received by the intermediaries with component j denoted by g_j, with $g_j \equiv \sum_{i=1}^{m} \sum_{l=1}^{2} q_{ijl}$
γ	n-dimensional vector of shadow prices associated with the intermediaries with component j denoted by γ_j
d	o-dimensional vector of market demands with component k denoted by d_k
$\rho_{3k}(d)$	the demand price (inverse demand) function at demand market k
V^i	the $(2n + o) \times (2n + o)$ dimensional variance-covariance matrix associated with source agent i
V^j	the $(2m + 2o) \times (2m + 2o)$ dimensional variance-covariance matrix associated with intermediary j
$c_{ijl}(q_{ijl})$	the transaction cost incurred by source agent i in transacting with intermediary j using mode l with the marginal transaction cost denoted by $\frac{\partial c_{ijl}(q_{ijl})}{\partial q_{ijl}}$

continued on next page

continued from previous page

Notation	Definition
$c_{ik}(q_{ik})$	the transaction cost incurred by source agent i in transacting with demand market k with marginal transaction cost denoted by $\frac{\partial c_{ik}(q_{ik})}{\partial q_{ik}}$
$c_{jkl}(q_{jkl})$	the transaction cost incurred by intermediary j in transacting with demand market k via mode l with marginal transaction cost denoted by $\frac{\partial c_{jkl}(q_{jkl})}{\partial q_{jkl}}$
$c_j(Q^1) \equiv c_j(g)$	conversion/handling cost of intermediary j with marginal handling cost with respect to g_j denoted by $\frac{\partial c_j}{\partial g_j}$ and the marginal handling cost with respect to q_{ijl} denoted by $\frac{\partial c_j(Q^1)}{\partial q_{ijl}}$
$\hat{c}_{ijl}(q_{ijl})$	the transaction cost incurred by intermediary j in transacting with source agent i via mode l with the marginal transaction cost denoted by $\frac{\partial \hat{c}_{ijl}(q_{ijl})}{\partial q_{ijl}}$
$\hat{c}_{jkl}(Q^2, Q^3)$	the unit transaction cost associated with obtaining the product at demand market k from intermediary j via mode l
$\hat{c}_{ik}(Q^2, Q^3)$	the unit transaction cost associated with obtaining the product at demand market k from source agent i

All vectors are assumed to be column vectors. The equilibrium solutions throughout this paper are denoted by $*$.

The m agents or sources of funds at the top tier of the financial network in Figure 1 seek to determine the optimal allocation of their financial resources transacted either physically or electronically with the intermediaries or electronically with the demand markets. Examples of source agents include: households and businesses. The financial intermediaries, in turn, which can include banks, insurance companies, investment companies, etc., transact with the source agents. In addition, they determine how to allocate the incoming financial resources among the distinct uses or financial products associated with the demand markets, which correspond to the nodes at the bottom tier of the financial network in Figure 1. Examples of demand markets are: the markets for real estate loans, household loans, business loans, etc. The transactions between the financial intermediaries and the demand markets can also take place physically or electronically via the Internet.

We denote a typical source agent by i; a typical financial intermediary by j, and a typical demand market by k. The mode of transaction is denoted by l with $l = 1$ denoting the physical mode and with $l = 2$ denoting the electronic mode.

We now describe the behavior of the decision-makers with sources of funds. We then discuss the behavior of the financial intermediaries and, finally, the consumers at the demand markets. Subsequently, we state the financial network equilibrium conditions and derive the variational inequality formulation governing the equilibrium conditions.

The Behavior of the Source Agents

The behavior of the decision-makers with sources of funds, also referred to as source agents is briefly recalled below (see Liu and Nagurney (2007)).

Since there is the possibility of non-investment allowed, the node $n + 1$ in the second tier in Figure 1 represents the "sink" to which the uninvested portion of the financial funds flows from the particular source agent or source node. We then have the following conservation of flow equations:

$$\sum_{j=1}^{n}\sum_{l=1}^{2} q_{ijl} + \sum_{k=1}^{o} q_{ik} \leq S^i, \quad i = 1, \ldots, m, \tag{1}$$

that is, the amount of financial funds available at source agent i and given by S^i cannot exceed the amount transacted physically and electronically with the intermediaries plus the amount transacted electronically with the demand markets. Note that the "slack" associated with constraint (1) for a particular source agent i is given by $q_{i(n+1)}$ and corresponds to the uninvested amount of funds.

Let ρ_{1ijl} denote the price charged by source agent i to intermediary j for a transaction via mode l and, let ρ_{1ik} denote the price charged by source agent i for the electronic transaction with demand market k. The ρ_{1ijl} and ρ_{1ik} are endogenous variables and their equilibrium values ρ_{1ijl}^* and ρ_{1ik}^*; $i = 1, \ldots, m$; $j = 1, \ldots, n$; $l = 1, 2$, $k = 1, \ldots, o$ are determined once the complete financial network model is solved. As noted earlier, we assume that each source agent seeks to maximize his net revenue and to minimize his risk. For further background on risk management, see Rustem and Howe (2002). We assume as in Liu and Nagurney (2007) that the risk for source agent i is represented by the variance-covariance matrix V^i so that the optimization problem faced by source agent i can be expressed as:

$$\text{Maximize} \quad U^i(q_i) = \sum_{j=1}^{n}\sum_{l=1}^{2} \rho_{1ijl}^* q_{ijl} + \sum_{k=1}^{o} \rho_{1ik}^* q_{ik} - \sum_{j=1}^{n}\sum_{l=1}^{2} c_{ijl}(q_{ijl})$$

$$- \sum_{k=1}^{o} c_{ik}(q_{ik}) - q_i^T V^i q_i \tag{2}$$

subject to:

$$\sum_{j=1}^{n}\sum_{l=1}^{2} q_{ijl} + \sum_{k=1}^{o} q_{ik} \leq S^i$$

$$q_{ijl} \geq 0, \quad \forall j, l,$$

$$q_{ik} \geq 0, \quad \forall k,$$

$$q_{i(n+1)} \geq 0.$$

The first four terms in the objective function (2) represent the net revenue of source agent i and the last term is the variance of the return of the portfolio, which represents the risk associated with the financial transactions.

We assume that the transaction cost functions for each source agent are continuously differentiable and convex, and that the source agents compete in a noncooperative manner in the sense of Nash (1950; 1951). The optimality conditions for all decision-makers with source of funds simultaneously coincide with the solution of the following variational inequality (Liu and Nagurney, 2007): determine $(Q^{1*}, Q^{2*}) \in \mathcal{K}^0$ such that:

$$\sum_{i=1}^{m} \sum_{j=1}^{n} \sum_{l=1}^{2} \left[2V_{z_{jl}}^i \cdot q_i^* + \frac{\partial c_{ijl}(q_{ijl}^*)}{\partial q_{ijl}} - \rho_{1ijl}^* \right] \times \left[q_{ijl} - q_{ijl}^* \right]$$

$$+ \sum_{i=1}^{m} \sum_{k=1}^{o} \left[2V_{z_{2n+k}}^i \cdot q_i^* + \frac{\partial c_{ik}(q_{ik}^*)}{\partial q_{ik}} - \rho_{1ik}^* \right] \times \left[q_{ik} - q_{ik}^* \right] \geq 0,$$

$$\forall (Q^1, Q^2) \in \mathcal{K}^0, \tag{3}$$

where $V_{z_{jl}}^i$ denotes the z_{jl}-th row of V^i and z_{jl} is defined as the indicator: $z_{jl} = (l-1)n + j$. Similarly, $V_{z_{2n+k}}^i$ denotes the z_{2n+k}-th row of V^i but with z_{2n+k} defined as the $2n+k$-th row, and the feasible set $\mathcal{K}^0 \equiv \{(Q^1, Q^2) | (Q^1, Q^2) \in R_+^{2mn+mo}$ and (1) holds for all $i\}$.

The Behavior of the Financial Intermediaries

The behavior of the intermediaries in the financial network model of Liu and Nagurney (2007) is recalled below.

Let the endogenous variable ρ_{2jkl} denote the product price charged by intermediary j with ρ_{2jkl}^* denoting the equilibrium price, where $j = 1, \ldots, n$; $k = 1, \ldots, o$, and $l = 1, 2$. We assume that each financial intermediary also seeks to maximize his net revenue while minimizing his risk. Note that a financial intermediary, by definition, may transact either with decision-makers in the top tier of the financial network as well as with consumers associated with the demand markets in the bottom tier. Noting the conversion/handling cost as well as the various transaction costs faced by a financial intermediary and recalling that the variance-covariance matrix associated with financial intermediary j is given by V^j (cf. Table 1), we have that the financial intermediary is faced with the following optimization problem:

$$\text{Maximize} \quad U^j(q_j) = \sum_{k=1}^{o} \sum_{l=1}^{2} \rho_{2jkl}^* q_{jkl} - c_j(Q^1) - \sum_{i=1}^{m} \sum_{l=1}^{2} \hat{c}_{ijl}(q_{ijl})$$

$$-\sum_{k=1}^{o}\sum_{l=1}^{2} c_{jkl}(q_{jkl}) - \sum_{i=1}^{m}\sum_{l=1}^{2} \rho_{1ijl}^* q_{ijl} - q_j^T V^j q_j \qquad (4)$$

subject to:

$$\sum_{k=1}^{o}\sum_{l=1}^{2} q_{jkl} \le \sum_{i=1}^{m}\sum_{l=1}^{2} q_{ijl}, \qquad (5)$$

$$q_{ijl} \ge 0, \quad \forall i,l,$$

$$q_{jkl} \ge 0, \quad \forall k,l.$$

The first five terms in the objective function (4) denote the net revenue, whereas the last term is the variance of the return of the financial allocations, which represents the risk to each financial intermediary. Constraint (5) guarantees that an intermediary cannot reallocate more of its financial funds among the demand markets than it has available.

Let γ_j be the Lagrange multiplier associated with constraint (5) for intermediary j. We assume that the cost functions are continuously differentiable and convex, and that the intermediaries compete in a noncooperative manner. Hence, the optimality conditions for all intermediaries simultaneously can be expressed as the following variational inequality (Liu and Nagurney, 2007): determine $(Q^{1*}, Q^{3*}, \gamma^*) \in R_+^{2mn+2no+n}$ satisfying:

$$\sum_{i=1}^{m}\sum_{j=1}^{n}\sum_{l=1}^{2}\left[2V_{z_{il}}^j \cdot q_j^* + \frac{\partial c_j(Q^{1*})}{\partial q_{ijl}} + \rho_{1ijl}^* + \frac{\partial \hat{c}_{ijl}(q_{ijl}^*)}{\partial q_{ijl}} - \gamma_j^* \right] \times [q_{ijl} - q_{ijl}^*]$$

$$+\sum_{j=1}^{n}\sum_{k=1}^{o}\sum_{l=1}^{2}\left[2V_{z_{kl}}^j \cdot q_j^* + \frac{\partial c_{jkl}(q_{jkl}^*)}{\partial q_{jkl}} - \rho_{2jkl}^* + \gamma_j^* \right] \times [q_{jkl} - q_{jkl}^*]$$

$$+\sum_{j=1}^{n}\left[\sum_{i=1}^{m}\sum_{l=1}^{2} q_{ijl}^* - \sum_{k=1}^{o}\sum_{l=1}^{2} q_{jkl}^* \right] \times [\gamma_j - \gamma_j^*] \ge 0,$$

$$\forall (Q^1, Q^3, \gamma) \in R_+^{2mn+2no+n}, \qquad (6)$$

where $V_{z_{il}}^j$ denotes the z_{il}-th row of V^j and z_{il} is defined as the indicator: $z_{il} = (l-1)m + i$. Similarly, $V_{z_{kl}}^j$ denotes the z_{kl}-th row of V^j and z_{kl} is defined as the indicator: $z_{kl} = 2m + (l-1)o + k$.

Additional background on risk management in finance can be found in Nagurney and Siokos (1997); see also the book by Rustem and Howe (2002).

The Consumers at the Demand Markets and the Equilibrium Conditions

By referring to the model of Liu and Nagurney (2007), we now assume, as given, the inverse demand functions $\rho_{3k}(d)$; $k = 1, \ldots, o$, associated with the demand markets at the bottom tier of the financial network. Recall that the demand markets correspond to distinct financial products. Of course, if the demand functions are invertible, then one may obtain the price functions simply by inversion.

The following conservation of flow equations must hold:

$$d_k = \sum_{j=1}^{n} \sum_{l=1}^{2} q_{jkl} + \sum_{i=1}^{m} q_{ik}, \quad k = 1, \ldots, o. \tag{7}$$

Equations (7) state that the demand for the financial product at each demand market is equal to the financial transactions from the intermediaries to that demand market plus those from the source agents.

The equilibrium condition for the consumers at demand market k are as follows: for each intermediary j; $j = 1, \ldots, n$ and mode of transaction l; $l = 1, 2$:

$$\rho_{2jkl}^* + \hat{c}_{jkl}(Q^{2*}, Q^{3*}) \begin{cases} = \rho_{3k}(d^*), & \text{if} \quad q_{jkl}^* > 0 \\ \geq \rho_{3k}(d^*), & \text{if} \quad q_{jkl}^* = 0. \end{cases} \tag{8}$$

In addition, we must have that, in equilibrium, for each source of funds i; $i = 1, \ldots, m$:

$$\rho_{1ik}^* + \hat{c}_{ik}(Q^{2*}, Q^{3*}) \begin{cases} = \rho_{3k}(d^*), & \text{if} \quad q_{ik}^* > 0 \\ \geq \rho_{3k}(d^*), & \text{if} \quad q_{ik}^* = 0. \end{cases} \tag{9}$$

Condition (8) states that, in equilibrium, if consumers at demand market k purchase the product from intermediary j via mode l, then the price the consumers pay is exactly equal to the price charged by the intermediary plus the unit transaction cost via that mode. However, if the sum of price charged by the intermediary and the unit transaction cost is greater than the price the consumers are willing to pay at the demand market, there will be no transaction between this intermediary/demand market pair via that mode. Condition (9) states the analogue but for the case of electronic transactions with the source agents.

In equilibrium, conditions (8) and (9) must hold for all demand markets. We can also express these equilibrium conditions using the following variational inequality (Liu and Nagurney, 2007): determine $(Q^{2*}, Q^{3*}, d^*) \in \mathcal{K}^1$, such that

$$\sum_{j=1}^{n} \sum_{k=1}^{o} \sum_{l=1}^{2} \left[\rho_{2jkl}^* + \hat{c}_{jkl}(Q^{2*}, Q^{3*}) \right] \times \left[q_{jkl} - q_{jkl}^* \right]$$

$$+ \sum_{i=1}^{m} \sum_{k=1}^{o} \left[\rho_{1ik}^* + \hat{c}_{ik}(Q^{2*}, Q^{3*}) \right] \times \left[q_{ik} - q_{ik}^* \right]$$

$$- \sum_{k=1}^{o} \rho_{3k}(d^*) \times \left[d_k - d_k^* \right] \geq 0, \quad \forall (Q^2, Q^3, d) \in \mathcal{K}^1, \tag{10}$$

where $\mathcal{K}^1 \equiv \{(Q^2, Q^3, d) | (Q^2, Q^3, d) \in R_+^{2no+mo+o} \text{ and (7) holds.}\}$

The Equilibrium Conditions for Financial Network with Electronic Transactions

In equilibrium, the optimality conditions for all decision-makers with source of funds, the optimality conditions for all the intermediaries, and the equilibrium conditions for all the demand markets must be simultaneously satisfied so that no decision-maker has any incentive to alter his or her decision. We recall the equilibrium condition in Liu and Nagurney (2007) for the entire financial network with intermediation and electronic transactions as follows.

Definition 1. Financial Network Equilibrium with Intermediation and with Electronic Transactions
The equilibrium state of the financial network with intermediation is one where the financial flows between tiers coincide and the financial flows and prices satisfy the sum of conditions (3), (6), and (10).

We now define the feasible set:

$$\mathcal{K}^2 \equiv \{(Q^1, Q^2, Q^3, \gamma, d) | (Q^1, Q^2, Q^3, \gamma, d) \in R_+^{m+2mn+2no+n+o}$$

and (1) and (7) hold}

and state the following theorem. For the proof of Theorem 1, please refer to the paper by Liu and Nagurney (2007).

Theorem 1. Variational Inequality Formulation
The equilibrium conditions governing the financial network model with intermediation are equivalent to the solution to the variational inequality problem given by: determine $(Q^{1*}, Q^{2*}, Q^{3*}, \gamma^*, d^*) \in \mathcal{K}^2$ *satisfying:*

$$\sum_{i=1}^m \sum_{j=1}^n \sum_{l=1}^2 \left[2V_{z_{jl}}^i \cdot q_i^* + 2V_{z_{il}}^j \cdot q_j^* + \frac{\partial c_{ijl}(q_{ijl}^*)}{\partial q_{ijl}} + \frac{\partial c_j(Q^{1*})}{\partial q_{ijl}} + \frac{\partial \hat{c}_{ijl}(q_{ijl}^*)}{\partial q_{ijl}} - \gamma_j^* \right]$$

$$\times \left[q_{ijl} - q_{ijl}^* \right]$$

$$+ \sum_{i=1}^m \sum_{k=1}^o \left[2V_{z_{2n+k}}^i \cdot q_i^* + \frac{\partial c_{ik}(q_{ik}^*)}{\partial q_{ik}} + \hat{c}_{ik}(Q^{2*}, Q^{3*}) \right] \times \left[q_{ik} - q_{ik}^* \right]$$

$$+ \sum_{j=1}^n \sum_{k=1}^o \sum_{l=1}^2 \left[2V_{z_{kl}}^j \cdot q_j^* + \frac{\partial c_{jkl}(q_{jkl}^*)}{\partial q_{jkl}} + \hat{c}_{jkl}(Q^{2*}, Q^{3*}) + \gamma_j^* \right] \times \left[q_{jkl} - q_{jkl}^* \right]$$

$$+ \sum_{j=1}^n \left[\sum_{i=1}^m \sum_{l=1}^2 q_{ijl}^* - \sum_{k=1}^o \sum_{l=1}^2 q_{jkl}^* \right] \times \left[\gamma_j - \gamma_j^* \right] - \sum_{k=1}^o \rho_{3k}(d^*) \times \left[d_k - d_k^* \right] \geq 0,$$

$$\forall (Q^1, Q^2, Q^3, \gamma, d) \in \mathcal{K}^2 . \tag{11}$$

The variables in the variational inequality problem (11) are: the financial flows from the source agents to the intermediaries, Q^1; the direct financial flows via electronic transaction from the source agents to the demand markets, Q^2; the financial flows from the intermediaries to the demand markets, Q^3; the shadow prices associated with handling the product by the intermediaries, γ, and the prices at demand markets ρ_3. The solution to the variational inequality problem (11), $(Q^{0*}, Q^{1*}, Q^{2*}, Q^{3*}, \gamma^*, d^*)$, coincides with the equilibrium financial flow and price pattern according to Definition 1.

3 The Financial Network Performance Measure and the Importance of Financial Network Components

In this section, we propose the novel financial network performance measure and the associated network component importance definition. For completeness, we also discuss the difference between our measure and a standard efficiency measure in economics.

3.1 The Financial Network Performance Measure

As stated in the Introduction, the financial network performance measure is motivated by the work of Qiang and Nagurney (2007). In the case of the financial network performance measure, we state the definitions directly within the context of financial networks, without making use of the transformation of the financial network model into a network equilibrium model with defined origin/destination pairs and paths as was done by Qiang and Nagurney (2007), who considered network equilibrium problems with a transportation focus (see also, Nagurney and Qiang (2007a; 2007b; 2007d) and Liu and Nagurney (2007).

Definition 2. The Financial Network Performance Measure
The financial network performance measure, $\mathcal{E}$, for a given network topology G, and demand price functions $\rho_{3k}(d)$ ($k = 1, 2, \ldots, o$), and available funds held by source agents S, is defined as follows:

$$\mathcal{E} = \frac{\sum_{k=1}^{o} \frac{d_k^*}{\rho_{3k}(d^*)}}{o}, \tag{12}$$

where o is the number of demand markets in the financial network, and d_k^* and $\rho_{3k}(d^*)$ denote the equilibrium demand and the equilibrium price for demand market k, respectively.

The financial network performance measure $\mathcal{E}$ defined in (12) is actually the average demand to price ratio. It measures the overall (economic) functionality of the financial network. When the network topology G, the demand price functions, and the available funds held by source agents are given, a financial network is considered performing better if it can satisfy higher demands at lower prices.

By referring to the equilibrium conditions (8) and (9), we assume that if there is a positive transaction between a source agent or an intermediary with a demand market at equilibrium, the price charged by the source agent or the intermediary plus the respective unit transaction costs is always positive. Furthermore, we assume that if the equilibrium demand at a demand market is zero, the demand market price (i.e., the inverse demand function value) is positive. Hence, the demand market prices will always be positive and the above network performance measure is well-defined.

In the above definition, we assume that all the demand markets are given the same weight when aggregating the demand to price ratio, which can be interpreted as all the demand markets are of equal strategic importance. Of course, it may be interesting and appropriate to weight demand markets differently by incorporating managerial or governmental factors into the measure. For example, one could give more preference to the markets with large demands. Furthermore, it would also be interesting to explore different functional forms associated with the definition of the performance measure in order to ascertain different aspects of network performance. However, in this paper, we focus on the definition in the form of (12) and the above issues will be considered for future research. Finally, the performance measure in (12) is based on the "pure" cost incurred between different tiers of the financial network. Another future research problem is the study of the financial network performance with "generalized costs" and multi-criteria objective functions.

3.2 Network Efficiency vs. Network Performance

It is worth pointing out further relationships between our network performance measure and other measures in economics, in particular, an *efficiency* measure. In economics, the total utility gained (or cost incurred) in a system may be used as an efficiency measure. Such a criterion is basically the underlying rationale for the concept of *Pareto efficiency*, which plays a very important role in the evaluation of economic policies in terms of social welfare. As is well-known, a Pareto efficient outcome indicates that there is no alternative way to organize the production and distribution of goods that makes some economic agent better off without making another worse off (see, e.g., Mas-Colell et al. (1995), Samuelson (1983)). Under certain conditions, which include that externalities are not present in an economic system, the equilibrium state assures that the system is Pareto efficient and that the social welfare is maximized. The concept of *Kaldor-Hicks efficiency*, in turn, relaxes the requirement of Pareto efficiency by incorporating the compensation principle: an outcome is efficient if those that are made better off could, in theory, compensate those that are made worse off and leads to a Pareto optimal outcome (see, e.g., Chipman (1987) and Buchanan and Musgrave (1999)).

The above economic efficiency concepts have important implications for the government and/or central planners such as, for example, by suggesting and enforcing policies that ensure that the system is running cost efficiently. For instance, in the transportation literature, the above efficiency concepts have been used to model the "system-optimal" objective, where the toll policy can be implemented to guarantee that the minimum total travel cost for the entire network (cf. Beckmann et al. (1956), Dafermos (1973), Nagurney (2000), and the references therein) is achieved. It is worth noting that the system-optimal concept in transportation networks has stimulated a tremendous amount of interest also, recently, among computer scientists, which has led to the study of the *price of anarchy* (cf. Roughgarden (2005) and the references therein). The price of anarchy is defined as the ratio of the system-optimization objective function evaluated at the user-optimized solution divided by that objective function evaluated at the system-optimized solution. It has been used to study a variety of noncooperative games on such networks as telecommunication networks and the Internet. Notably, the aforementioned principles are mainly used to access the tenability of the resource allocation policies from a societal point of view. However, we believe that in addition to evaluating an economic systems in the sense of optimizing the resource allocation, there should also be a measure that can assess the network performance and functionality. Although in such networks as the Internet and certain transportation networks, the assumption of having a central planner to ensure the minimization of the total cost may, in some instances, be natural and reasonable, the same assumption faces difficulty when extended to the larger and more complex networks as in the case of financial networks, where the control by a "central planner" is not realistic.

The purpose of this paper is not to study the efficiency of a certain market mechanism or policy, which can be typically analyzed via the Pareto criterion and the Kaldor-Hicks test. Instead, we want to address the following question: given a certain market mechanism, network structure, objective functions, and demand price and cost functions, how should one evaluate the performance and the functionality of the network? In the context of a financial network where there exists noncooperative competition among the source agents as well as among the financial intermediaries, if, on the average and across all demand markets, a large amount of financial funds can reach the consumers, through the financial intermediaries, at low prices, we consider the network as performing well. Thus, instead of studying the efficiency of an economic policy or market mechanism, we evaluate the functionality and the performance of a financial network in a given environment. The proposed performance measure of the financial network is based on the equilibrium model outlined in Section 2. However, our measure can be applied to other economic networks, as well, and has been done so in the case of transportation networks and other critical infrastructure networks (see Nagurney and Qiang (2007a; 2007b; 2007d)). Notably, we believe that such a network equilibrium model is general and relevant and, moreover, it also has deep theoretic foundations (see, for example, Judge and Takayama (1973)).

Furthermore, three points merit discussion as to the need of a network performance measure besides solely looking at the total cost of the network. First, the function of an economic network is to serve the demand markets at a reasonable price. Hence, it is reasonable and important to have a performance measure targeted at the functionality perspective. Secondly, when faced with network disruptions with certain parts of the network being destroyed, the cost of providing services/products through the dysfunctional/disconnected part reaches infinity. Therefore, the total cost of the system is also equal to infinity and, hence, becomes undefined. However, since the remaining network components are still functioning, it is still valid to analyze the network performance in this situation. Finally, it has been shown in the paper of Qiang and Nagurney (2007) that the total system cost measure is not appropriate as a means of evaluating the performance of a network with elastic demands and, hence, a unified network measure is needed.

Based on the discussion in this section, we denote our proposed measure as the "financial network performance measure" to avoid confusion with efficiency measures in economics and elsewhere.

3.3 The Importance of a Financial Network Component

The importance of the network components is analyzed, in turn, by studying the impact on the network performance measure through their removal. The financial network performance is expected to deteriorate when a critical network component is eliminated from the network. Such a component can include a link or a node or a subset of nodes and links depending on the financial network problem under investigation. Furthermore, the removal of a critical network component will cause more severe damage than that caused by the removal of a trivial component. Hence, the importance of a network component is defined as follows (cf. Qiang and Nagurney (2007)):

Definition 3. Importance of a Financial Network Component
The importance of a financial network component $g \in G$, $I(g)$, is measured by the relative financial network performance drop after g is removed from the network:

$$I(g) = \frac{\triangle\mathcal{E}}{\mathcal{E}} = \frac{\mathcal{E}(G) - \mathcal{E}(G - g)}{\mathcal{E}(G)} \tag{13}$$

where $G - g$ is the resulting financial network after component g is removed from network G.

It is worth pointing out that the above importance of the network components is well-defined even in a financial network with disconnected source agent/demand market pairs. In our financial network performance measure, the elimination of a transaction link is treated by removing that link from the

Sources of Financial Funds

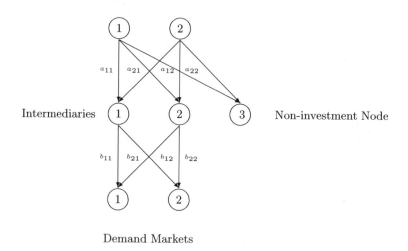

Demand Markets

Fig. 2. The financial network structure of the numerical examples.

network while the removal of a node is managed by removing the transaction links entering or exiting that node. In the case that the removal results in no transaction path connecting a source agent/demand market pair, we simply assign the demand for that source agent/demand market pair to an abstract transaction path with an associated cost of infinity. The above procedure(s) to handle disconnected agent/demand market pairs, will be illustrated in the numerical examples in Section 4, when we compute the importance of the financial network components and their associated rankings.

4 Numerical Examples

In order to further demonstrate the applicability of the financial network performance measure proposed in Section 3, we, in this section, present two numerical financial network examples. For each example, our network performance measure is computed and the importance and the rankings of links and the nodes are also reported.

The examples consist of two source agents, two financial intermediaries, and two demand markets. These examples have the financial network structure depicted in Figure 2. For simplicity, we exclude the electronic transactions. The transaction links between the source agents and the intermediaries are denoted by a_{ij} where $i = 1, 2$; $j = 1, 2$. The transaction links between the intermediaries and the demand markets are denoted by b_{jk} where $j = 1, 2$; $k = 1, 2$. Since the non-investment portions of the funds do not participate in the actual transactions, we will not discuss the importance of the links and the nodes related to the non-investment funds. The examples below were solved using the Euler method (see, Nagurney and Zhang (1996; 1997), Nagurney and Ke (2003), and Nagurney et al. (2006)).

Example 1

The financial holdings for the two source agents in the first example are: $S^1 = 10$ and $S^2 = 10$. The variance-covariance matrices V^i and V^j are identity matrices for all the source agents $i = 1, 2$. We have suppressed the subscript l associated with the transaction cost functions since we have assumed a single (physical) mode of transaction being available. Please refer to Table 1 for a compact exposition of the notation.

The transaction cost function of source agent 1 associated with his transaction with intermediary 1 is given by:

$$c_{11}(q_{11}) = 4q_{11}^2 + q_{11} + 1.$$

The other transaction cost functions of the source agents associated with the transactions with the intermediaries are given by:

$$c_{ij}(q_{ij}) = 2q_{ij}^2 + q_{ij} + 1, \quad \text{for } i = 1, 2; j = 1, 2$$

while i and j are not equal to 1 at the same time.

The transaction cost functions of the intermediaries associated with transacting with the sources agents are given by:

$$\hat{c}_{ij}(q_{ij}) = 3q_{ij}^2 + 2q_{ij} + 1, \quad \text{for } i = 1, 2; j = 1, 2.$$

The handling cost functions of the intermediaries are:

$$c_1(Q^1) = 0.5(q_{11} + q_{21})^2, \quad c_2(Q^1) = 0.5(q_{12} + q_{22})^2.$$

We assumed that in the transactions between the intermediaries and the demand markets, the transaction costs perceived by the intermediaries are all equal to zero, that is,

$$c_{jk} = 0, \quad \text{for } j = 1, 2; k = 1, 2.$$

The transaction costs between the intermediaries and the consumers at the demand markets, in turn, are given by:

$$\hat{c}_{jk} = q_{jk} + 2, \quad \text{for } j = 1, 2; k = 1, 2.$$

The demand price functions at the demand markets are:

$$\rho_{3k}(d) = -2d_k + 100, \quad \text{for } k = 1, 2.$$

The equilibrium financial flow pattern, the equilibrium demands, and the incurred equilibrium demand market prices are reported below.

For Q^{1*}, we have:

$$q_{11}^* = 3.27, \ q_{12}^* = 4.16, \ q_{21}^* = 4.36, \ q_{22}^* = 4.16.$$

For Q^{2*}, we have:

$$q_{11}^* = 3.81, \ q_{12}^* = 3.81, \ q_{21}^* = 4.16, \ q_{22}^* = 4.16.$$

Also, we have:

$$d_1^* = 7.97, \ d_2^* = 7.97,$$

$$\rho_{31}(d^*) = 84.06, \ \rho_{32}(d^*) = 84.06.$$

The financial network performance measure (cf. (12)) is:

$$\mathcal{E} = \frac{\frac{7.97}{84.06} + \frac{7.97}{84.06}}{2} = 0.0949.$$

The importance of the links and the nodes and their ranking are reported in Table 2 and 3, respectively.

Table 2. Importance and ranking of the links in example 1.

Link	Importance Value	Ranking
a_{11}	0.1574	3
a_{12}	0.2003	2
a_{21}	0.2226	1
a_{22}	0.2003	2
b_{11}	0.0304	5
b_{12}	0.0304	5
b_{21}	0.0359	4
b_{22}	0.0359	4

Discussion

First note that, in Example 1, both source agents choose not to invest a portion of their financial funds. Given the cost structure and the demand price functions in the network of Example 1, the transaction link between source agent 2 and intermediary 1 is the most important link because it carries a large amount of financial flow, in equilibrium, and the removal of the link causes the highest performance drop assessed by the financial network performance measure. Similarly, because intermediary 2 handles the largest amount of financial input from the source agents, it is ranked as the most important node in the above network. On the other hand, since the transaction links between intermediary 1 to demand markets 1 and 2 carry the least amount of equilibrium financial flow, they are the least important links.

Table 3. Importance and ranking of the nodes in example 1.

Node	Importance Value	Ranking
Source Agent 1	0.4146	4
Source Agent 2	0.4238	3
Intermediary 1	0.4759	2
Intermediary 2	0.5159	1
Demand Market 1	0.0566	5
Demand Market 2	0.0566	5

Example 2

In the second example, the parameters are identical to those in Example 1, except for the following changes.

The transaction cost function of source agent 1 associated with his transaction with intermediary 1 is changed to:

$$c_{11}(q_{11}) = 2q_{11}^2 + q_{11} + 1$$

and the financial holdings of the source agents are changed, respectively, to $S_1 = 6$ and $S_2 = 10$.

The equilibrium financial flow pattern, the equilibrium demands, and the incurred equilibrium demand market prices are reported below.

For Q^{1*}, we have:

$$q_{11}^* = 3.00, \ q_{12}^* = 3.00, \ q_{21}^* = 4.48, \ q_{22}^* = 4.48.$$

For Q^{2*}, we have:

$$q_{11}^* = 3.74, \ q_{12}^* = 3.74, \ q_{21}^* = 3.74, \ q_{22}^* = 3.74.$$

Also, we have:

$$d_1^* = 7.48, \ d_2^* = 7.48,$$

$$\rho_{31}(d^*) = 85.04, \ \rho_{32}(d^*) = 85.04.$$

The financial network performance measure (cf. (12)) is:

$$\mathcal{E} = \frac{\frac{7.48}{85.04} + \frac{7.48}{85.04}}{2} = 0.0880.$$

The importance of the links and the nodes and their ranking are reported in Table 4 and 5, respectively.

Table 4. Importance and ranking of the links in example 2.

Link	Importance Value	Ranking
a_{11}	0.0917	2
a_{12}	0.0917	2
a_{21}	0.3071	1
a_{22}	0.3071	1
b_{11}	0.0211	3
b_{12}	0.0211	3
b_{21}	0.0211	3
b_{22}	0.0211	3

Table 5. Importance and ranking of the nodes in example 2.

Node	Importance Value	Ranking
Source Agent 1	0.3687	3
Source Agent 2	0.6373	1
Intermediary 1	0.4348	2
Intermediary 2	0.4348	2
Demand Market 1	-0.0085	4
Demand Market 2	-0.0085	4

Discussion

Note that, in Example 2, the first source agent has no funds non-invested. Given the cost structure and the demand price functions, since the transaction links between source agent 2 and intermediaries 1 and 2 carry the largest amount of equilibrium financial flow, they are ranked the most important. In addition, since source agent 2 allocates the largest amount of financial flow in equilibrium, it is ranked as the most important node. The negative importance value for demand markets 1 and 2 is due to the fact that the existence of each demand market brings extra flows on the transaction links and nodes and, therefore, increases the marginal transaction cost. The removal of one demand market has two effects: first, the contribution to the network performance of the removed demand market becomes zero; second, the marginal transaction cost on links/nodes decreases, which decreases the equilibrium prices and increases the demand at the other demand markets. If the performance drop caused by the removal of the demand markets is overcompensated by the improvement of the demand-price ratio of the other demand markets, the removed demand market will have a negative importance value. It simply implies that the "negative externality" caused by the demand market has a larger impact than the performance drop due to its removal.

5 Summary and Conclusions

In this paper, we proposed a novel financial network performance measure, which is motivated by the recent research of Qiang and Nagurney (2007) and Nagurney and Qiang (2007a; 2007b; 2007d) in assessing the importance of network components in the case of disruptions in network systems ranging from transportation networks to such critical infrastructure networks as electric power generation and distribution networks. The financial network measure examines the network performance by incorporating the economic behavior of the decision-makers, with the resultant equilibrium prices and transaction flows, coupled with the network topology. The financial network performance measure, along with the network component importance definition, provide valuable methodological tools for evaluating the financial network vulnerability and reliability. Furthermore, our measure is shown to be able to evaluate the importance of nodes and links in financial networks even when the source agent/demand market pairs become disconnected.

We believe that our network performance measure is a good starting point from which to begin to analyze the functionality of an economic network, in general, and a financial network, in particular. Especially in a network in which agents compete in a noncooperative manner in the same tier and coordinate between different tiers without the intervention from the government or a central planner, our proposed measure examines the network on a functional level other than in the traditional Pareto sense. We believe that the proposed measure has natural applicability in such networks as those studied in this paper. Specifically, with our measure, we are also able to study the robustness and vulnerability of different networks with partially disrupted network components (Nagurney and Qiang, 2007c). In the future, additional criteria and perspectives can be incorporated to analyze the network performance more comprehensively. Moreover, with a sophisticated and informative network performance measure, network administrators can implement effective policies to enhance the network security and to begin to enhance the system robustness.

Acknowledgements

This chapter is dedicated to Professor Manfred Gilli, a true scholar and gentleman.

The authors are grateful to the two anonymous reviewers and to the Guest Editor, Professor Peter Winker, for helpful comments and suggestions on two earlier versions of this paper.

This research was supported, in part, by NSF Grant No.: IIS-0002647, under the Management of Knowledge Intensive Dynamic Systems (MKIDS) program. The first author also acknowledges support from the John F. Smith Memorial Fund at the University of Massachusetts at Amherst. The support provided is very much appreciated.

References

Beckmann, M.J., McGuire, B.C. and Winsten, B.C.: 1956, *Studies in the Economics of Transportation*, Yale University Press, New Haven, Connecticut.

Boginski, V., Butenko, S. and Pardalos, P.M.: 2003, On structural properties of the market graph, *in* A. Nagurney (ed.), *Innovations in Financial and Economic Networks*, Edward Elgar Publishing, Cheltenham, England, pp. 29–45.

Buchanan, J.M. and Musgrave, R.A.: 1999, *Public Finance and Public Choice: Two Contrasting Visions of the State*, MIT Press, Boston, MA.

Caldarelli, G., Battiston, S., Garlaschelli, D. and Catanzaro, M.: 2004, Emergence of complexity in financial networks, *Lecture Notes in Physics* **650**, 399–423.

Charnes, A. and Cooper, W.W.: 1967, Some network characterizations for mathematical programming and accounting approaches to planning and control, *The Accounting Review* **42**, 24–52.

Chipman, J.S.: 1987, Compensation principle, *in* J. Eatwell, M. Milgate and P. Newman (eds), *The New Palgrave: A Dictionary of Economics, Vol. 1*, The Stockton Press, New York, NY.

Christofides, N., Hewins, R.D. and Salkin, G.R.: 1979, Graph theoretic approaches to foreign exchange operations, *Journal of Financial and Quantitative Analysis* **14**, 481–500.

Copeland, M.A.: 1952, *A Study of Moneyflows in the United States*, National Bureau of Economic Research, New York, NY.

Crum, R.L. and Nye, D.J.: 1981, A network model of insurance company cash flow management, *Mathematical Programming Study* **15**, 86–101.

Dafermos, S.C.: 1973, Toll patterns for multi-class user transportation networks, *Transportation Science* **7**, 211–223.

Doherty, N.A.: 1997, Financial innovation in the management of catastrophe risk, *Journal of Applied Corporate Finance* **10**, 84–95.

Fei, J.C.H.: 1960, The study of the credit system by the method of the linear graph, *The Review of Economics and Statistic* **42**, 417–428.

Geunes, J. and Pardalos, P.M.: 2003, Network optimization in supply chain management and financial engineering: An annotated bibliography, *Networks* **42**, 66–84.

Gilli, M. and Këllezi, E.: 2006, An application of extreme value theory for measuring financial risk, *Computational Economics* **27**, 207–228.

Jackson, M.O. and Wolinsky, A.: 1996, A strategic model of social and economic networks, *Journal of Economic Theory* **71**, 44–74.

Judge, G.G. and Takayama, T.: 1973, *Studies in Economic Planning Over Space and Time*, North-Holland, Amsterdam, The Netherlands.

Kim, D.H. and Jeong, H.: 2005, Systematic analysis of group identification in stock markets, *Physical Review E* **72**, Article No. 046133.

Kontoghiorghes, E.J., Rustem, B. and Siokos, S.: 2002, *Computational Methods in Decision-Making, Economics and Finance, Optimization Models,* Kluwer Academic Publishers, Boston, MA.

Latora, V. and Marchiori, M.: 2001, Efficient behavior of small-world networks, *Physical Review Letters* **87**, Article No. 198701.

Latora, V. and Marchiori, M.: 2002, Is the Boston subway a small-world network?, *Physica A: Statistical Mechanics and its Applications* **314**, 109–113.

Latora, V. and Marchiori, M.: 2003, Economic small-world behavior in weighted networks, *The European Physical Journal B* **32**, 249–263.

Latora, V. and Marchiori, M.: 2004, How the science of complex networks can help developing strategies against terrorism, *Chaos, Solitons and Fractals* **20**, 69–75.

Liu, Z. and Nagurney, A.: 2007, Financial networks with intermediation and transportation network equilibria: A supernetwork equivalence and computational management reinterpretation of the equilibrium conditions with computations, *Computational Management Science* **4**, 243–281.

Loubergé, H., Këllezi, E. and Gilli, M.: 1999, Using catastrophe-linked securities to diversify insurance risk: A financial analysis of cat-bonds, *The Journal of Insurance Issues* **2**, 125–146.

Mas-Colell, A., Whinston, M. and Green, J.R.: 1995, *Microeconomic Theory,* Oxford University Press, New York, NY.

Mulvey, J.M.: 1987, Nonlinear networks in finance, *Advances in Mathematical Programming and Financial Planning* **20**, 187–217.

Nagurney, A.: 2000, *Sustainable Transportation Networks,* Edward Elgar Publishers, Cheltenham, England.

Nagurney, A.: 2007, Networks in finance, *to appear,* in D. Seese, C. Weinhardt and F. Schlottmann (eds), *Handbook on Information Technology and Finance,* Springer, Berlin, Germany.

Nagurney, A. and Cruz, J.: 2003a, International financial networks with electronic transactions, in A. Nagurney (ed.), *Innovations in Financial and Economic Networks,* Edward Elgar Publishing, Cheltenham, England, pp. 136–168.

Nagurney, A. and Cruz, J.: 2003b, International financial networks with intermediation: modeling, analysis, and computations, *Computational Management Science* **1**, 31–58.

Nagurney, A., Dong, J. and Hughes, M.: 1992, Formulation and computation of general financial equilibrium, *Optimization* **26**, 339–354.

Nagurney, A. and Hughes, M.: 1992, Financial flow of funds networks, *Networks* **22**, 145–161.

Nagurney, A. and Ke, K.: 2001, Financial networks with intermediation, *Quantitative Finance* **1**, 309–317.

Nagurney, A. and Ke, K.: 2003, Financial networks with electronic transactions: modeling, analysis, and computations, *Quantitative Finance* **3**, 71–87.

Nagurney, A. and Qiang, Q.: 2007a, A network efficiency measure for congested networks, *Europhysics Letters* **79**, Article No. 38005.

Nagurney, A. and Qiang, Q.: 2007b, A network efficiency measure with application to critical infrastructure networks, *Journal of Global Optimization (in press)* .

Nagurney, A. and Qiang, Q.: 2007c, Robustness of transportation networks subject to degradable links, *Europhysics Letters* **80**, Article No. 68001.

Nagurney, A. and Qiang, Q.: 2007d, A transportation network efficiency measure that captures flows, behavior, and costs with applications to network component importance identification and vulnerability, *Proceedings of the 18th Annual POMS Conference*, Dallas, Texas.

Nagurney, A. and Siokos, S.: 1997, *Financial Networks: Statics and Dynamics*, Springer-Verlag, Heidelberg, Germany.

Nagurney, A., Wakolbinger, T. and Zhao, L.: 2006, The evolution and emergence of integrated social and financial networks with electronic transactions: A dynamic supernetwork theory for the modeling, analysis, and computation of financial flows and relationship levels, *Computational Economics* **27**, 353–393.

Nagurney, A. and Zhang, D.: 1996, *Projected Dynamical Systems and Variational Inequalities with Applications*, Kluwer Academic Publishers, Boston, MA.

Nagurney, A. and Zhang, D.: 1997, Projected dynamical systems in the formulation, stability analysis, and computation of fixed demand traffic network equilibria, *Transportation Science* **31**, 147–158.

Nash, J.F.: 1950, Equilibrium points in n-person games, *Proceedings of the National Academy of Sciences*, USA, pp. 48–49.

Nash, J.F.: 1951, Noncooperative games, *Annals of Mathematics* **54**, 286–298.

Newman, M.: 2003, The structure and function of complex networks, *SIAM Review* **45**, 167–256.

Niehaus, G.: 2002, The allocation of catastrophe risk, *Journal of Banking and Finance* **26**, 585–596.

Odell, K. and Phillips, R.J.: 2001, Testing the ties that bind: financial networks and the 1906 San Francisco earthquake, *working paper*, Department of Economics, Colorado State University, Fort Collins, Colorado.

Onnela, J.P., Kaski, K. and Kertész, J.: 2004, Clustering and information in correlation based financial networks, *The European Physical Journal B* **38**, 353–362.

Qiang, Q. and Nagurney, A.: 2007, A unified network performance measure with importance identification and the ranking of network components, *Optimization Letters (in press)* .

Quesnay, F.: 1758, Tableau economique. Reproduced in facsimile with an introduction by H. Higgs by the British Economic Society, 1895.

Robinson, C.P., Woodard, J.B. and Varnado, S.G.: 1998, Critical infrastructure: interlinked and vulnerable, *Issues in Science and Technology* **15**, 61–67.

Roughgarden, T.: 2005, *Selfish Routing and the Price of Anarchy*, MIT Press, Cambridge, Massachusetts.

Rustem, B. and Howe, M.: 2002, *Algorithms for Worst-Case Design and Risk Management*, Princeton University Press, Princeton, New Jersey.

Samuelson, P.A.: 1983, *Foundations of Economic Analysis, Enlarged Edition*, Harvard University Press, Boston, MA.

Sheffi, Y.: 2005, *The Resilient Enterprise: Overcoming Vulnerability for Competitive Advantage*, MIT Press, Cambridge, Massachusetts.

Thore, S.: 1969, Credit networks, *Economica* **36**, 42–57.

Thore, S.: 1980, *Programming the Network of Financial Intermediation*, Universitetsforlaget, Oslo, Norway.

Thore, S. and Kydland, F.: 1972, Dynamic flow-of-funds networks, *in* S. Eilon and T. R. Fowkes (eds), *Applications of Management Science in Banking and Finance*, Gower Press, Epping, England, pp. 259–276.

An Analysis of Settlement Risk Contagion in Alternative Securities Settlement Architectures

Giulia Iori[1] and Christophe Deissenberg[2]

[1] Department of Economics, City University, Northampton Square, London, EC1V 0HB, United Kingdom. g.iori@city.ac.uk.
[2] Université de la Méditerranée and GREQAM, Château Lafarge, Route des Milles, 13290 Les Milles, France. christophe.deissenberg@univmed.fr.

Summary. The so-called gross and net architectures for securities settlement are compared. The settlement risk arising from exogenous operational delays is studied and the importance of settlement failures under the two architectures is investigated as a function of the length of the settlement cycle and of different market conditions. Under both architectures, settlement failures are non-monotonically related to the length of the settlement cycle. There is no evidence that continuous time settlement provides always higher stability. Gross systems appear to be more stable than net systems.

Key words: Security clearing and settlement, gross and net systems, contagion.

Preamble

Manfred Gilli is a fine man. Not only is he an excellent scientist and a person of striking culture. He also inherited the physical leanness, the moral robustness, the humbleness of the mountaineers of Südtirol, his birthplace. His unobtrusive sense of humor is second to none. Most importantly perhaps, he is a profoundly gentle and kind person.

It is with a profound respect and friendship that we dedicate this paper to Manfred. It has been a pleasure and a privilege to know him for many years. May life bring us together for many more years to come.

1 Introduction

Securities Settlement Systems (SSSs) are institutional arrangements for the confirmation, clearance and settlement of securities trades and for the safe-keeping of securities. They involve three steps. The first one, *trade confirmation*, aims at ensuring that the buyer and the seller (typically both banks) agree on the terms of the trade. To that purpose, following a trade each party sends an advisory message identifying the counterpart, the security, the number of shares, the invoice price, and the settlement date. After confirmation comes *clearance*, i.e., the computation of the obligations of the counterparts to make deliveries or to make payments at the settlement date. Finally, *settlement* consists of the operations by which the shares are transferred from the seller to the buyer and the payments from the buyer to the seller.

Settlement systems may operate in one tier or two tiers. In a one-tier securities settlement models all end-investor security accounts are within the Central Security Depository (CSD). This model can be found for example in the Nordic countries, Slovenia and Greece. In a one-tier system participants settle mainly their customers' transactions. In a two-tier system the CSD keeps accounts only for the participating banks (or custodians) and only the inter-participant transactions are booked on these accounts while the end-investor accounts are with the participants/custodians. See Holthausen and Tapking (2007) for an analysis of competition between CSD and custodians. The participanting banks can settle the transactions of the end-investor internally.

The banks face a variety of risks, see Committee on Payment and Settlement Systems (2001b). Among them, there is the risk that creditors do not pay back a loan (*credit risk*) or that settlement is delayed because of shortage of cash and securities (*liquidity risk*). There is the risk that securities are delivered but payment not received, and vice-versa (*principal risk*). Other risks arise from mistakes and deficiencies in information and control (*operational risk*), from the safekeeping of securities by third parties (*custody risk*), and from potential failures of the legal system that supports the rules and procedures of the settlement system (*legal risk*).

A financial or operational problem during the settlement process may make a clearing bank unable to meet its obligations. Default by a bank, in turn, may render other banks unable to meet their own obligations, triggering a chain of defaults within the SSS. In that sense, a SSS is susceptible to *systemic risk*. In addition, SSSs are critical components of the infrastructure of global financial markets. Serious dysfunction at their level have the potential to propagate to other payment systems used by or using the SSS to transfer collateral. Thus, problems in the settlement process may induce systemic risk not only for the SSS but also for the financial system and the economy as a whole. See de Bandt and Hartmann (2000) for a review on systemic risk.

Crucial characteristics of a SSS the timing and modalities of settlement. The transactions can be settled in real time or in batches. In the case of batches, the settlement can be conducted gross or net. Under gross settlement, the clearing house settles the trades in the order they have been inputed in the system by the participants. Real time settlements can only be conducted gross as *Real Time Gross Settlements* (RTGS), where payments are executed continuously via transfers of central bank funds from the account of the paying bank to the account of the receiving bank. Under net settlement, each party delivers at batch time the net amount it sold (or receives the net amount it purchased) since the last batch. Netting is appealing because it results in a very significant reduction of the amount of cash and security that needs to be available to the banks during the batch. However, a failure to settle a trade leads to an *unwind*, i.e., to the deletion of some or all of the trades in which the defaulting bank are involved, and in the re-calculation of the settlement obligations of the other banks. An unwind imposes liquidity pressures and replacement costs on the non-defaulting banks that have traded with the defaulting one. Under these conditions, the system will almost surely fail to settle if one or more of the initially non-defaulting banks proves unable to cover the shortfalls and default. It is then likely that both the securities markets and the payment system will be disrupted. To mitigate this, a partial net settlement is implemented in some markets. In that case, only the transactions that cannot be settled are deleted in such a way as to reduce the overall disruption. Likewise, real markets may implement a hybrid RTGS with partial netting queuing. Such hybrid mechanisms make use of the so called upper bound and lower bound on liquidity needs. The lower bound liquidity is the net amount of sent and received transactions. The upper bound liquidity is the amount of liquidity needed to settle immediately without queuing. The difference between upper bound and lower bound liquidity is the difference in liquidity needs between (1) a RTGS system without queuing; and (2) a deferred net settlement system with queuing until the end of the day. If the RTGS system includes a queuing facility and a netting facility at the end of the day, it can use the lower bound liquidity efficiently during the day and thereby settle some of the transactions with the same amount of liquidity as the net system, but possibly earlier during the day. See, e.g., Leinonen and Soramäki (1999) and Committee on Payment and Settlement Systems (2001a). The usual wisdom is that reducing as much as possible the delay between trade and settlement improves the system's stability. The rational is the following. The longer is the lag between the date of trade and the date of settlement, the greater is the risk that one of the parties will default on the trade. The greater is also the possibility that the security current price will move away from the contract price, i.e., the greater is the replacement costs risk. Both the default and the replacement costs risks can be lowered by reducing the lag between trade and settlement. Thus, the G30 recommended in 1989 that the final settlement of cash transactions should occur on $T + 3$, i.e., three business days after trade date, and noted that same day settlement, $T + 0$, should be the final goal to minimize counterpart risk and market exposure. See also Leinonen (2003). Similarly, the International *Organization of Securities Commissions* (IOSCO) created, in December 1999, the *Task Force on Securities Settlement Systems*. Amongst others, the Task Force has recommended that $T + 3$ settlement be retained as a *minimum* standard, but strongly suggests that each market assesses whether a shorter interval than $T + 3$ is needed as a function of the transaction volume, the price volatility, and the financial strength of the banks, among others.

Thus, there is a need to understand how the various type of risks are affected by a shortening of the time between trade and settlement. Consider first the principal risk. It is typically taken care of by the *Delivery versus payment* (DVP) practice that links securities transfers to funds transfers. The settlement of securities transactions on a DVP basis reduces the likelihood that the failure of a participant bank could result in systemic disruptions, but does not eliminate it. If one party fails to deliver, the counterpart still needs to replace the transaction at the current market price. The magnitude of the *replacement cost risk* depends on the volatility of the security price and on the time that elapses between the trade and the settlement dates. As this time becomes shorter, the replacement cost risk becomes less and less important. Moreover, the replacement cost risk has little systemic implication.

The credit and liquidity risks are mitigated in some markets by using a *central counterpart* (CCP) that acts as the buyer to the seller and the seller to the buyer. As previously discussed, most markets have also established central securities depositories that immobilize physical securities and transfer ownership by means of book entries to electronic accounting systems. Because of this mechanism, liquidity is usually not a problem on the security leg of the transaction if short selling is not allowed. The cash leg of the transactions is typically settled through the central bank payment system, as this has the advantage of eliminating the credit risk to the seller. Furthermore, as mentioned earlier, intraday credit is typically available (possibly against provision of collateral) to the banks.

In moving from $T+n$ to $T+0$, which is the current policy target, the liquidity risk becomes particularly important on the payments side because the in- and out-coming flows of payments are not known long in advance by the cash managers. Gridlock may occur if the flow of payments is disrupted because banks are waiting to receive payments before sending them.[3] By contrast, liquidity is not a problem on the securities side because the custodians already have the securities at the time the trade is conducted. Nonetheless, in some markets the rate of settlement falls significantly short of 100% because of human errors or operational problems on the security side. Errors or delays may result from the incomplete or inaccurate transmission of information or documentation,

---

[3] Angelini (1998) studies RTGS systems under payment flow uncertainty. He shows that uncertainty, together with a costly daylight liquidity, may induce participants to postpone payment. This affects the quality of information available to the counterpart for cash management purposes and may induce higher than socially optimal levels of end-of-day reserve holding.

and from system deficiencies or interruptions. Thus, a move to $T + 0$ and to real time settlement could increase settlement failures on the security side.

The previous discussion hints that there are complex interactions between the diverse characteristics of a SSS that have a number of implications at the bank as well as, arguably more importantly, at the overall settlement level. The Bank of Finland has recently developed a powerful tool for simulating SSSs in encompassing, realistic settings, see e.g. Koponen and Soramäki (1998), Leinonen (2005), Leinonen (2007). Nonetheless, additional insight on specific questions can be gained by investigating highly stylized, simple models. In particular, such simple models may help recognize salient aspects of settlement contagion in Securities Settlement Systems.

The only attempt to study contagion in a simple model of a SSS we are aware of is that of Devriese and Mitchell (2005). The authors focus on liquidity risk on the payment side and study contagion effects triggered by the default of the largest player in a gross settlement system. They show that large settlement failures may occur even if ample liquidity is provided.

By contrast, we investigate in this paper the implications of operational risks on the security side which, as noted above, may become particularly important as one goes towards $T + 0$. We are not interested in the replacement risk which, as noted previously, is not very important if the settlement interval is short. Likewise, we do not consider the credit and liquidity risks, and do not allow short selling. We use numerical simulations to compare, in a one tier framework, the performance of pure net and gross settlement architectures as a function of the length of the settlement batches. Specifically, we study the effects of increasing the number of intraday settlement batches when exogenous random delays affect the settling process. The delays are intrinsic to the system and do not depend on the length of the batches or on the gross/net arrangement. Then, a decrease in the length of the batches increases the likelihood that delays will lead to settlement failures. We compare the implications of these failures under net and gross architectures.

While the results presented here do not amount to an exhaustive study of the behavior of a SSS under alternative architectures they give first, and deep, insights on the forces at work and on the complexity of their interaction. In particular, they make clear that there is no simple monotonic relation between the length of settlement cycles and failures, and that shorter batch lengths do not necessarily improve the performance of a SSS. They suggest that the gross architecture is almost always more stable than the net one.

2 The Basic Framework

The stylized situation we consider is the following. A system of M tier one banks trade S shares of a security among themselves. Short selling is not allowed. Let's T be the length of a trading day. Settlement operates at $T + 0$. Trades are registered in a queueing system and settled for by a clearing house in N intraday batches that occur at regular intervals. The N batches define N *settlement cycles* of length $T_n = T/N$. Note that, for N large, the model approximates real time settlement. The number of possible trades over a settlement cycle is $N_T = MT_n$ (M banks can trade on each time step).

Our analysis concentrates on what might happen during a settlement cycle. The settlement cycle is sub-divided in time steps, smaller than T_n, that represent the shortest time necessary for concluding a trade. At any time step $t = 1, ..., N_T$ we randomly select a buyer and a seller. The seller transfers an uniformly distributed random fraction of the shares in its virtual position, that is, of the shares it would have if all previous transactions (since the last settlement batch) had settled properly.

We assume that liquidity is always available. Thus, we consider only one of two possible reasons for default, the incapacity of the selling bank to deliver the shares on time, i.e., the security leg. The other possible cause, the inability of the buying bank to pay, or cash leg, cannot arise. Within this framework we investigate, out of all possible, one particular mechanism that could lead to a default: The existence of an unpredictable delay τ between the conclusion of a trade and the moment the trade is confirmed and cleared. This delay may be due e.g. to human or technical failure. Such a delay is always possible, even if the all actors involved acts diligently and follows best business practice. A delay τ such that the trade cannot be settled within the current cycle generates a *triggering default* as it can trigger a chain of subsequent defaults. Indeed, assume that the trade that defaults was between seller A and buyer B. The buyer B is not yet aware that the trade with A will default and may sell the security to another bank C. Bank C, in turn, may sell the security to another bank D, and so on, possibly generating a chain of defaults. These induced defaults will be called *contagious defaults*.

Note that in our scenario a default results, directly or indirectly, from involuntary causes only. It is not the consequence of a strategic behavior, such as short selling for speculative reasons.

We use simulations to compare the two settlement architectures, net and gross, previously introduced. In the case of gross settlement each trade is settled at batch time in the order it has occurred. If a trade concluded at time t does not settle, the buyer may be unable to settle a trade concluded in some $t' > t$, since the later was made assuming that the trade in t would be properly settled. In the case of net settlement, all trades concluded during the settlement period are settled together at batch time by netting the banks' positions.

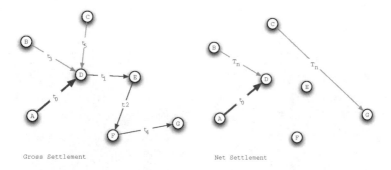

Gross Settlement   Net Settlement

Fig. 1. Comparison of contagious defaults, triggered by an initial operational problem at time t_0, in the gross and net systems.

Only the bank's net position at the end of the settlement period needs to be actually settled. Thus, a default that under the gross system would induce the buyer to default on another trade may not do so under the net system.

The functioning of these two alternative systems, gross on the left and net on the right, is illustrated in Figure 1 with the help of a simple example. The dots represent the banks, the links the trades, with the arrows indicating the trade direction (from the seller to the buyer). The symbols t_j indicate the time when the trade took place, with $t_j < t_k$ for $j < k$. Originally, the banks A, B, and C have one unit of the security and the other banks have none. Under both systems the trade in t_0, plotted in both cases as a thick red link[4] *initially* defaults. In the case of gross settlement, this initial default triggers further defaults touching the trades concluded at times t_1, t_2, and t_4, for a total of four defaults (red arrows). In the netting case, by contrast, the initial default at time t_0 does not propagate through the system. Indeed, thanks to the trades at times t_3 and t_5, the overall netting position of bank D remains positive in spite of the default at time t_0. Only one default, the initial one, occurs.

In the case of net settlement, however, default is followed by an unwind. The unwind causes (a) the deletion of all the trades involving the bank that could not fulfill its commitments to sell, that is, the defaulting bank; and (b) the recalculation of the settlement obligations of the non-defaulting banks. Because the trades of the defaulting bank are deleted, it is possible that other traders will find themselves unable to settle after (b). This may trigger more failures and thus more unwinding. The settlement process is completed only when all remaining banks, if any, can settle, possibly after many unwinding cycles. The effects of the unwinding process are illustrated in the Figures 2 and

---

[4] All figures can be downloaded in color and high definition under http://www.giuliaiori.com/SettlementFigures

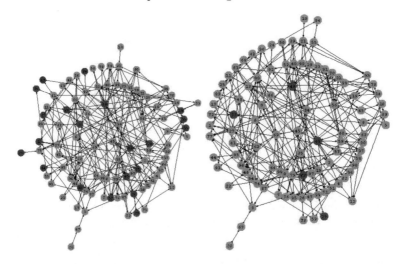

Fig. 2. Defaulting banks under gross arrangements (left) and under netting arrangements (right) as the unwinding process develops.

3, where the red dots indicate the banks that *initially* default, the green dots those who can *initially* settle[5]. As could be expected, the number of banks that initially default (Figure 2) is lower under the net than under the gross system. Nonetheless, as the unwinding process takes place (Figure 3) and defaulting banks are removed from the system, more and more banks default under the netting arrangements. *In fine*, the number defaults is lower under the gross system (75 out of 1319) than under the net system (356 out of 1319).

The advantage of the gross settlement system is that it is always possible to identify and delete the exact trade that defaults. The disadvantage is that the order of the trades matter. The advantage of net settlement is that the order of the trades up to the batch time does not matter. This may reduce the initial number of contagious defaults. Nonetheless, via the unwinding mechanism, even trades that could otherwise settle are cancelled. This may generate new rounds of contagious defaults. We study in the next section the interplay between these two effects.

3 Numerical Analysis

In the numerical analysis presented here, $N = 100$ banks and $S = 1000$ shares. The trading day has length $T = 65536$ time steps. Initially, the S shares are distributed randomly among the N firms. Let $\Pi_i(t)$ be the actual position of

---

[5] The graphs were generated by simulating trading accordingly to the rules described in section 3.2.

Fig. 3. Defaults in the net system as the unwinding progresses (from left to right). The final graph shows only the trades that can ultimately settle.

bank i at time t, and $\tilde{\Pi}_i(t)$ its virtual position. Initially, the actual and the virtual positions coincide, $\Pi_i(0) = \tilde{\Pi}_i(0)$.

At each time step t, a trade occurs with a probability λ. That is, a high value of λ indicates a very liquid market. The trade is defined in the following way. A bank i sells a integer random number of shares $s_i(t) \sim U\left[1, \tilde{\Pi}_i(t)\right]$ to buyer j. The buyer is randomly chosen. However, the seller is with probability $p \geq 0$ the buyer at time $t-1$. This introduces a structure in the sequence of trades, unless $p = 0$, in which case the choice of the seller is purely random and does not depend on past activity.

Note that $s_i(t) \leq \tilde{\Pi}_i(t)$ does not warrant $s_i(t) \leq \Pi_i(t)$. In other words, the bank sells shares that belongs to its virtual portfolio. However, should a default occur, its actual portfolio may be smaller than the virtual one and it may be unable to settle the trade it realized.

A random delay τ between the conclusion of the trade and its confirmation and clearing occurs with probability $\mu \geq 0$. If $t + \tau > T_n$, the trade is not communicated on time to the clearing house and thus automatically defaults. Otherwise it is communicated and the clearing house will *try* to settle it.

Following a trade, the positions of buyer and seller are updated in the following, self-explanatory way:

1a. If $t + \tau \leq T_n$ and $s_i(t) \leq \Pi_i(t)$,

$$\Pi_i(t) \longrightarrow \Pi_i(t) - s_i(t), \ \tilde{\Pi}_i(t) \longrightarrow \tilde{\Pi}_i(t) - s_i(t),$$
$$\Pi_j(t) \longrightarrow \Pi_j(t) + s_i(t), \ \tilde{\Pi}_j(t) \longrightarrow \tilde{\Pi}_j(t) + s_i(t).$$

1b. If $t + \tau \leq T_n$ and $s_i(t) > \Pi_i(t)$,

$$\Pi_i(t) \longrightarrow \Pi_i(t), \; \tilde{\Pi}_i(t) \longrightarrow \tilde{\Pi}_i(t) - s_i(t),$$
$$\Pi_j(t) \longrightarrow \Pi_j(t), \; \tilde{\Pi}_j(t) \longrightarrow \tilde{\Pi}_j(t) + s_i(t).$$

2. If $t + \tau > T_n$,

$$\Pi_i(t) \longrightarrow \Pi_i(t), \; \tilde{\Pi}_i(t) \longrightarrow \tilde{\Pi}_i(t) - s_i(t),$$
$$\Pi_j(t) \longrightarrow \Pi_j(t), \; \tilde{\Pi}_j(t) \longrightarrow \tilde{\Pi}_j(t) + s_i(t).$$

In the cases 1b. and 2. $\Pi_i(t)$ respectively $\Pi_j(t)$ are not updated because the trade will not take place due to a delay or to an insufficient portfolio.

The trades concluded up to time step t for which $t + \tau \leq T_n$ are stored in a matrix $J(t)$. The element $J_{i,j}(t)$ gives the net position between agent i and j at time t, as it will be known to the clearing house. If $J_{i,j}(t) > 0$ then i is a net buyer from j. If $J_{i,j}(t) < 0$ then i is a net seller to j. Obviously $J_{i,j}(t) = -J_{j,i}(t)$.

We simulate the system, under both gross and net architectures, for different values of N, and average the resulting default rates over 10000 simulations[6].

Gross Settlement: Under gross settlement, the clearing house checks all the trades in order of arrival. If at any t a bank i has committed to sell a security it did not have in its portfolio at that time, i.e. if $\Pi_i(t) < 0$, the corresponding trade cannot be settled. Accordingly, any occurrence of a negative position $\Pi_i(t) < 0$ is a contagious default. In addition, a triggering default occurs every time a trade is not communicated to the clearing house, $t + \tau > T_n$, even if $\Pi_i(t) \geq 0$.

Net Settlement: Under net settlement, the clearing house is only concerned with the final net position $n_i(T_n)$ of trader i at the batch time. Not counting the trades that were not communicated because of delays, the net position of bank i is given by

$$n_i(T_n) := \sum_j J_{i,j}(T_n).$$

If $n_i(T_n)$ is positive (negative), trader i has to receive (transfer) $n_i(T_n)$ stocks. If $n_i(T_n) < 0$ and $\Pi_i(0) > |n_i(T_n)|$ the trade can settle. However, the trade cannot settle if $n_i(T_n) < 0$ and $\Pi_i(0) < |n_i(T_n)|$. There is a contagious default. Banks who cannot settle are eliminated from the system and all their trades are canceled (unwinding mechanism). Accordingly, if bank i e.g. cannot settle, we set $J_{ij} = J_{ji} = 0$ for all j. Settlement is then attempted without the defaulting banks by recalculating the new net position of surviving banks.

---

[6] The C code for the simulations is available on request.

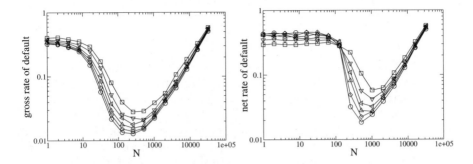

Fig. 4. Default rate r_d as a function of N for $\lambda = 1.0$, $N_a = 100$, $S = 1000$, and $p = 0$ (circle), $p = 0.2$ (diamond), $p = 0.4$ (triangle up), $p = 0.6$ (triangle left), $p = 0.8$ (triangle down) $p = 1$ (square) under gross (left) and net (right) arrangements.

3.1 Experiments I

In a first batch of experiments, we set the probability μ of a delay equal to 0, construct a single default at the beginning of the simulation, and study the induced contagious effects as trading goes on. We slowly increase the value of p (the probability that the buyer of one period is the seller of the next period), starting with $p = 0$. For each value of p, we compute the average rate of default r_d, defined as the ratio of the number of trades that fail to settle,

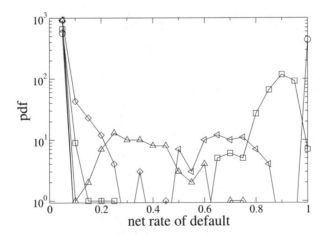

Fig. 5. Probability distribution of defaults in the netting system for $p = 0.2$, $\lambda = 1.0$, $N_a = 100$ and $S = 1000$ and $N = 1$ (circle), $N = 128$ (square), $N = 1024$ (diamond), $N = 8192$ (triangle up), $N = 32768$ (triangle left).

n_f, over the total number of trades that occurred during the trading day, n_T,

$$r_d := \frac{n_f}{n_T}.$$

Figure 4 shows the outcome of these experiments as p and N increase in the case of gross (left side of the figure) and net settlement (right side). One recognizes that under gross settlement the rate of default increases with p for any value of N. This reflects the fact that a larger p augments the likelihood of long chains of trades. Under the gross architecture, this in turn raises the likelihood of default contagion. Under the net architecture, by contrast, one might think that the trade structure and thus the value of p are irrelevant. Nonetheless, we find that when N is large an increase in p leads to a higher rate of default. Indeed, when N is large, the interval between batches is short so that few trades occur between two batches, leaving little scope for netting. Netting and gross architecture then behave almost identically. Note that for $N = 32768$ only one trade can follow the initial one within a batch, for a total of two trades. In this case the rate of default approaches 50%, as should be expected given that the initial trade defaults by construction. When N is small, a higher level of p implies a lower rate of default. That is, netting seems to perform better when the trading bank acts both as a buyer and a seller – a point that needs further study.

Under both architectures, the rate of default depends non-monotonically on N. It reaches a clear minimum in the range $N = 100$ to 1000. In the net case this can be explained in the following way. Since many trades take place between two batches when N is small, the netting mechanism is very effective. However, if a bank defaults in spite of netting, the number of transactions that are deleted via the unwinding mechanism may be very high. This is illustrated in Figure 5 where the distribution of defaults is plotted for various N given $p = 0.2$. For $N = 1$ (circle, black) the distribution is bimodal with two peaks at $r_d = 0$ and $r_d = 1$. That is, either defaults do not occur or, if they do occur, contagious effects may affect the whole settlement process. This explains the abrupt increase of r_d to about 44% as N decreases below $N = 100$. For N large, on the other hand, netting is not effective anymore. The default rate r_d increases with N since there is always at least one default, the initial one, although the number of trades becomes very small. A similar effect is found in gross systems. In that case, the increase of r_d as N decreases is due to an increase in the length of the chain of trading. For N large, netting and gross arrangements produce very similar results.

The average rate of default is always higher for netting systems than for gross systems except for $p \geq 0.8$ and N small. However, this does not tell the whole story. Consider for example the case $N = 1$ and $p = 0.2$. In the netting system, no defaults occur in about 60% of the simulations. But $r_d = 100\%$ in the remaining 40% of the simulations. In the gross system, no defaults occur in only about 17% of the simulations. In more than 80% of the simulations

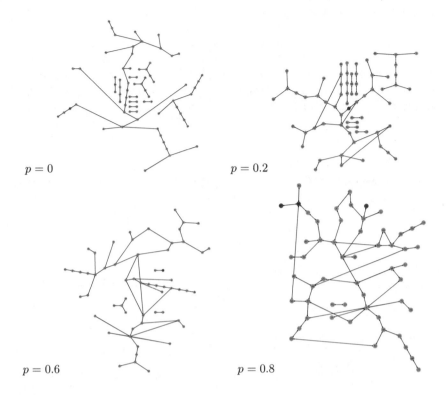

Fig. 6. Transaction network for $\lambda = 1.0$, $N = 1024$, $S = 1000$, $N_a = 100$ and $p = 0$ (top, left), $p = 0.2$ (top, right), $p = 0.6$ (bottom, left), $p = 0.8$ (bottom, right).

about half of the trades cannot be settled. Hence there is no clear cut ranking of which architecture delivers greater stability.

In Figure 6, we show the transaction networks for $N = 1024$ and different values of p. For small values of p the network is disconnected in several smaller components. At the same time, the rate of default is lower in the gross case. This is in line with the predictions of Allen and Gale (2000) who suggest that systemic risk increases with the network connectivity. Nonetheless the network structure is lost when netting multilaterally merges the positions of each bank. In fact, in the netting case we find that contagion decreases with p at low N and increases with p at high N.

3.2 Experiments II

In this second batch of simulations we set the probability p that a buyer becomes seller equal to 0, and assume that a random delay τ can occur with probability μ. We take $\tau = \tau_M \epsilon$ where $\epsilon \sim |N(0,1)|$ and τ_M is a positive constant (qualitatively similar results were obtained with a uniform distribution).

A default occurs at time t if $t + \tau > T_n$.

We study the dependence of the default rate on the number N of intraday batches. Reducing the batch length has the advantage of reducing the number of parties exchanging any given security between two settlement cycles, and hence, of lowering contagion. However, a higher N decreases T_n without affecting τ. Thus, a higher N increases the rate of triggering defaults generated by random delays.

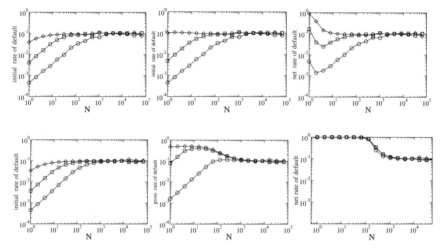

Fig. 7. Initial default rate (left), total default rate in gross systems (center) and total default rate in net systems (right), as a function of N, for $M = 100$ $S = 1000$, $\mu = 0.1$, and for $\tau_M = T$ (diamond), $T/10$ (square), $T/100$ (circle). Top: $\lambda = 0.01$. Bottom: $\lambda = 1$.

We compare the average rate of default r_d in the gross and net cases under different market conditions. Specifically, we let vary λ, which is a proxy for liquidity; μ and τ_M, which measure the likelihood of operational problems. Figure 7 shows the rate of triggering events (defined as the total number of triggering events divided by the total number of trades, left), the rate of default under the gross architecture (center), and the rate of default under the net architecture (right) for $\mu = 0.1$, $M = 100$, $S = 1000$ and $\lambda = 0.01$ (top) or $\lambda = 1$ (bottom). The curves correspond to $\tau_M = T$ (diamond), $\tau_M = T/10$ (square), and $\tau_M = T/100$ (circle). One observes that the number of triggering events increases with N both for $\lambda = 0.01$ and $\lambda = 1$. The *gross* rate of default increases with N when λ is small. For large λ, it decreases with N for $\tau_M = T$, increases with N for $\tau_M = T/100$, and for $\tau_M = T/10$ it is first increasing then decreasing in N with a maximum around $N \approx 16$. The *net* rate of default decreases with N when λ is large. For small λ it decreases with N for $\tau_M = T$, but reaches a minimum and then increases with N for $\tau_M = T/10$ and $\tau_M = T/100$.

Figure 7 reveals that gross systems are little sensitive to contagion when markets are illiquid. In this case the average rate of default is almost identical to the initial rate of default (top, left and center). However, contagious effects can be significative in the gross case when markets are liquid, due to the long chains of trade that can arise in this context (center, bottom). By contrast, in net systems contagion can be very important even when there is limited trading activity.

Further investigation is required to explain this rich variety of behavior.

4 Conclusions

In this paper we examined some issues related to the performance of different securities settlement architectures under the assumption of exogenous random delays in settlement. In particular, we focused on the impacts of the length of settlement cycles on default under different market conditions. Factors such as the market liquidity, the trading volume, the frequency and length of delays, and to some extend the trade structure, were taken into account.

We found that the length of settlement cycles has a non-monotonic effect on failures under both gross and net architectures. This reflects to a large extend the interplay between (a) the stabilization resulting from a decrease in system defaults due to a shorter settlement cycle involving fewer parties; and (b) the destabilization resulting from an increase in triggering defaults due to the greater likelihood that a delay will impair a transfer. We also showed that, contrary to a common wisdom, real time settlement does not improve the performance of the settlement process under all market conditions. Finally, under the scenarios we studied, the gross architecture appears to be more stable than the net one.

The susceptibility of a SSS to contagion depends of many factors in addition to the institutional arrangements governing the exchanges. These factors, that were not taken into account in this paper, include among others the topology of the transaction network and the distribution of portfolio among the banks. The impact of factors of this kind have been in part investigated in the financial literature using techniques from graph theory, for example in the analysis of the interbank market. We are considering to do a similar analysis for SSSs. First steps in this direction would be introduce heterogeneity on the bank size, to simulate a two tier system, and to link the trading activity of each bank to its size. In that way, bigger banks would possibly become hubs of the transaction network, with important implications for contagion.

Another interesting extension would to take into account the possibility of strategic default, i.e., of rational decisions by some participants not to settle in response to movements in securities prices. Doing so may be particularly challenging since, ideally, one should also take into account the behavior of the operator of the SSS. Indeed, the operator typically attempts to discourage strategic default by imposing a fine aiming at taxing away any potential gain from defaulting.

Albeit further investigation is needed to fully understand the complex mechanisms underlying settlement risk in SSSs, we trust that the existing results convincingly show that the approach pursued in this paper sheds light on the behavior of SSSs that could not be gained otherwise, and may potentially help improve their architecture.

Acknowledgements

This paper has crucially benefited from the Lamfalussy Fellowship Program sponsored by the European Central Bank. It expresses the views of the authors and not necessarily represents those of the ECB or the Eurosystem.

The authors are grateful to Mark Bayle, Philipp Hartmann, Cornelia Holthausen, Cyril Monnet, Thorsten Koeppl, Jens Tapking and Saqib Jafarey for valuable comments and interesting discussions. The paper was written within the ESF–Cost Action P10 "Physics of Risk". It benefited from the valuable comments of three anonymous referees. All errors and omissions are the authors sole responsibility.

References

Allen, F. and Gale, D.: 2000, Financial contagion, *Journal of Political Economy* **108**, 1–33.

Angelini, P.: 1998, An analysis of competitive externalities in gross payment systems, *Journal of Banking and Finance* **21**(1), 1–18.

Committee on Payment and Settlement Systems: 2001a, Core principles for systemically important payment systems, CPSS publication No. 43, Basel.

Committee on Payment and Settlement Systems: 2001b, Recommendations for securities settlement systems, CPSS publication No. 46, Basel.

de Bandt, O. and Hartmann, P.: 2000, Systemic risk: A survey, *Working Paper 35*, European Central Bank, Frankfurt.

Devriese, J. and Mitchell, J.: 2005, Liquidity risk in securities settlement, *CEPR Discussion Paper No. 5123* .

Holthausen, C. and Tapking, J.: 2007, Raising rival's costs in the securities settlement industry, *Journal of Financial Intermediation* **16**(1), 91–116.

Koponen, R. and Soramäki, K.: 1998, Intraday liquidity needs in a modern interbank payment system – a simulation approach, *Expository Studies 14*, Bank of Finland, Helsinki.

Leinonen, H.: 2003, Restructuring securities systems processing – a blue print proposal for real-time/$t + 0$ processing, *Discussion Papers 7*, Bank of Finland.

Leinonen, H.: 2005, Liquidity, risks and speed in payment and settlement systems – a simulation approach, *Expository Studies 31*, Bank of Finland, Helsinki. Editor.

Leinonen, H.: 2007, Simulation studies of liquidity needs, risks and efficiency in payment networks, *Expository Studies 39*, Bank of Finland, Helsinki. Editor.

Leinonen, H. and Soramäki, K.: 1999, Optimizing liquidity usage and settlement speed in payment systems, *Discussion Paper 16*, Bank of Finland, Helsinki.

Integrated Risk Management: Risk Aggregation and Allocation Using Intelligent Systems

Andreas Mitschele[1,2], Frank Schlottmann[2], and Detlef Seese[1]

[1] Institute AIFB, Universität Karlsruhe (TH), D-76128 Karlsruhe, Germany
   mitschele|seese@aifb.uni-karlsruhe.de
[2] GILLARDON AG financial software, Research Department, Edisonstr. 2,
   D-75015 Bretten, Germany
   frank.schlottmann@gillardon.de

Summary. In recent years integrated approaches have become state-of-the-art practice for risk management in financial institutions. Contrary to the still common silo-based approach where risk categories and business lines are predominantly analyzed separately, an integrated risk management system adopts an enterprise-wide perspective to appropriately account for cross-sectional dependencies between all significant banking risks. In this contribution an application of intelligent systems that provides management with risk-return efficient bank-wide asset allocation strategies is outlined. It is based on multi-objective evolutionary algorithms and considers different market risks and credit risk as well as position volume constraints. The presented novel approach is not only able to integrate the differing goals concerning the risk management function but also to partly overcome the obstacles for risk integration and aggregation. Using real market data a sample portfolio analysis is performed and possible conclusions for a bank risk manager are drawn. The approach is extendable concerning for instance advanced risk measurement methodologies, correlation assumptions, different time horizons and additional risk types. Further real-world constraints, such as regulatory capital, portfolio or P&L restrictions can also be easily integrated into the model.

Key words: Integrated risk management, risk aggregation, asset allocation, multi-objective evolutionary algorithms.

1 Introduction

The integrated management of financial risks has drawn increasing attention of banking supervisors as well as of banks themselves in recent years. Within the context of the supervisory review process of Basel II the Basel Committee on Banking Supervision (2006) postulates that *"Institutions should have a process for assessing their overall capital adequacy in relation to their risk profile and a strategy for maintaining their capital levels"*. Furthermore, in recently published guidelines for the application of the supervisory review process the Committee of European Banking Supervisors (2006) requires European banks to *"develop sound risk management processes that adequately identify, measure, aggregate and monitor their risks."*.

These recent revisions have been preceded by the worldwide study *Trends in risk integration and aggregation* by the Working Group on Risk Assessment and Capital of the Basel Committee (2003) which reports about two major trends in financial risk management. Firstly, the study has identified a strong emphasis on the management of risk on an integrated firm-wide basis. The second emerging trend comprises rising efforts to aggregate risks through mathematical models. At the end of the day, not only banking supervisors but also banks themselves are highly motivated to approximate their required capital base, that serves as a buffer against unexpected losses, increasingly accurate (Kuritzkes and Schuermann, 2007).

While some banks undertake high endeavors to obtain an integrated sight of their entire risk profile, this aim in reality usually still rather resembles a mere vision. In real-world applications different types of risk are commonly assessed and controlled in a more silo-based manner (Pézier, 2003; Kuritzkes et al., 2002), that is, market risk is measured and managed separately from credit risk etc. When an aggregate risk measure is desired, the resulting risk numbers for the different risk categories are often simply added up, assuming comonotone dependence between them. As such crude methods completely neglect potential diversification benefits, they can only mark a first step towards true integrated risk management.

The remainder of this contribution is organised as follows: In the next section we describe the still widely adopted silo-based approach to risk management, including a characterisation of the main banking risks. After an overview of integrated risk management we outline our multi-objective approach that enables bank managers to obtain an integrated view on their bank portfolio's risk and return, taking common real-world constraints into account. The approach fits easily into current risk management practice and avoids some of the implementation issues discussed before. Finally, for a sample bank the application of two different MOEAs (multi-objective evolutionary algorithms) in our setting is illustrated. We use the popular NSGA-II and the more recent ϵ-MOEA with a set of current market data and discuss the conclusions that the (integrated) risk manager can draw from the results.

2 Current Silo-Based Approach to Risk Management

Financial risks are inherent in financial markets and their management represents one of the main tasks in the business of financial institutions. In this context risk is generally perceived as *"potential for deviation from expected results"* (Kuritzkes and Schuermann, 2007). Among the different goals that are pursued with risk management and with its integration in particular two main objectives may be denoted according to Cumming and Hirtle (2001), namely the view of owners and investors (*optimal capital allocation*) and the view of bank supervisors and creditors (*financial solvency*). While for the supervisory motivated solvency concern adverse deviations are particularly important, the owners additionally take advantageous deviations into account.

The Basel Committee on Banking Supervision (2006) has identified market, credit and operational risk as the main risk sources faced by banks. However, the individual exposure to these risk categories may be very different among institutions and depends on factors such as the strategic orientation of management. Consequently, individual institutions may be exposed to entirely different mixes of significant risks, comprising also new risk categories, e.g., high legal risks. In the following, we describe the three key risk types according to the Basel Committee.

In general, **market risk** arises through adverse movements in the market prices of financial instruments. There exist a number of subcategories depending on the considered market factor, for instance *interest rate, equity* or *currency risk*. Compared to other risk types market risk measurement and management is rather well developed as extensive research has been carried out in this area (Cumming and Hirtle, 2001; Kuritzkes and Schuermann, 2007). For most of the instruments that are traded on financial markets long historical data sets are available. However, this does not apply to newly introduced financial instruments (e.g., new stock issues) and to assets without an actively traded market (e.g., real estate). These are generally valued using peer instruments as approximation or by modeling return distributions based on certain assumptions (cf. also to Monte Carlo simulation). Additionally, returns of market instruments are often assumed to resemble the standard normal distribution. While this assumption facilitates the handling in the context of market risk measurement, it has to be noted that empirical return distributions usually contain *fat tails* (Rachev et al., 2005). This means that the outer quantiles of the observed market distributions usually possess a higher probability density than assumed by the standard normal distribution.

In the area of **credit risk** strongly intensified efforts have been undergone, both in research and practice. Credit risk concerns possible losses through unfavorable changes in a counterparty's credit quality. This category comprises *default risk* in case a contractual partner is not capable to repay his debt anymore. Possible depreciations in the bond value of an obligor through changes in his individual credit spread (*spread risk*, cf. Crouhy et al. (2001)) may also lead to credit losses. Apart from the Basel Accords issued by the Basel Committee on Banking Supervision in 1988; 1996, and 2006, competitive forces to establish adequate credit risk pricing systems have recently further increased the focus of financial institutions on credit risk. Nowadays there exist a number of credit risk models that have emerged as common practice. Even though credit risk data availability is still lagging far behind market risk data due to comparably rare events (default or rating deterioration) it has improved substantially in recent years. Also risk management possibilities have increased. *Credit default swaps* that pay off the outstanding credit amount in case an obligor defaults are an example for such credit portfolio insurance. A detailed description of the prevailing approaches is however beyond the scope of this contribution (Bluhm et al., 2003).

Particularly through the new Basel Capital Accord **operational risk** has recently come into focus. The Basel Committee on Banking Supervision (2006) defines operational risk as *"the risk of loss resulting from inadequate or failed internal processes, people and systems or from external events. This definition includes legal risk, but excludes strategic and reputational risk."* As there are high uncertainties concerning its quantification and there is a considerable lack of differentiation in the Basel Standardised Approach provided by the Basel Committee (2006), we have focussed on credit risk and a set of different market risks in our integrated risk management example which is described later. In this respect we deviate from our model presented in Schlottmann et al. (2005). Further risks such as *reputational, strategic* or *business risks* have been intentionally excluded from the Basel operational risk definition and from the main supervisory risk focus in general. However, we like to mention that the flexibility of our model setup presented in section 3.3 facilitates the inclusion of even sophisticated and complicated further risk types.

The presented state-of-the-art practices for risk management in financial institutions are typically performed within a silo-based environment which means that different risks are assessed and managed in their separate "silos" (Kuritzkes et al., 2002). For instance a bank would calculate a VaR figure for the trading book to account for market risk and a Credit-VaR figure for the banking book. Usually, these figures are calculated based on differing methodological prerequisites, for instance different holding periods, confidence levels and time spans for the historical model input data. In less sophisticated banks, i.e. banks that do not employ advanced methods such as copulas for instance (cf. 3.1), these two entirely different figures would often simply be added up to obtain a total bank risk number, assuming comonotone dependence. This rather inappropriate methodology that neglects potential diversification benefits between the silos is partly due to the organisational structure of banks and partly due to historical regulatory constraints. Integrated risk management which is introduced in the following section aims at overcoming these drawbacks.

3 An Integrated Approach to Risk Management

After an introduction to the integration of risk management we present our multi-objective framework for the retrieval of risk-return-efficient portfolio allocations. Research overviews in the context are provided in Saita (2004) and in the more recent contribution of Rosenberg and Schuermann (2006).

3.1 Integrated Risk Management and Measurement

For *integrated risk management* (IRM) there exist different roughly inter-changeable terms, which include *consolidated (financial) risk management* and *enterprise(-wide) risk management*. The Joint Forum of the Basel Committee on Banking Supervision (2003) proposes the following definition for an IRM system:

> *"An integrated risk management system seeks to have in place man-agement policies and procedures that are designed to help ensure an awareness of, and accountability for, the risks taken throughout the fi-nancial firm, and also develop the tools needed to address these risks."*

Even though there sometimes is an intensive discussion about its necessity in this context,[3] the core of such an integrated risk management system is typ-ically represented by a risk aggregation methodology, which we call *integrated risk measurement*. It involves a set of quantitative risk measures that incor-porate multiple types or sources of risk. *Economic capital* (EC) has emerged as state-of-the-art to assess how much capital is needed to cover potential losses from the considered risks, making it a common 'currency of risk' that allows the direct comparison of different risk types. EC is usually computed using Value-at-Risk (VaR) or a similar technique. However, to appropriately aggregate the different risk figures it is important to harmonise the former presented heterogenous approaches across risk categories concerning param-eters such as confidence level (e.g., 99.9%) and time horizon (e.g., 1 year) (Kuritzkes and Schuermann, 2007). *Copulas* represent a promising advanced concept for the task to aggregate risk measures across different risk types, as they make it possible to convolve the different distributions of credit, market and operational risk for instance. However, a big drawback stems from the fact that extensive data input is necessary to calibrate copulas adequately, making them prone to model risk. Thus, copulas cannot be seen as general problem solver for the integration of risk measurement.[4] It should be noted that, while Alexander (2003) proclaims the accurate aggregation of banking risks as *"one of the main challenges facing risk managers today"* she also warns that *"crude risk aggregation techniques"* represent the major source of risk for bank capi-tal models. Thus, in section 3.3 we opt for an integrated approach that does not aggregate the different risk types prior to the optimisation run.

The adequate provision for diversification benefits, giving the bank man-agement a realistic estimate of the total bank risk, and the possibility to appropriately allocate bank capital to the different business lines are among the striking advantages of an IRM system. They enforce the promotion of sound economic incentives throughout the different management levels and ultimately improve the competitive position of the financial institution.

---

[3] Cf. Cumming and Hirtle (2001) and the Joint Forum survey (2003).

[4] For a comprehensive introduction to the copula method and its applications to finance and risk management cf. Embrechts (2000) and McNeil et al. (2005).

However, there are a number of ongoing issues that impede the straightforward implementation of such a system. These include the heterogenuous risk types with different and inconsistent risk metrics, insufficient data availability, different time horizons for risk assessment and inhomogenous distributional assumptions (c.f. also section 2). Additionally, correlation estimates as crucial component for adequate integrated risk modeling, often have to be based on relatively small data sets. Thus, they tend to be rather unstable, causing reluctance from regulators to recognise the corresponding diversification benefits. Even though in the future it may well be expected that the use of proprietary correlation estimates for regulatory capital computation will become more flexible, much more experience with correlations is still required today to avoid under- as well as overestimation. From the general setup point of view, for true IRM a *top-down process* seems by far superior to the prevailing *bottom-up approaches* that are still common in many financial institutions. Other technological, legal and regional problems can further aggravate the successful implementation of an integrated system.[5]

It becomes obvious that the bank-wide integration of risks still turns out to be highly complicated. Our latter presented approach can handle the discussed weaknesses of today's widespread structures in financial institutions and provides an integrated view while being able to build right upon the bottom-up risk management as it is still commonly performed in practice.

3.2 Asset Allocation within Integrated Risk Management

Generally speaking, asset allocation is classified into *Tactical Asset Allocation (TAA)* and *Strategic Asset Allocation (SAA)*. While the former is commonly used to optimise the specific composition of a predefined portfolio, e.g., to select individual stock names within an emerging market trading portfolio, the latter is intended to support strategic decisions about the fundamental investment orientation of an entire portfolio. For instance, an SAA analysis may lead the bank management to establish the following overall investment strategy for the banks' assets: 25% in European stocks, 35% in government bonds and the remaining 40% in retail business, e.g., mortgages. In a possible second step the trading department that manages the stock market portfolio would then start a TAA investigation to determine the individual European stock names that are to be invested in. In line with the above-mentioned term "integrated risk measurement" the process of finding a risk-return efficient investment mix for the entire bank portfolio, in consideration of all significant risk types, could well be denoted *integrated asset allocation*.

With his seminal work on portfolio selection Markowitz (1952) set the foundation for modern portfolio theory. He realised that a portfolio may be diversified by adding more and more investment opportunities. This lead to

---

[5] Cf. Cumming and Hirtle (2001) for a comprehensive introduction into benefits and challenges of an IRM system.

the conclusion that there exist *Markowitz efficient portfolios* that dominate other portfolios with the same risk but smaller return or with the same return but higher risk. Markowitz had to make numerous strong assumptions (Chang et al., 2000; Maringer, 2005), e.g., the perfectness of markets and the multi-variate normal distribution to describe asset returns, to be able to formulate a quadratic programming (QP) optimisation problem which can be solved in reasonable computing time. However, in real-world applications generally these assumptions do not hold, making the problem considerably harder to solve.

Consequently, different authors have proposed to use modern heuristic algorithms in portfolio optimisation. Chang et al. (2000) have incorporated cardinality constraints and proportional asset limits into the optimisation problem and solved it using different intelligent systems (genetic algorithms, tabu search and simulated annealing). Further successful applications in this context using threshold accepting[6] were presented by Maringer (2005) and Gilli et al. (2006).

In this contribution, we employ evolutionary algorithms (EAs) that mimic their strategies from the omnipresent evolution process in nature. EAs perform the natural evolution steps *selection, recombination* and *mutation* in an algorithmic way. Their adaptivity and flexibility makes them very successful problem solvers in hard problem instances and thus ideal for the constrained portfolio optimisation problem. As we have to deal with numerous contradicting objectives, i.e. risk (risks respectively) and return, we propose a *multi-objective* approach using multi-objective evolutionary algorithms (MOEAs).[7]

3.3 Multi-Objective Model Framework

In the remainder, we consider a universe of $n \in \mathbf{N}$ investment opportunities for a certain bank. While in Schlottmann et al. (2005) we focused more on tactical asset allocation,[8] in our example presented here (cf. sections 3.5 and 3.6) we concentrate on the efficient allocation of the bank's entire capital from a more strategic point of view, i.e. strategic asset allocation.

When running a tactical asset allocation analysis in particular, the investment opportunities $n \in \mathbf{N}$ may well be single assets, for instance VOLKSWAGEN AG stocks or certain DEUTSCHE TELEKOM AG bonds. In the case of strategic asset allocation, which we have tailored our sample bank for, the investments will rather comprehend asset classes. These asset classes are typically represented by indices, for example the German DAX Performance stock index as a proxy for investments in large German companies. Any portfolio consisting of a subset of these assets or asset classes, respectively, is specified by an n-dimensional vector

---

[6] Cf. Winker (2001) for an introduction to threshold accepting and applications.

[7] Cf. Deb (2001) for a very comprehensive introduction to MOEAs.

[8] In Schlottmann et al. (2005) the general framework has been first introduced. This approach will be described in further detail in 3.4.

$$X = (x_1, x_2, ..., x_n),$$ (1)

which satisfies the conditions

$$\sum_{i=1}^{n} x_i = 1 \wedge \forall i \in \{1, \ldots, n\} : x_i \in [0, 1].$$ (2)

Each decision variable x_i represents the percentage of the bank's current wealth which is to be invested in an investment opportunity $i \in \{1, \ldots, n\}$.

Remark. Due to the following reasoning we neither admit credit financed long positions nor short positions in our model. Assuming a considered bank is highly liquid, it can thus virtually always raise sufficient and inexpensive funds through the interbank market. Consequently, after a decision for a certain allocation – excluding financing and short positions – has been reached, the bank management can quite easily leverage the bank portfolio through further interbank credit. Hereby the manager can increase the portfolio expected return while simultaneously increasing the risk exposure through this leverage. Additionally, the sample bank manager is not willing to become involved in direct short sales of stocks and thus restricts the optimization to long positions. Through the inclusion of certificates from hedge funds, that generally hold large short positions themselves, into the investment decision process, opportunities similar to short positions can in principle also be taken. Certainly, under these assumptions different optimal portfolios might be obtained compared to those obtained when allowing external financing and short positions in the optimisation run.

Objective Functions and Optimisation Problem

The following target functions reflect the usual objectives in a bank's general silo-based approach to integrated risk and return management (Schlottmann et al., 2005). The first objective is the expected rate of return from the portfolio, given by

$$ret(X) := \sum_{i=1}^{n} x_i r_i,$$ (3)

where r_i is the expected rate of return of investment opportunity i. This objective is to be maximised.

For the second and third objective function Value-at-Risk (VaR), a prevalent method to measure market, credit and operational risk exposures,[9] is used. The formal definition by Frey and McNeil (2002) which is pursuant to Artzner et al. (1999) is as follows: Given a loss L with probability distribution **P**, the Value-at-Risk of a portfolio at the given confidence level $\alpha \in \;]0, 1[$ is represented by the smallest number l such that the probability that the loss L exceeds l is no larger than $(1 - \alpha)$. Formally,

---

[9] For a comprehensive introduction to the Value-at-Risk concept cf. Jorion (2006).

$$VaR_\alpha = \inf \{l \in \mathbf{R}, \mathbf{P}(L > l) \leq 1 - \alpha\} \, . \tag{4}$$

Even though VaR does not represent a *coherent risk measure* (Artzner et al., 1999) it has still become a widespread measure of choice for risk management in financial institutions. Furthermore in a recent risk aggregation study Rosenberg and Schuermann (2006) found that the explanatory power of alternative risk measures, like expected shortfall,[10] does not deviate strongly from the conclusions that can be drawn from a VaR-based analysis. Due to these results and as we intend to present a real-world application we have decided to build our analyses upon the common risk measure VaR.

Hence, the second objective function is the Value-at-Risk of the portfolio due to changes of market prices (market risk), denoted by

$$mr(X) := VaR_{market\ risk}(X) \, , \tag{5}$$

where $VaR_{market\ risk}(X)$ is determined by one of the common calculation methods, historical simulation, variance-covariance approach or Monte Carlo simulation. Usually, this objective is short-term oriented, e.g., measured on a time horizon of one or ten trading days. It is to be minimised.

The third objective function is the Value-at-Risk of the portfolio due to credit risk, i.e. defaults of obligors or other losses resulting from changing credit qualities of the obligors. It is denoted by

$$cr(X) := VaR_{credit\ risk}(X) \, . \tag{6}$$

As mentioned in the first section, the $VaR_{credit\ risk}$ is commonly calculated using a portfolio model, such as CreditMetrics, CreditRisk+, CreditPortfolioView or similar approaches.[11] A common time horizon for the credit risk calculations is one year, and this risk measure should also be minimised.

A fourth objective which is relevant to the context is the required capital for operational risk compensation which can, for instance, be calculated according to the Basel Committee on Banking Supervision's Standardised Approach (Basel Committee on Banking Supervision (2006), p. 137ff). This approach yields a target function

$$or(X) := \sum_{i=1}^{n} x_i \beta_i \, , \tag{7}$$

where β_i is specific for the business line in the bank which is affected by the investment $x_i > 0$ in opportunity i. This objective is also to be minimised.

Summarizing the above definitions and restrictions as well as converting maximisation of the $ret(X)$-function into minimisation of $-ret(X)$, we obtain the following general optimisation problem setting with four objective functions $f_1(X)$ to $f_4(X)$ that are to be minimised:

---

[10] Also known as Tail-VaR or Conditional VaR (CVaR), cf. Artzner et al. (1999) and Rockafellar and Uryasev (2002).

[11] Cf., e.g., Bluhm et al. (2003) for an overview of these models.

$$f_1(X) := -ret(X) \tag{8}$$
$$f_2(X) := mr(X) \tag{9}$$
$$f_3(X) := cr(X) \tag{10}$$
$$f_4(X) := or(X) \tag{11}$$
$$X := (x_1, \ldots, x_n) \tag{12}$$
$$\forall i \in \{1, \ldots, n\} : x_i \in [0, 1] \tag{13}$$
$$\sum_{i=1}^{n} x_i = 1 \tag{14}$$

Further Model Assumptions and Remarks

For the derivation of efficient portfolios we use the following definition, that is compatible to both the usual definition of dominated portfolios in the finance context and the common definition of dominated points in the context of multi-objective optimisation:

A portfolio X_2 is (weakly) *dominated* by a portfolio X_1 if the following condition is met:

$$\forall j \in \{1, \ldots, 4\} : f_j(X_1) \le f_j(X_2) \wedge \exists k \in \{1, \ldots, 4\} : f_k(X_1) < f_k(X_2). \tag{15}$$

We assume that the bank is a *rational investor*, i.e. the bank is not going to invest in a dominated portfolio (Markowitz, 1952). Moreover, we assume that the bank's management prefers to choose from a whole set of individually optimal solutions, particularly by evaluating the trade-off between the desired expected rate of return and the different risks which have to be accepted for the respective portfolio. Hence, we search for a set of non-dominated portfolios well-distributed in the four-dimensional objective function space $f_1(X)$ to $f_4(X)$ over the feasible search space which is specified by conditions (12) to (14).

To justify the use of heuristic approaches let us consider the complexity of the problem. The problem of finding even a single feasible non-dominated point is **NP**-*hard* if the decision variables are restricted to integer values. This can be proved by reducing the standard *KNAPSACK* setting to a discrete version of our problem.[12] Without this restriction the complexity of the problem seems to be open. Although we consider the problem without restriction to integer variables the mathematical properties of the objective functions also indicate its high complexity. According to Artzner et al. (1999) and Gaivoronski and Pflug (2002) the Value-at-Risk risk measure is a nonlinear and non-convex function and has usually many local optima. In the presented general setup f_2 and f_3 share this property which is hence delicate for conventional

---

[12] Cf., e.g., Seese and Schlottmann (2003) for a formal analysis in the two-objective function case which can be generalised to more than two objectives.

optimisation approaches.[13] If we had assumed a Value-at-Risk measure for operational risk then this would also apply to f_4. Furthermore, through the simultaneous optimisation of the four objective functions we are able to provide the decision maker with a choice of unbiased optimal allocations without superimposing a certain utility function prior to the optimisation. Compared to the common procedure, where a utility function has to be employed before the optimisation run, this is a clear advantage. It should also be noted in this context that it is very hard (if at all possible) to determine specific utility functions for the investor in practical applications (Elton et al., 2007). Last but not least, additional constraints such as min/max position volume limits, that are usually imposed on the portfolio positions in practice, considerably aggravate the use of standard portfolio opimisation algorithms or even make them intractable.

Thus, we opt for a heuristic approach to compute approximated solutions. More details about implementation considerations concerning the algorithms are described in the following section. Thereafter, sample bank portfolios are illustrated.

Solution Using Multi-Objective Evolutionary Algorithms

To solve the optimisation problem we employ multi-objective evolutionary algorithms (MOEAs) which are appropriate here since we search for a well-distributed approximation set in a restricted four-dimensional objective function space. The striking advantage of the MOEA approach is that it computes the entire efficient frontier in one single step of the algorithm. The frontier is then represented through the final best population.[14]

In the literature, different algorithms which implement a specific MOEA scheme are discussed, see, e.g., Deb (2001), Coello Coello et al. (2002) and many other theoretical and empirical comparisons between the alternative approaches to evolutionary multi-objective optimisation. In general, most of these MOEAs should be appropriate in our problem setting. It has to be pointed out here that it is not the goal of our work to propose a specific MOEA in this context as the best one out of all these algorithms.

To illustrate the successful application of MOEAs in our real-world problem of integrated risk management, we chose the NSGA-II by Deb et al. (2000) that has successfully been applied to many problem contexts in general, and more specifically, to other constrained portfolio optimisation problems using less than four objective functions.[15] Additionally, for our presented study in section 3.5 and 3.6 we picked the more recently published ϵ-MOEA by Deb

---

[13] In our presented strategic asset allocation sample in Sections 3.5 and 3.6 VaR is used in all objective risk functions (f_2 to f_4).

[14] A very comprehensive introduction into MOEA's is given in Deb (2001).

[15] Cf. Schlottmann and Seese (2004b) for a general and Schlottmann and Seese (2005) for a more specific overview of such studies.

et al. (2003) that comes with a superior ability to preserve diversity in the solution set using the so-called ϵ-*dominance principle*. As we intend to deliver a well organised set of pareto-optimal portfolio allocations to the bank decision maker this special feature of the ϵ-MOEA can be of particular importance. We apply the standard NSGA-II and ϵ-MOEA implementations, which are both provided by Kalyanmoy Deb, to our problem instances. Using the genetic variation operators made available in these implementations (simulated binary crossover and a corresponding mutation operator for real-coded genes), we set the crossover probability to 0.8 and the mutation rate to $\frac{1}{n}$.

For the restriction of the decision variables specified in formula (2), we set the bounds for each real-coded gene in the NSGA-II to $[0, 1]$. In addition, we have to ensure that each portfolio X satisfies $\sum_{i=1}^{n} x_i = 1$. Since we have observed a worse empirical convergence of the algorithms when using an objective function penalty for infeasible individuals, we opt for a simple repair algorithm (in accordance to other studies in different application contexts): Immediately after performing crossover and mutation, every allele value x_j of an offspring individual is re-normed according to

$$\widetilde{x_j} = \frac{x_j}{\sum_{i=1}^{n} x_i},\qquad(16)$$

and only the re-normed individuals $\widetilde{X} = (\widetilde{x_1}, \ldots, \widetilde{x_n})$ are considered in the succeeding steps of the NSGA-II.

3.4 Sample Bank 1 (Schlottmann et al., 2005)

Concerning our sample applications, we firstly want to outline the setup that was used by Schlottmann et al. (2005) where we considered a more tactical asset allocation for a sample bank with $n = 20$ investment opportunities with the characteristics shown in Table 1. The historical market data range covered closing prices for the ten traded instruments from 15-MAY-2003 to 30-SEP-2004. For the 10 loans with annual net interest payments, we assumed that the bank was hedged against market risk changes, i.e. interest rate risk was not relevant to these instruments. All calculations were based on a decision to be made by the bank's integrated risk manager on 30-SEP-2004.

For any given portfolio X, the expected rate of return $ret(X)$ was estimated from the historical time series for the ten traded instruments and from the expected annual net interest payment by the respective loan obligor. Moreover, we assumed the bank used historical simulation for the calculation of $mr(X)$ using a confidence level of 99% and a time horizon of 1 trading day in the general optimisation problem (cf. (8) to (14)). Furthermore, we assumed the bank applied the CreditMetrics model by Gupton et al. (1997) in the two-state variant described by Gordy (2000) to determine $cr(X)$ for a confidence level of 99.9% and a one-year horizon. The function value $or(X)$ was calculated according to the Basel Standardised Approach as specified in section 3.3.

Table 1. Investment opportunities for sample bank 1.

QTY	Category	Issuer/Obligor	Coupon	Maturity	Rating
1	Bond	German government (BUND)	6.250%	26-APR-2006	AAA
1	Bond	German government (BUND)	4.500%	04-JUL-2009	AAA
1	Bond	German government (BUND)	5.625%	20-SEP-2016	AAA
1	Bond	Deutsche Telekom (Corporate)	8.125%	29-MAY-2012	BBB+
1	Bond	Volkswagen (Corporate)	4.125%	22-MAY-2009	A-
1	Equity	BASF AG	-	-	-
1	Equity	Deutsche Bank AG	-	-	-
1	Equity	DaimlerChrysler AG	-	-	-
1	Equity	SAP AG	-	-	-
1	Equity	Siemens AG	-	-	-
2	Loan	Private Obligor	8.000%	30-SEP-2005	BB
2	Loan	Private Obligor	8.000%	30-SEP-2005	BB-
2	Loan	Private Obligor	8.000%	30-SEP-2005	B+
2	Loan	Private Obligor	8.000%	30-SEP-2005	B
2	Loan	Private Obligor	8.000%	30-SEP-2005	B-

For the solution of the optimisation problem we applied the standard NSGA-II implementation provided by Kalyanmoy Deb, with the parameter setting and extensions that were pointed out in section 3.3. Thus, we obtained a four-dimensional solution space with non-dominated efficient portfolio allocations with respect to the 20 investment opportunities. At this point it has to be emphasized that it is an important advantage of our multi-objective model framework, concerning real-world applications, that different risk measures, distinct confidence levels and varying time horizons can be used within the search for non-dominated portfolios without adversely affecting the results.

Discussion of results. In Figure 1 the trade-off between the three considered risk factors (market , credit and operational risk) can be analysed.[16] Clearly, an increase of the market risk exposure boosts the operational risk for the total bank portfolio which is due to the fact that market risk investments, such as stocks, demand a higher operational risk capital charge according to the Basel II standardised approach. Contrarily, a higher concentration on the private loans reduces operational risk while at the same time market risk falls and credit risk rises considerably. The decision maker will have to take her individual risk appetite into account to choose an efficient allocation from this presented set of portfolio compositions. Cf. Schlottmann et al. (2005) for further discussion of the results.

---

[16] Note that we obtain a 4-dimensional objective space. Further advanced visualisation techniques will be introduced in 3.5 and 3.6.

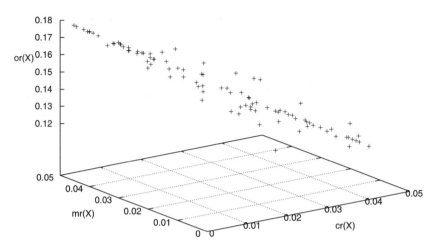

Fig. 1. Projection of mr(X), cr(X) and or(X) in Schlottmann et al. (2005).

3.5 Sample Bank 2 (Strategic Asset Allocation)

Table 2. Investment opportunities for the sample bank.

#	Index	Type	Country	Currency	Start Date
1	DAX30	Stock index (Performance)	Germany	-	03-MAR-1997
2	ATX	Stock index (Price)	Austria	EUR	03-MAR-1997
3	S&P500	Stock index (Price)	USA	USD	03-MAR-1997
4	RTX	Stock index (Price)	Russia	USD	03-MAR-1997
5	REXP	Bond index (Performance)	Germany	-	06-MAY-1992
6	iBoxx overall	Bond index (Performance)	Europe	-	10-MAY-2001

For our here presented sample bank, we consider $n = 6$ investment opportunities with their specific characteristics shown in Table 2. We have chosen these instruments as reasonable proxies for investments in the respective markets and also in a pragmatic way as we want to deliver a sound proof of concept for our optimisation model in the context of strategic bank asset allocation that is based on real-world data. While all historical data sets run until 28-FEB-2007, their entire time spans differ widely as it is often the case in the risk management of heterogenous financial instruments. For the stock index instruments we have chosen a 10 year time span which includes several stock market crashes, e.g., the Russian crisis in 1998 and the burst of the "dot-com bubble" in the year 2000. The credit risk bearing iBoxx time series starts in 2001 while the REXP bond index starts in 1992. For any given portfolio X, the expected rate of return $ret(X)$ is estimated from the historical time series of the indices.[17] In Figure 2 the cumulative logarithmic returns of the four stock indices are shown. The Russian Traded Index (RTX) has obviously had the best performance over the 10 year time horizon, however, also with the highest volatility. All calculations are performed to support the allocation decision of the bank's integrated risk manager on 28-FEB-2007.

Fig. 2. Performance of stock indices.

Deviating from the approach in Schlottmann et al. (2005), presented in section 3.4, we calculate the credit risk measure $cr(x)$ (and credit return) solely based on the historical iBoxx time series and not using a portfolio model as we prefer to use a benchmark market instrument in the context of strategic asset allocation. Additionally, we do not account for operational risk capital as the simple Basel standardised approach lacks differentiation concerning the risk capital charges. Instead, we consider two market risk objective functions, namely $mr_1(X)$ as aggregated risk measure for the four stock market indices and $mr_2(X)$ for the REXP index that represents almost risk free bond investments. Notably, in many banks the stock market portfolio is very likely managed separately from the interest bearing bond portfolio. Thus, our approach fits perfectly into the organisational structure of many financial institutions. Moreover, we assume the bank uses historical simulation for the calculation of the function values of $mr_1(X)$, $mr_2(X)$ and $cr(X)$, using a confidence level of 99% and a time horizon of 1 trading day in each separate case. As we employ differing lengths of historical time series we have annualised the re-

---

[17] Note: Possible distortions through the differences between price and performance indices as well as currency effects have been intentionally neglected.

turn figures for each specific portfolio / objective function. Hereby, we ensure that the contributions of each individual instrument to the overall portfolio return are adequately accounted for. Thus, our general optimisation problem as introduced in (8) - (14) is slightly modified by replacing (9) - (11) with

$$f_2(X) := mr_1(X) \tag{17}$$
$$f_3(X) := mr_2(X) \tag{18}$$
$$f_4(X) := cr(X) \tag{19}$$

Results for Sample Bank 2. Due to our four objective functions the results cannot be easily visualised in standard coordinate representation, which is limited to three dimensions. Thus, we have projected our four dimensional results into three dimensions, omitting one dimension each time. Figure 3 displays such a 3-dimensional projection of the objective function values of the final individuals after $1,000$ generation steps in the NSGA-II with 100 individuals per population. The same parameter setting for the ϵ-MOEA delivers a final archive of 103 pareto-optimal solutions which are depicted in Figure 4. It can be clearly seen that the ϵ-MOEA produces much more evenly spread solutions over the objective space, due to the above mentioned ϵ-dominance principle.

In Figure 4 the three components concerning stock index risk $mr_1(X)$, interest rate risk $mr_2(X)$ and expected return $ret(X)$ of the respective approximated portfolio set X are shown. The three single outlier points to the right of the figure represent portfolios with heavy investment in the RTX as these combine the highest returns with by far the highest values of $mr_1(X)$. The bank's risk manager can use this information straightforward to verify the current portfolio allocation of the bank against the drawn portfolios: Assume the bank has a current portfolio status Y. If the risk manager computes $f_i(Y)$ for $i = 1, .., 4$ he can immediately check the bank's current position in the three-dimensional plot. Of course, he also has to check the value $f_4(Y)$ against the objective function values in an additional figure (omitted here) plotting credit risk $cr(X)$ and, e.g., $mr_1(X)$ against the expected return. If the bank's current position Y is dominated by a portfolio X he can directly observe the outcome of possible improvements and derive corresponding managing decisions to move the bank's risk-return profile into an improved position.

It is a striking advantage of the multi-objective view that the risk manager can see the consequences of decision alternatives concerning different risk sources and the corresponding expected rate of return, simultaneously. For instance, in Figure 4 he can choose a portfolio that has high stock market risk $mr_1(X)$ while having very low interest rate risk $mr_2(X)$. Such portfolios are located towards the upper right corner of the figure and yield a high expected rate of return. If he does not desire high stock market risk (e.g., if the bank's trading limits are low), he can choose a portfolio in the left area of the figure, having higher interest rate risk but lower expected return.

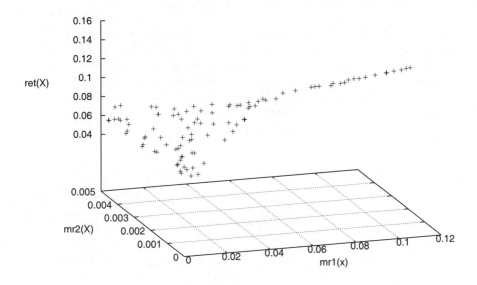

Fig. 3. Projection of $mr_1(X)$, $mr_2(X)$ and $ret(X)$ with NSGA-II.

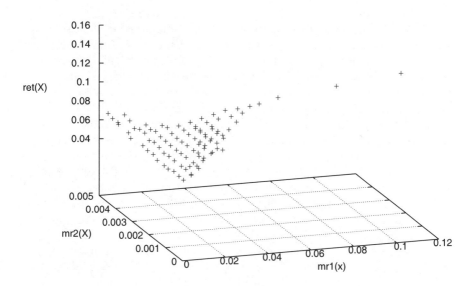

Fig. 4. Projection of $mr_1(X)$, mr_2 and $ret(X)$ with ϵ-MOEA.

3.6 Sample Bank 2 with Volume Constraints

To demonstrate the flexibility of MOEAs in the context of integrated risk management we extend the optimisation problem from section 3.5 with specific constraints on the minimum and maximum holding size of each individual considered asset, representing *one* common limitation in practical applications. The respective limits that may, for instance, be imposed by the regulatory authority or due to bank management preference are shown on the left hand side of Table 3 (*Input Constraints*).

Table 3. Holding size constraints for Sample Bank 2.

#	Index	Input Constraints		Optimisation Results	
		Minimum	Maximum	Minimum	Maximum
1	DAX30	0.05	0.50	**0.05**	0.20
2	ATX	0.00	0.30	0.01	**0.30**
3	S&P500	0.05	0.50	**0.05**	0.31
4	RTX	0.00	0.10	0.00	**0.10**
5	REXP	0.20	1.00	**0.20**	0.77
6	iBoxx overall	0.10	0.80	**0.10**	0.66

For the optimisation we had to modify the simple repair algorithm that was presented in section 3.3 and that was used for the optimisation of the problems in 3.4 and 3.5: In each optimisation run the individual constraints are checked and non-feasible individuals are repaired in a way similar to the simple algorithm. Hereby it is possible to almost completely preserve the speed characteristics of the algorithm without volume constraints.

Results for Sample Bank 2 with Volume Constraints. In Table 3 we have indicated in **bold** which constraints represent true restrictions for the optimisation problem (*Optimisation Results*). Clearly, the maximum volume limits on ATX and RTX as well as the minimum constraints for DAX30, S&P500, REXP and iBoxx both prohibited the retrieval of efficient portfolios with higher expected returns. Subsequently, Figure 5 gives an interesting insight into the trade-off between different sources of risk for the given investment opportunities. It can be observed that when the risk figure $mr_1(X)$ is maximally boosted through stock index investments in the constrained problem this goes in parallel with bond market investments, represented by $mr_2(X)$. Also, we realise a multitude of possible efficient trade-offs between the three risk functions $mr_1(X)$, $mr_2(X)$ and $cr(X)$. Note that in current real-world applications, this trade-off is usually not analyzed in such detail.

To further improve the visualisation of our objective function values and to increase the intuition behind our results we have also illustrated the efficient

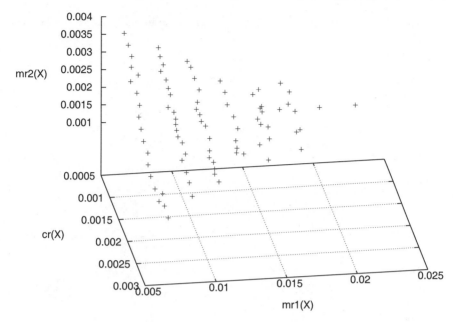

Fig. 5. Projection of $mr_1(X)$, $mr_2(X)$ and $cr(X)$ with ϵ-MOEA and constraints.

allocations in parallel coordinate representation which has been introduced by Inselberg (1985). In the evolutionary multi-objective optimisation context the method has subsequently been applied by Fonseca and Fleming (1993) to visualise trade-offs in their objectives. Purshouse and Fleming (2003) discuss the meaning of parallel coordinates when identifying conflicting goals. Referring to our problem these findings will be discussed in more detail below.

Parallel coordinates enable us to visualise high-dimensional data, i.e. with more than three dimensions, in two dimensions. The different parallel coordinate axes (one for each dimension) are set up equidistantly along one single horizontal axis, with the different objective functions being interchangeable within the representation. According to Purshouse and Fleming (2003) we have normalised each single objective to the interval $[0; 1]$. In parallel coordinates, single solution vectors are represented by a crestfallen line spanning from market risk 1 over market risk 2 and credit risk to the return axis.

Figure 6 consequently shows the parallel coordinate representation for all four objective functions, namely $mr_1(X)$, $mr_2(X)$, $cr(X)$ and $ret(X)$. Given a grayscale spectrum that is based on the return function values, and with the knowledge that conflicting objectives cause many crossing lines and harmonious objectives exhibit more parallel lines (Purshouse and Fleming, 2003), it can be clearly seen that all objectives are exposed to trade-offs in our optimisation problem due to the fact that the lines heavily intersect here. For instance, the grayscale spectrum signals noticeably through the light gray lines that investments into the stock indices (high value for $mr_1(X)$) stimulate the returns.

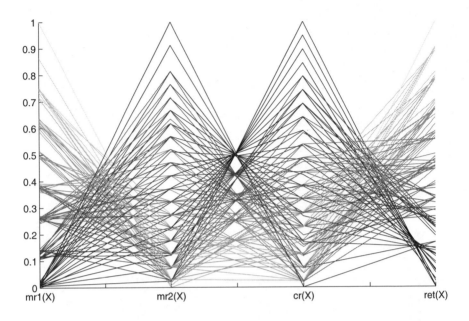

Fig. 6. Parallel coordinate representation for constrained problem.

Finally, we present heatmaps which are a visualisation technique that has recently been introduced to the multi-objective optimisation community (Pryke et al., 2007). In Figure 7 for each individual optimal portfolio all parameter values, i.e. all distinct allocation percentages x_i, as well as all objective function values $ret(X)$, $mr_1(X)$, $mr_2(X)$ and $cr(X)$ are visualised in a grayscale spectrum, ranging from black (0.0) to white (1.0).[18] The allocations have been sorted according to the expected return $ret(X)$ column. The heatmap gives a very good insight into the dependencies within the optimal portfolio structures. Obviously, portfolios with high expected return comprise investments in ATX and RTX. The worst return is obtained through an overweighting of credit risk investments (iBoxx), followed by investments in riskless bonds (REXP) and mixtures of the two. Even in this constrained problem with a minimum holding size of 5%, the DAX still plays a relatively insignificant role as can be derived from the predominantly black column above DAX. Interestingly, but probably due to its high risk, the Russian index comes only very lately into play despite its very good return characteristics. Furthermore, from the bird's eye view on the grayscale spectrum it becomes apparent that $ret(X)$ and stock index risk $mr_1(X)$ exhibit high dependence.

---

[18] Note: All parameters and objective function values are normalised to [0; 1].

The preceding considerations represent a novel approach compared to the current state-of-the-art within the financial industry and the integrated risk management literature. Moreover, it has to be pointed out that the 6 asset class sample might seem small at first glance, however, the risk manager can use more global risk categories representing more asset classes and the MOEA can of course process larger problems. Thus, our approach can be applied even to large portfolios in a top-down approach over different asset classes.

If an aggregate risk number is desired the aggregation of different risk types from the finance point of view can also be integrated into a refined MOEA approach due to its flexibility, for instance, by introducing a further objective function or by replacing formerly disaggregated functions in the optimisation problem. Thus, the multi-objective approach benefits from the development of new financial tools while still being attractive for analyzing the trade-off between different sources of risk and the expected rate of return as pointed out above. In addition, more objective functions of the bank, such as liquidity or reputational risk, which do not necessarily need to possess convenient mathematical properties as well as complex real-world constraints, such as P&L limits, might be incorporated quite easily into our MOEA-based approach in the future.

4 Summary and Outlook

The integrated management of different sources of risk is one of the largest challenges to the financial industry. Regulatory authorities have increased their focus on this matter and will oblige financial institutions to deliver more appropriate aggregate risk numbers in the future. Additionally, an increasingly competitive environment will drive banks to continuously improve their entire business mix, taking all significant risks under reasonable economic considerations into account. Thus, it is decisive to analyse the current bank portfolio adequately and in a way that accurately incorporates institution-wide risk interdependencies.

Concerning optimal strategic asset allocation, which is – next to the sustainment of solvability – one of the two main objectives in the context, we proposed a multi-objective approach that enables banks to maintain their silo-based approach to risk management. In real-world applications each risk category has its own specific properties and is typically measured on a distinct time horizon under an individual confidence level. Our approach does not require the aggregation of incompatible risk figures into a single number. Moreover, it does not necessarily require dependence assumptions between different risk types which are difficult to estimate in real-world applications due to the lack of data. Instead, the risk manager is provided with an unbiased number of solutions which he can use for an analysis of the trade-off between the different risk types and the expected rate of return. Thus, the

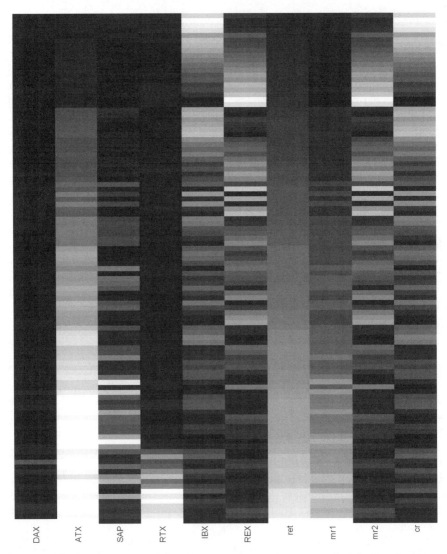

Fig. 7. Heatmap representation for parameter values x_i and objective functions $ret(X)$, $mr_1(X)$, $mr_2(X)$ and $cr(X)$ - ordered by objective $ret(X)$.

manager can either stick to the still widely applied silo-based approach for the integrated management of risk and return, or alternatively, he can employ more advanced integrated risk management techniques, such as copulas, within our framework. Moreover, due to the use of MOEAs in the search of non-dominated portfolios, Value-at-Risk which is commonly used in real-world applications, can be kept as a risk measure in the respective category. To illustrate a real-world application of our approach, we have provided an empirical example using the NSGA-II and the ϵ-MOEA to find approximations of non-dominated portfolios with and without common volume constraints.

For a thorough analysis of the MOEA algorithm performance in this area of application, a more detailed empirical study is necessary. Furthermore, our multi-objective approach to integrated risk management might be an adequate real-world application for an empirical comparison between even more alternative MOEA schemes. A potential improvement of the approximation algorithm in terms of convergence speed could hybridise MOEAs and problem-specific knowledge, confer to the ideas used within the two-objective function approach presented in Schlottmann and Seese (2004a).

Acknowledgements

The authors would like to thank GILLARDON AG financial software for support of their work. Nevertheless, the views expressed in this contribution reflect the personal opinion of the authors and are neither official statements of GILLARDON AG financial software nor of its partners or its clients.

We are also grateful to Kalyanmoy Deb for providing the NSGA-II and ϵ-MOEA source code on the WWW.

References

Alexander, C.: 2003, The present, future and imperfect of financial risk management, *Discussion papers in finance*, ICMA Centre, University of Reading, UK.

Artzner, P., Delbaen, F., Eber, J.M. and Heath, D.: 1999, Coherent measures of risk, *Mathematical Finance* **9**(3), 203–228.

Basel Committee on Banking Supervision: 1988, *Internal Convergence of Capital Measurement and Capital Standards*.

Basel Committee on Banking Supervision: 1996, *Amendment to the Capital Accord to Incorporate Market Risks*.

Basel Committee on Banking Supervision: 2006, *Basel II: International Convergence of Capital Measurement and Capital Standards: A Revised Framework*.

Bluhm, C., Overbeck, L. and Wagner, C.: 2003, *An Introduction to Credit Risk Modeling*, Chapman-Hall, Boca Raton.

Chang, T.-J., Meade, N., Beasley, J.E. and Sharaiha, Y.M.: 2000, Heuristics for cardinality constrained portfolio optimisation, *Computers and Operations Research* **27**, 1271–1302.

Coello Coello, C., Van Veldhuizen, D. and Lamont, G.: 2002, *Evolutionary Algorithms for Solving Multi-objective Problems*, Kluwer, New York.

Committee of European Banking Supervisors: 2006, *Guidelines on the Application of the Supervisory Review Process under Pillar 2*.

Crouhy, M., Mark, R. and Galai, D.: 2001, *Risk Management*, McGraw-Hill.

Cumming, C.M. and Hirtle, B.J.: 2001, The challenges of risk management in diversified financial companies, *FRBNY Economic Policy Review* **7**(1), 1–17.

Deb, K.: 2001, *Multi-Objective Optimisation Using Evolutionary Algorithms*, John Wiley & Sons, Chichester.

Deb, K., Agrawal, S., Pratap, A. and Meyarivan, T.: 2000, A fast elitist non-dominated sorting genetic algorithm for multi-objective optimisation: NSGA-II, *in* K. Deb, E. Lutton, G. Rudolph, H. Schwefel and X. Yao (eds), *Parallel Problem Solving from Nature, LNCS 1917*, Springer, Berlin, pp. 849–858.

Deb, K., Mishra, S. and Mohan, M.: 2003, A fast multi-objective evolutionary algorithm for finding well-spread pareto-optimal solutions, *Techn. report*, KanGAL, Kanpur, India.

Elton, E.J., Gruber, M.J., Stephen, S.J. and Goetzmann, W.N.: 2007, *Modern Portfolio Theory and Investment Analysis*, John Wiley & Sons.

Embrechts, Paul (ed.): 2000, *Extremes and Integrated Risk Management*, Risk-Books, London.

Fonseca, C.M. and Fleming, P.J.: 1993, Genetic algorithms for multiobjective optimization: Formulation, discussion, generalization, *Genetic Algorithms: Proceedings of the Fifth International Conference*, Morgan Kaufmann, pp. 416–423.

Frey, R. and McNeil, A.: 2002, Var and expected shortfall in portfolios of dependent credit risks: conceptual and practical insights, *Journal of Banking and Finance* **26**(7), 1317–1334.

Gaivoronski, A. and Pflug, G.: 2002, Properties and computation of value-at-risk efficient portfolios based on historical data, *Working paper*, Trondheim University.

Gilli, M., Hysi, H. and Këllezi, E.: 2006, A data-driven optimization heuristic for downside risk minimization, *Journal of Risk* **8**(3), 1–19.

Gordy, M.: 2000, A comparative anatomy of credit risk models, *Journal of Banking and Finance* **24**, 119–149.

Gupton, G., Finger, C. and Bhatia, M.: 1997, CreditMetrics, *Techn. report*, JP Morgan & Co., New York.

Inselberg, A.: 1985, The plane with parallel coordinates, *The Visual Computer* **1**(2), 69–91.

Joint Forum of the Basel Committee on Banking Supervision: 2003, *Trends in risk integration and aggregation*.

Jorion, P.: 2006, *Value at risk*, 2 edn, McGraw-Hill, USA.

Kuritzkes, A. and Schuermann, T.: 2007, What we know, don't know and can't know about bank risk, *in* F.X. Diebold and R.J. Herring (eds), *The Known, The Unknown and The Unknowable in Financial Risk Management*, Princeton University Press, p. forthcoming.

Kuritzkes, A., Schuermann, T. and Weiner, S.: 2002, Risk measurement, risk management and capital adequacy in financial conglomerates, *Working paper*, Wharton Financial Institutions Center.

Maringer, D.: 2005, *Portfolio Management with Heuristic Optimization*, Vol. 8 of *Advances in Computational Management Science*, Springer.

Markowitz, H.: 1952, Portfolio selection, *Journal of Finance* **7**(1), 77–91.

McNeil, A., Frey, R. and Embrechts, P.: 2005, *Quantitative Risk Management: Concepts, Techniques, and Tools*, Princeton University Press.

Pézier, J.: 2003, Application-based financial risk aggregation methods, *Discussion Papers in Finance 11*, ICMA Centre, University of Reading, UK.

Pryke, A., Mostaghim, S. and Nazemi, A.: 2007, Heatmap visualization of population based multi objective algorithms, *Proceedings of the Evolutionary Multi-Criterion Optimization (EMO) 2007 Conference*, Vol. 4403 of *LNCS*, Springer, pp. 361–375.

Purshouse, R.C. and Fleming, P.J.: 2003, Conflict, harmony, and independence: Relationships in evolutionary multi-criterion optimisation, *in* K. Deb, P.J. Fleming, C.M. Fonseca, L. Thiele and E. Zitzler (eds), *Evolutionary Multi-Criterion Optimization, Second International Conference*, Lecture Notes in Computer Science, Springer, Faro, Portugal, pp. 16–30.

Rachev, S.T., Menn, C. and Fabozzi, F.J.: 2005, *Fat-Tailed and Skewed Asset Return Distributions: Implications for Risk Management, Portfolio Selection, and Option Pricing*, Wiley, Chichester.

Rockafellar, R.T. and Uryasev, S.: 2002, Conditional value-at-risk for general loss distributions, *Journal of Banking & Finance* **26**(7), 1443–1471.

Rosenberg, J.V. and Schuermann, T.: 2006, A general approach to integrated risk management with skewed, fat-tailed risks, *Journal of Financial Economics* **79**(3), 569–614.

Saita, F.: 2004, Risk capital aggregation: the risk managers perspective, *Working paper*, Newfin Research Center and IEMIF.

Schlottmann, F., Mitschele, A. and Seese, D.: 2005, A multi-objective approach to integrated risk management, *in* A.H. Aguirre, C. Coello Coello and E. Zitzler (eds), *Evolutionary Multi-Criterion Optimization, Third International Conference*, Lecture Notes in Computer Science, Springer, Guanajuato, Mexico, pp. 692–706.

Schlottmann, F. and Seese, D.: 2004a, A hybrid heuristic approach to discrete portfolio optimization, *Computational Statistics and Data Analysis* **47**(2), 373–399.

Schlottmann, F. and Seese, D.: 2004b, Modern heuristics for finance problems: a survey of selected methods and applications, *in* C. Marinelli and S.T. Rachev (eds), *Handbook on Numerical Methods in Finance*, Springer, pp. 331–360.

Schlottmann, F. and Seese, D.: 2005, Financial applications of multi-objective evolutionary algorithms: Recent developments and future research, *in* C. Coello Coello and G. Lamont (eds), *Handbook on Applications of Multi-*

Objective Evolutionary Algorithms, World Scientific, Singapore, pp. 627–652.

Seese, D. and Schlottmann, F.: 2003, *The Building Blocks of Complexity: A Unified Criterion and Selected Applications in Risk Management.* Complexity 2003: Complex behaviour in economics, Aix-en-Provence.

Winker, P.: 2001, *Optimization Heuristics in Econometrics: Applications of Threshold Accepting*, Wiley&Sons, Chichester.

A Stochastic Monetary Policy Interest Rate Model

Claudio Albanese[1] and Manlio Trovato[2]

[1] Level3finance, London, United Kingdom `claudio@level3finance.com`
[2] Merrill Lynch, 2 King Edward Street, London EC1A 1HQ and Imperial College, Department of Mathematics, London SW7 2AZ, United Kingdom `manlio_trovato@ml.com`

Summary. A new stochastic monetary policy interest rate term structure model is introduced within an arbitrage-free framework. The 3-month spot LIBOR rate is taken as modelling primitive and the model is constructed with local volatility, asymmetric jumps, stochastic volatility regimes and stochastic drift regimes. This can be done in an arbitrage-free framework as the chosen modelling primitive is not an asset price process and therefore its drift is not constrained by the no-arbitrage condition. The model is able to achieve a persistent smile structure across maturities with a nearly time homogeneous parameterisation and explain steep, flat and inverted yield curves, consistently with historical data. It is shown that the drift process, which is made correlated to the underlying LIBOR rate, is the main driver for long time horizons, whilst jumps are predominant at short maturities and stochastic volatility has the greatest impact at medium maturities. As a consequence, the drift regime process provides a powerful tool for the calibration to the historical correlation structure and the long dated volatility smile. The model is solved by means of operator methods and continuous-time lattices, which rely on fast and robust numerical linear algebra routines. Finally, an application to the pricing of callable swaps and callable CMS spread range accruals is presented and it is shown that this modelling framework allows one to incorporate economically meaningful views on central banks' monetary policies and, at the same time, provides a consistent arbitrage-free context suitable for the pricing and risk management of interest rate derivative contracts.

Key words: Interest rate models, monetary policy, stochastic drift, stochastic volatility, operator methods, continuous-time lattice, Markov generator, volatility smile, callable swap, callable CMS spread.

1 Introduction

The view on central banks' monetary policies has a crucial impact on the risk taking and hedging strategies of financial institutions. Surprisingly, most commonly used pricing models do not allow for a direct specification of such a view. In this work, we present a novel approach to interest rate modelling, which incorporates stochastic monetary policy.

Taking inspiration on short rate models, previously discussed by Vasicek (1977); Cox et al. (1985); Black and Karasinski (1991); Hull and White (1993), we choose a modelling primitive which is not an asset price process. The main consequence of this choice is that the drift term is not implied by the specification of the volatility term and the no-arbitrage condition. We exploit this property and introduce an additional degree of freedom by making the drift a stochastic process which can be directly specified.

Furthermore, similarly to what is done in the so called *LIBOR Market Models* by Brace et al. (1996), we choose to work with a simple compounding rate, rather than an instantaneous rate, as this is a market observable quantity. We take the 3-month LIBOR spot rate L_t as our modelling primitive. The process for L_t can formally be written as follows:

$$dL_t = \mu_{a_t}(L_t)dt + \sigma_{b_t}(L_t)dW_t + \text{jumps}, \tag{1}$$

where W_t is a one-dimensional Brownian motion and the drift and the volatility terms are stochastic and driven by the processes a_t, b_t correlated to the rates themselves.

We enforce a nearly time-homogeneous parameterisation and show that the model is able to generate a persistent smile across tenors and maturities, in good qualitative agreement with the EUR market. We also show that the model is able to explain steep, flat and inverted yield curve shapes and we find that these shapes are considerably affected by the drift regime process, as one would intuitively expect. Although we find that a limitation of the model is that it can not obtain humped yield curve shapes, our model compares favourably with time-homogeneous short rate models, which have much greater limitation in the shapes that they can reproduce. For example, Brigo and Mercurio (2006) show that in the Vasicek model it is not possible to obtain inverted curves, no matter the values of the model parameters. We show how the drift regimes can be used as a direct control for the calibration of the long dated portion of the volatility smile in high and low strike regions without considerably affecting the nature of the process at earlier maturities. In fact, Konikov and Madan (2001) explain that jumps are predominant at short maturities and it is well known that stochastic volatility persists for longer time span. Our finding is that the drift process, associated with the first moment of L_t, has a predominant impact at long maturities. The three different components can therefore be effectively used to drive the process at different time spans whilst retaining a time homogeneous parameterisation.

Our model is solved by means of continuous-time finite-state Markov chains. We use Markov chains in two conceptually distinct ways: as an approximation to the continuum limit and as a modelling framework in its own right. In particular, the first approach is used for the approximation of the conditional local volatility diffusion process. Whereas the drift and volatility regime processes are constructed with a direct specification of the transition probabilities themselves. The combined process is therefore the result of an hybrid approach.

Discretisation schemes of the first type have been previously discussed in Albanese and Kuznetsov (2005) and Albanese and Kuznetsov (2003), and applied to models for which the spectrum of the Markov generator can be computed in analytically closed form. Whilst retaining the insights provided by their spectral analysis treatment, here we choose to do without analytic solvability and follow instead a non-parametric approach, similarly to what was proposed by Albanese and Trovato (2005). The idea is to leverage not on the ability to evaluate special functions but instead on numerical linear algebra routines in order to compute a matrix exponentiation. The discretisation scheme is based on operator methods and level-3 numerical linear algebra (i.e., matrix-matrix multiplication), which can be performed with robust and efficient third party libraries, available with both freeware and commercial licensing. A popular and well established freeware for numerical linear algebra is ATLAS (Automatically Tuned Linear Algebra Software), available from the *sourceforge* web site and discussed by Whaley and Petitet (2005). Markov chain modelling of the second type has recently been explored by Albanese and Trovato (2005), in the context of interest rates, and Di Graziano and Rogers (2006) and Albanese and Vidler (2007) in the context of credit modelling.

The introduction of stochastic monetary policy in interest rate modelling is the main technical contribution of this work. Furthermore, the use of discretization schemes based on continuous-time finite-state Markov chains and operator methods also represents a distinctive feature of our approach.

The rest of the paper is organised in two main parts: the first part sets out the mathematical framework for continuous-time lattices. Here we describe the fundamental concepts for the construction of approximating Markov chains. We take a jump diffusion process as case study and discuss the construction of the Markov generator, the numerical method for matrix exponentiation, the computation of the transition probability kernel, and the lattice convergence properties. In the second part we describe the interest rate model and its calibration to the implied volatility smile and correlation. Finally we show an application to callable swaps and callable CMS spreads range accruals and analyse the impact of drift regimes on the price functions and exercise boundaries.

2 Continuous-time Lattices

In the construction of our interest rate model we work with continuous-time Markov chains with finitely many states. In Appendix A, we recall the basic concepts of Markov processes and Markov generators, which play a pivotal role in the specification of our model.

In this section, we describe the construction of the approximating Markov chain for jump diffusion processes with continuous-time lattices. We start with a diffusion process and proceed with the construction of its discretized Markov generator. We then overlay jumps on the original process by means of Bochner subordinators, and show how to construct the discretized Markov generator which incorporates both the diffusion and jump parts. Finally, we show how to compute the transition probability matrix on the lattice by means of numerical matrix exponentiation, and discuss convergence properties.

2.1 Local Volatility Diffusion Processes

We consider a generic asset price process F_t as a modelling primitive and assume that it satisfies the following SDE for a diffusion process:

$$dF_t = \mu(F_t)dt + \sigma(F_t)dW_t, \tag{2}$$

where W is a one dimensional Brownian motion and both $\mu(F_t)$ and $\sigma(F_t)$ are assumed to be continuous.

We consider the probability density function $u(f, t; F, T)$ of the diffusion process given by eq. (2), where $f, F \in \mathbb{R}$. If $\sigma^2 > 0$, then the probability density function $u(f, t; F, T)$ is known to be the unique solution of the backward Kolmogorov equation, as described in Karatzas and Shreve (1991). In particular, as a function of the *backward* variables f and t, the density satisfies:

$$\frac{\partial u}{\partial t} + \mathcal{L}u = 0 \qquad\qquad \text{for } t < T, \tag{3}$$

$$u(f, t; F, T) = \delta(f - F), \qquad\qquad \text{at } t = T, \tag{4}$$

where the Markov generator $\mathcal{L}$, defined on a continuum state space, acts on a twice differentiable function u in the following way, (see Friedman (1975, 1976); Karatzas and Shreve (1991)):

$$(\mathcal{L}u)(F_t) = \mu(F_t)\frac{\partial u}{\partial F}(F_t) + \frac{1}{2}\sigma^2(F_t)\frac{\partial^2 u}{\partial F^2}(F_t). \tag{5}$$

The solution to eq. (3), subject to the terminal condition (4) can be formally written as:

$$u(f, t; F, T) = e^{\mathcal{L}(T-t)}(f, F), \tag{6}$$

where in this notation we mean that the operator $e^{\mathcal{L}(T-t)}$ is applied to the variables (f, F). It is therefore obvious that, in order to solve for the probability density function u, also referred to as pricing kernel, all we need is the specification of the Markov generator $\mathcal{L}$. We first discuss the specification of the Markov generator, then in section (2.3) we describe the computation of the transition probability matrix.

Markov Generator Construction

We proceed with the construction of the approximating finite-state Markov chain that discretises the local volatility diffusion process (2). The construction is based on the specification of the local properties of the Markov chain, and in particular on the specification of the Markov generator. Kushner and Dupuis (2001) prove that, if the Markov generator satisfies certain local consistency conditions, then the resulting Markov chain is a correct approximation of the original process, meaning that the discretised process converges to the continuous limit process in a weak or distributional sense. Once the Markov generator is correctly specified, the transition probabilities between any two points in the continuous-time lattice can be computed by calculating the exponential of the Markov generator matrix itself (eq. 6).

We define a discrete domain for the forward rate process F_t. Let the state space be $\Omega = \{0, 1, 2, ..., N\}$ and $F : \Omega \to \mathbb{S}$ be a monotone mapping, where $\mathbb{S}$ is a countable set of $\mathbb{R}$. The map F defines the discrete grid for all possible values that the discretized version $F_{\Omega t}$ of the process F_t can take on the lattice Ω. We note that the grid is not necessarily uniform, therefore it is possible to adopt a judicious and efficient node placing scheme. In particular, for a given total number of nodes N, one can improve the convergence of the discretized process by reducing the node spacing in the region close to a guessed expected value of F_t, or alternatively around the spot value f, and making use of a sparser grid in regions of very high and low levels in F_t.

It can be shown, as discussed in Albanese and Mijatovic (2006) and Kushner and Dupuis (2001), that an appropriate Markov chain approximation of the diffusion process can be constructed by imposing the following local consistency conditions:

$$\mathbb{E}_t \left[F_{t+dt} - F_t \right] = \mu(F_t)dt, \tag{7}$$

$$\mathbb{E}_t \left[(F_{t+dt} - F_t)^2 \right] = \sigma^2(F_t)dt. \tag{8}$$

By satisfying the above two equations in our discretization scheme we enforce that the first two moments in the discrete system match those in the continuum state space.

Theorem 1. *Let* $\mathcal{L}_\Omega$ *be the following discretization of* $\mathcal{L}$:

$$\mathcal{L}_\Omega := \mu \nabla_h + \frac{\sigma^2}{2} \Delta_h, \tag{9}$$

where ∇_h *and* Δ_h *are the discrete finite difference and Laplace operator respectively, defined for a lattice spacing* $h = F(x) - F(x-1)$, *with* $x \in \Omega - \{0\}$ *as follows:*

$$\nabla_h = \frac{F(x+1) - F(x-1)}{2h}, \tag{10}$$

$$\Delta_h = \frac{F(x+1) + F(x-1) - 2F(x)}{h^2}, \tag{11}$$

Then one can write the local consistency conditions (7) and (8) in terms of the discretised Markov generator $\mathcal{L}$ as follows:

$$\sum_{y \in \Omega} \mathcal{L}_\Omega(x,y)(F(y) - F(x)) = \mu(F(x)), \tag{12}$$

$$\sum_{y \in \Omega} \mathcal{L}_\Omega(x,y)(F(y) - F(x))^2 = \sigma^2(F(x)). \tag{13}$$

Proof. A sketch of the proof is reported in Appendix B.

Equations (12) and (13), together with the probability conservation condition

$$\sum_{y \in \Omega} \mathcal{L}_\Omega(x,y) = 0, \tag{14}$$

uniquely identify the Markov generator for the diffusion process. This follows from the fact that the matrix $\mathcal{L}_\Omega$ is tridiagonal and can thus be computed as the solution of the system of linear equations (12, 13, 14). This is also a consequence of the fact that, locally, we have imposed first and second moment matching conditions only.

2.2 Adding Jumps by Subordination

At this stage of the construction one can add jumps by means of Bochner subordinators. In this section we show the numerical procedure for the construction of the discretized subordinated Markov generator, having computed the discretized Markov generator for a diffusion process. But, before we proceed, we recall a fundamental theorem of functional calculus, which we are going to need in our algorithm.

Theorem 2. *If $\mathcal{L}_\Omega \in \mathbb{C}^{n \times n}$, with $\mathcal{L}_\Omega = U\Lambda U^{-1}$ and $\Lambda = diag(\lambda_1, \lambda_2, .., \lambda_n)$, then a function φ, defined on the spectrum of $\mathcal{L}_\Omega$, can be applied to the matrix $\mathcal{L}_\Omega$ by means of the following formula:*

$$\varphi(\mathcal{L}_\Omega) = U\varphi(\Lambda)U^{-1}. \tag{15}$$

As Λ is a diagonal matrix, the calculation of $\varphi(\Lambda)$ is a very simple task:

$$\varphi(\Lambda) = \begin{pmatrix} \varphi(\lambda_1) & 0 & \cdots & 0 & 0 \\ 0 & \varphi(\lambda_2) & \cdots & 0 & 0 \\ 0 & 0 & \ddots & 0 & 0 \\ \vdots & \vdots & \cdots & \ddots & \vdots \\ 0 & 0 & \cdots & 0 & \varphi(\lambda_n) \end{pmatrix}. \tag{16}$$

The formula also extends to the continuum case, as described by Phillips (1952).

Bochner subordinators are a special type of time changed processes, first introduced by Bochner (1955), where the time changed process is independent of the original process being subordinated. The technique of time changes have been extensively used in finance both for describing jump diffusions and stochastic volatility, and it has been discussed by Geman et al. (2003, 2001). A comprehensive survey on time changes used in finance can be found in Geman (2005). We restrict our attention to subordinators.

Definition 1. *The process T_t is called a subordinator if it is a right-continuous non decreasing process with values in $\mathbb{R}^+$ such that $T_0 = 0$ and it has independent and homogeneous increments.*

If we choose a subordinator T_t with discontinuous paths, then we can construct a new process $Y_t = F_{T_t}$, which can be naturally interpreted as a jump diffusion process. Possible choices for T_t could be, for example, the Poisson or the Gamma process.

In our modelling framework, we work with the local properties of the process, so we are interested in the construction of the Markov generator for Y_t. The relation between the transition probability densities of F_t and Y_t allows one to compute the relationship between the Markov generators for F_t and Y_t.

If the Markov generator $\mathcal{L}$ of the diffusion process F_t can be diagonalised, then $\mathcal{L}(f, F) = \sum_n \lambda_n u_n(f) v_n(F)$, where $f, F \in \mathbb{R}$ and λ_n, u_n and v_n are the eigenvalues, the right eigenvectors and the left eigenvectors of $\mathcal{L}$ respectively. In this case the transition probability density function for the process F_t can be formally written as follows:

$$u(f, 0; F, \tau) = e^{\mathcal{L}\tau}(f, F) = \sum_n e^{\lambda_n \tau} u_n(f) v_n(F), \qquad (17)$$

where $\tau = T - t$. Also, the transition probability u^j of the subordinated process Y_t is given by:

$$u^j(f, 0; F, \tau) = \mathbb{E}\left[\sum_n e^{\lambda_n T_\tau} u_n(f) v_n(F)\right] = \sum_n \mathbb{E}\left[e^{\lambda_n T_\tau}\right] u_n(f) v_n(F), \quad (18)$$

where the first equality follows from the subordinating procedure and the second from the linearity of the expectation operator. For particular choices of the process T_t, Bochner (1955) shows that it is possible to find a function $\phi(\lambda)$ of the form:

$$\phi(\lambda) = c\lambda + \int_0^\infty \left(1 - e^{\lambda x}\right) d\nu(x), \qquad (19)$$

such that:

$$\mathbb{E}\left[e^{-\lambda T_t}\right] = e^{-t\phi(\lambda)}, \qquad (20)$$

where $c > 0$ and ν is a positive measure with $\int_0^\infty x\nu(dx) < \infty$. A function $\phi(\lambda)$ of the form (19) that satisfies eq. (20) is called the *Bernstein* function of T_t. If T_t admits a *Bernstein* function $\phi(\lambda)$, then the transition probability density of the jump diffusion process Y_t can be simply calculated as follows:

$$u^j(f, 0; F, \tau) = \sum_n e^{-\tau\phi(-\lambda_n)} u_n(f) v_n(F). \tag{21}$$

It follows from the fundamental result of functional calculus (2) that the Markov generator $\mathcal{L}^j$ of the subordinated process Y_t is simply given by:

$$\mathcal{L}^j = -\phi(-\mathcal{L}). \tag{22}$$

Markov Generator Construction

Eq.(22) allows one to calculate, in the continuum limit, the Markov generator of the jump diffusion process from the Markov generator of the diffusion process and the *Bernstein* function associated to the chosen Bochner subordinator. In a finite state space Ω we can apply the following numerical algorithm. First, the matrix $\mathcal{L}_\Omega$ is diagonalised by means of numerical routines:

$$\mathcal{L}_\Omega = U\Lambda U^{-1}, \tag{23}$$

where Λ is a diagonal matrix containing the eigenvalues of $\mathcal{L}_\Omega$ on its main diagonal and U is the matrix having as columns the right eigenvectors. We note that this computation is stable as $\mathcal{L}_\Omega$ is symmetric by construction (see 2.4). Then we apply the *Bernstein* function to the matrix $\mathcal{L}_\Omega$ by taking advantage of theorem (2):

$$\mathcal{L}_\Omega^j = -\phi(-\mathcal{L}_\Omega) = -U\phi(-\Lambda)U^{-1}. \tag{24}$$

The matrix $\mathcal{L}_\Omega^j$ is a Markov generator for a diffusion process with jumps. This describes the transition of the process between both contiguous and non contiguous states. Hence the matrix $\mathcal{L}_\Omega^j$ has non zero entries also in the elements off the three leading diagonals. Although in this case it is not possible to take advantage of the numerical optimisations that may have been used when dealing with the tridiagonal matrices of diffusion processes, the dimensionality of the problem has not increased and therefore the overall performance of the discretization scheme is only marginally affected.

Having constructed a Markov generator for a diffusion process with jumps, our next task is the computation of the transition probability matrix.

2.3 Transition Probabilities

It follows form eq. (6) that the discretized transition probability kernel defined on the lattice can be formally obtained by computing the exponential of the Markov generator matrix itself:

$$u_\Omega(x, t; y, T) = e^{\mathcal{L}_\Omega(T-t)}(x, y). \tag{25}$$

where $x, y \in \Omega$. The numerical algorithm for the computation of the matrix exponential is described in the next section (2.4).

Also, for the interested reader, it may be instructive to see a simple example on how to use continuous-time lattices in practice. We do this in Appendix C, where we discuss the pricing of call options with a one dimensional continuous-time lattices approximating a diffusion process.

2.4 Matrix Exponentiation

In the discretisation schemes discussed in Albanese and Kuznetsov (2003), the computation of the exponential of the Markov generator matrix was performed uniquely by an analytical spectral decomposition of the Markov generator. This represented a limitation, as the analytical spectral decomposition is only possible for a limited range of processes. Instead we follow a non parametric approach and propose to perform the exponentiation numerically. This simple observation has an important impact for the modeller, as it allows for a considerably enhanced modelling flexibility. Obviously, the drawback being the fact that the method relies on fast and accurate numerical methods.

A comprehensive review on available methods for the computation of the exponential of a matrix is Moler and Van Loan (2003). Unfortunately there is no single method that appears to work well in all situations and the classification criterion for selecting the most appropriate method is not straightforward. We have experimented two particular methods: the diagonalisation method and the scaling and squaring method. Consistently with what reported in Moler and Van Loan (2003), we have verified that the former can only be reliably used for symmetric matrices. Instead, when dealing with asymmetric matrices, the scaling and squaring method appears more robust. For applications of our interest, the calculation of the transition probability kernel for models based on diffusion processes can be safely solved with the diagonalisation technique; however for more complex dynamics the scaling and squaring method is more stable and therefore represents our preferred choice.

We discuss our chosen numerical libraries and review the above mentioned two methods in the following three sections.

Numerical Libraries

The solution of the exponential of the Markov generator matrix relies on fast and robust numerical libraries. Our choice has been the popular BLAS and LAPACK libraries.

The BLAS (Basic Linear Algebra Subprograms) are routines that provide standard building blocks for performing basic vector and matrix operations. The Level 1 BLAS perform scalar, vector and vector-vector operations, the

Level 2 BLAS perform matrix-vector operations, and the Level 3 BLAS perform matrix-matrix operations.

LAPACK stands for Linear Algebra PACKage, it is written in Fortran77 and provides routines for solving systems of simultaneous linear equations, least-squares solutions of linear systems of equations, eigenvalue problems, and singular value problems. The associated matrix factorizations (LU, Cholesky, QR, SVD, Schur, generalized Schur) are also provided, as are related computations such as reordering of the Schur factorizations and estimating condition numbers. Dense and banded matrices are handled, but not general sparse matrices. In all areas, similar functionality is provided for real and complex matrices, in both single and double precision.

Both BLAS and LAPACK ara available from the *netlib* web site.

Diagonalisation

Given a discretized Markov generator $\mathcal{L}_\Omega$ defined on a finite lattice Ω containing N points, we consider the following pair of eigenvalue problems:

$$\mathcal{L}_\Omega u_n = \lambda_n u_n, \qquad \mathcal{L}_\Omega{}^T v_n = \lambda_n v_n, \qquad (26)$$

where the superscript ^{T} denotes matrix transposition, u_n and v_n are the right and left eigenvectors of $\mathcal{L}_\Omega$, respectively, whereas λ_n are the corresponding eigenvalues. Except for the simplest cases, the Markov generator $\mathcal{L}_\Omega$ is not a symmetric matrix, hence u_n and v_n are different. Also, in general, the eigenvalues are not real. We are only guaranteed that their real part is non-positive $\mathcal{R}e\lambda_n \leq 0$ and that complex eigenvalues occur in complex conjugate pairs, in the sense that if λ_n is an eigenvalue then its conjugate $\bar{\lambda}_n$ is also an eigenvalue.

The diagonalization problem can be rewritten in the following matrix form:

$$\mathcal{L}_\Omega = U \Lambda U^{-1}, \qquad (27)$$

where U is the matrix having as columns the right eigenvectors and Λ is the diagonal matrix having the eigenvalues λ_i as elements. The numerical implementation of the diagonalization problem can be efficiently carried out by means of routines such as geev in LAPACK, however care must be taken in analysing the stability and accuracy of the spectrum. The method works well for symmetric matrices, as the eigenvectors can be chosen orthogonal. Theoretically the method breaks down when $\mathcal{L}_\Omega$ does not admit a complete set of linearly independent eigenvectors. However, numerically, the method becomes unstable when the conditioning number $cond(U) = \|U\| \|U^{-1}\|$ of the matrix of eigenvector is large. Assuming that $\mathcal{L}_\Omega$ can be decomposed in eigenvalues and eigenvectors, its exponential can be trivially computed by making use of theorem (2) and nothing that:

$$e^{t\mathcal{L}_\Omega} = U e^{t\Lambda} U^{-1}. \qquad (28)$$

Fast Exponentiation

The scaling and squaring method, also known as *fast exponentiation method*, is based on the following fundamental property of the exponential function:

$$e^{t\mathcal{L}_\Omega} = \left(e^{\delta t \mathcal{L}_\Omega}\right)^{\frac{t}{\delta t}}. \tag{29}$$

We choose $\delta t > 0$ so that it is possible to correctly compute the elementary propagator by means of a Taylor approximation:

$$u_{\delta t}(x,y) = \delta_{xy} + \delta t \mathcal{L}(y,y). \tag{30}$$

We then compute $e^{t\mathcal{L}_\Omega}$ by evaluating in sequence $u_{2\delta t} = u_{\delta t} \cdot u_{\delta t}$, $u_{4\delta t} = u_{2\delta t} \cdot u_{2\delta t}$, ... $u_{2^n \delta t} = u_{2^{n-1}\delta t} \cdot u_{2^{n-1}\delta t}$. We find that a suitable choice of δt is the largest positive time interval for which both of the following properties are satisfied:

$$\min_{y \in \Omega}(1 + \delta t \mathcal{L}(y,y)) \geq 1/2, \tag{31}$$

$$\log_2 \frac{t}{\delta t} = n \in \mathbb{N}. \tag{32}$$

Assuming a naive triple loop implementation, a matrix-matrix multiplication requires $\mathcal{O}(N^3)$ floating point operations, where N is the dimension of the matrix. Therefore a naive implementation of the numerical calculation of the matrix exponential would require $\mathcal{O}(\log_2 \frac{t}{\delta t} N^3)$ operations. However, matrix-matrix multiplication can be accelerated both by means of clever algorithms and more efficient hardware and software architecture. In particular, it has been shown that matrix-matrix multiplication algorithm scales no slower than $\mathcal{O}(N^{2.376})$, if this is performed by means of the asymptotic acceleration technique described in Coppersmith and Winograd (1987). Also, by using efficient numerical libraries of Level-3 BLAS, one can take advantage of processor-specific optimisation and clever cache management. What's more, it is possible to further improve code execution performance by means of the emergent technology of hardware acceleration on GPUs (Graphic Processing Unit), which are designed to optimise linear algebra operations on low cost hardware. Finally, it should be noted that, the advantage of encapsulating the bottleneck of the numerical algorithm in the repeated matrix-matrix multiplication is that it is possible to considerably improve the overall performance of the discretization scheme by optimising only a small section of the code.

2.5 Convergence Properties

Diffusions

For the case of diffusion processes, two types of convergence properties have been shown to hold: convergence of the finite dimensional distribution and weak convergence, or convergence in distribution.

For the former case, it has been shown that the transition probability kernel of the discretised process converges to the probability density function of the diffusion process, when the discretisation step h tends to zero (see Albanese and Mijatovic (2006)). The convergence rate has been shown to be $O(h^2)$.

For a slight different Markov chain approximation, where both the state space and time are discretised, Kushner and Dupuis (2001) show that the discretised Markov chain converges *in distribution* to the diffusion process. This is a stronger result, as it implies the convergence between probability measures, however in this case convergence rate estimations are not easily obtainable.

In both cases, it has been shown that, if a central difference approximation is used for the discrete first difference operator, then the following condition must hold:

$$\inf_x \left[\sigma^2(F(x)) - h\left|\mu(F(x))\right| \right] \geq 0, \tag{33}$$

where h is the lattice spacing and $x \in \Omega$.

Jump Diffusions

For the case of jump diffusions of section (2.2), to the best of our knowledge, convergence results have not been derived yet. However we note that an alternative construction of the jump diffusion approximating Markov chain is considered in Kushner and Dupuis (2001), where it is shown that the discretised process admits a continuous limit and that the resulting discretised process converges to this limit in distributional sense.

3 The Interest Rate Model

3.1 Desiderata for an Interest Rate Term Structure Smile Model

The large variety of payoffs available in the interest rate derivatives market means that a trading desk builds up exposure not only to interest rate movements, but also to several other derived state variables, some of which are not easily observable on the market. For consistent pricing and robust hedging one would want to try and make use of the same model across all products, and retain efficient numerical solubility and stable hedge ratios computation. Whilst we do not engage in a discussion on what should be the selection criterion for a good interest rate model, which has been an active debate in the literature, we note that an interest rate derivatives desk of a major financial institution is likely to build up exposure to:

- interest rates movements;

- at-the-money implied volatility movements;
- movements of the liquid portion of the volatility smile, i.e., strike levels around the at-the-money forward;
- movements of the asymptotic volatility smile, i.e., in regions of very high and very low strike levels;
- forward volatility movements;
- forward rates correlation movements.

Interest rates and at-the-money implied volatilities are observable in the market, as is the liquid portion of the smile. The latter is driven by European swaption contracts and, as an example, for the current EUR market it spans a strike range between 2% and 8% approximately. The asymptotic smile spans extreme strike levels and is driven primarily by constant maturity swaps (CMS). These exotic structures receive a floater, the coupon being a spread over LIBOR, and pay the equilibrium swap rate of a fixed tenor prevailing at each coupon date. An analysis of the CMS leg coupon structure leads to the conclusion that CMS contracts are particularly sensitive to the asymptotic behavior of implied volatilities for very large strikes. Therefore market CMS rates are an average indication of the asymptotic smiles and they suggest that implied volatilities flatten out and converge asymptotically to a constant. Forward volatility is not directly observable in the interest rates market, apart from some fairly illiquid products, and implied forward rates correlation is only observable for a few forward rate pairs over a limited range of strikes.

A consistent model aims to recover all market prices and observables and can be used to extrapolate the values for those quantities that, although not directly observable on the market, play an important role in the pricing and risk management of some interest rate derivative products. Also, ideally, a consistent model should be able to describe the market over time with a parameter set that can stay almost constant.

In the construction of our model we make every effort to retain the economic interpretation of the assumptions imposed on the process, hoping that this would lead to an economically meaningful extrapolation of the model derived quantities.

The first goal of our model construction is to try and describe consistently both the liquid and the asymptotic portion of the smile, in good qualitative agreement with the market. The rich model parameterization and the ability to control the transition probability of the spot rate to span and stay in different value regions allow us to achieve this goal.

The second goal is to ensure that the model is qualitatively consistent with historical data. In our calibration exercise we ensure that the yield curve shapes obtained by the model, as a function of the initial condition, are consistent with historical market scenarios. The third goal is to ensure that the model implies realistic values for the implied forward volatility. For this we use time-homogeneity as a guiding criterion of our methodology. In fact, in a time-homogeneous model the implied forward smile is necessarily similar to the spot smile. This is a desirable feature for an interest rate model.

The fourth goal is to recover the historical correlation structure between forward rates. The drift regime process provides the necessary control for this.

3.2 Historical Time Series Analysis

A qualitative analysis of the historical time series for the 3-month LIBOR spot rate of four major currencies (EUR, USD, GBP and JPY) gives us the necessary insight for the specification of our stochastic monetary policy model. From Figure 1 it is obvious that the 3-month LIBOR spot rate is characterised by trends which may last for periods that span multiple years. One can easily identify falling, stable and rising trends, in the EUR, GBP and USD time series. In our model, these trends are associated with drift regimes corresponding to falling, stable and rising rates. The JPY time series is interesting because it spans regions not explored by the EUR, GBP and USD time series, and in particular regions where the rate is close to zero. When the rate approaches these extremely low levels, it is subject to deflationary pressure and it may be constrained to stay at such low levels for an extended period of time, as it has happened for the Japanese economy between 2000 and 2005. In our model, we include a deflationary drift regime which is associated to this phenomenon.

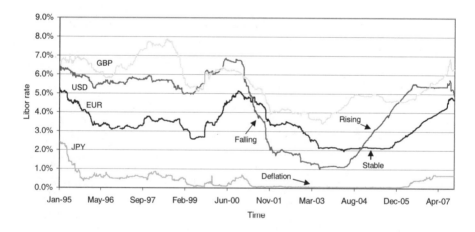

Fig. 1. Spot 3m LIBOR rate historical time series.

3.3 The Modelling Framework

We proceed with the specification of our modelling framework. We define a tenor structure t_i, $i \in \{0, 1, ..., M\}$ such that $\tau = t_j - t_{j-1} = 0.25$ for each

$j \in \{1, ..., M\}$, and take the 3-month LIBOR spot rate L_t as our modelling primitive. The term structure of discount bonds is recovered through a model dependent formula:

$$Z(t_i, t_j) = \mathbb{E}_{t_i} \left[\prod_{k=i}^{j-1} \frac{1}{1 + \tau L_{t_k}} \right], \tag{34}$$

where the expectation is taken in the 3-month spot LIBOR measure, $Z(t_i, t_j)$ is the value at t_i of a zero coupon bond paying one currency unit at t_j and L_{t_k} is the prevailing 3-month LIBOR spot rate at time t_k. The process for L_t can formally be written as follows:

$$dL_t = \mu_{a_t}(L_t)dt + \sigma_{b_t}(L_t)dW_t + \text{jumps}, \tag{35}$$

where the drift and the volatility terms are stochastic and are driven by the processes a_t, b_t correlated to the rates themselves. We note that, as L_t is not a fixed asset price process, the drift and the volatility terms can be independently specified.

In our modelling framework the process for the LIBOR rate is driven by a conditional local volatility diffusion process with asymmetric jumps, a volatility regime variable process and a drift regime variable process. The conditional local volatility is specified as a CEV process, where the exponent is made a function of the rates themselves. Asymmetric jumps are also introduced, in order to describe the pronounced and asymmetric smile at very short maturities. The volatility regime variable is constrained to take three possible values: low, medium and high volatility. The drift regime variable can take values corresponding to regimes of rising, stable, falling rates and deflation.

In this modelling framework, the 3-month LIBOR spot rate, and therefore the yield curve dynamics, are driven by three factors: the conditional local volatility diffusion process, the process for the volatility regime variable and the process for the drift regime variable.

3.4 The Model Construction

We introduce $N_L = 70$ states for the conditional local volatility diffusion process, $N_a = 4$ states for the drift regime variable and $N_b = 3$ states for the volatility regime variable. Let Ω_L be a finite set $\{0, 1, ..., N_L - 1\}$ containing the first $N_L - 1$ integers together with zero, Ω_a the finite set $\{0, 1, 2, 3\}$ and Ω_b the finite set $\{0, 1, 2\}$. Let $L : \Omega_L \to \mathbb{S}$ be an increasing positive function such that $L(0) = 0$, where $\mathbb{S}$ is some countable set of $\mathbb{R}$. Also, let $y_i = (x_i, a_i, b_i)$ be a generic element of the lattice $\Omega = \Omega_L \times \Omega_a \times \Omega_b$, where $x_i \in \Omega_L, a_i \in \Omega_a$, and $b_i \in \Omega_b$.

The Conditional Local Volatility Diffusion Process

Conditional on a drift regime state a_i and a volatility regime state b_i, we define an approximating Markov chain associated with the conditional local volatility diffusion process for the 3-month LIBOR spot rate. The Markov generator is constructed by imposing probability conservation and the local consistency properties, in the sense of eq. (14), (12) and (13) respectively.

Drift specification

The drift specification is an important ingredient in the construction of our model. In classical short rate modelling, the functional form for the drift is specified according to qualitative economic assumptions, which are derived from an analysis of historical data in the real world measure. In particular, the drift is typically made mean reverting to a long time mean reverting level at a given mean reverting speed, a choice adopted by Vasicek (1977); Hull and White (1993); Cox et al. (1985). In these models, the level of the rates is then fitted in the risk neutral measure in order to match the term structure of discount bonds. This procedure is based on the implicit assumption that the qualitative behaviour observed in the real world measure remains valid in the pricing measure, although the actual values are then computed in the pricing measure. This is not unreasonable if one believes that the market risk premium does not alter the qualitative behaviour of the drift functions. However, not all drift parameters are calibrated to the discount curve, as they have different economical meanings. In particular the degree of freedom provided by mean reversion speed is typically used in order to generate meaningful correlation dynamics. In fact, it is well known that the mean reversion speed has a crucial impact on the correlation of interest rates at different times, as described in Hunt and Kennedy (2004). For example, Hunt and Kennedy (2004) show that, in the Vasicek-Hull-White model where the short rate satisfies the following SDE

$$dr_t = (\theta_t - a_t r_t)dt + \sigma_t dW_t, \qquad (36)$$

with θ_t, a_t and σ_t deterministic function of times, the correlation structure for the short rate is given by:

$$\rho(r_{t_i}, r_{t_j}) = e^{a(t_j - t_i)} \sqrt{\frac{var(r_{t_i})}{var(r_{t_j})}}. \qquad (37)$$

Thus, the higher the mean reversion speed the lower is the correlation between the short rate at different times. For other short rate models it may not be possible to recover an analytical formula for the correlation structure, but the principle still applies. For this reason, the mean reversion speed is typically given in input in such a way to recover the desired correlation structure.

Our model allows for greater flexibility, as we have four drift functions, one for each drift regime. We follow the same approach of traditional short rate modelling and assume that the qualitative behaviour observed in the real world measure remains valid in our pricing measure. Although the yield curve dynamics are generated by the interplay of the different factors built into the model, we ensure that the drift mean reversion levels are primarily fit to the term structure of discount bonds and the mean reversion speeds to the correlation term structure.

The estimation of the drift functions from historical time series is a very active research field in econometrics. Andersen et al. (2004) consider parametric jump diffusions with stochastic volatility, whereas Stanton (1997) discusses a fully non parametric approach for one factor diffusion processes and show that the drift is substantially non linear. In particular, they show that the mean reversion effect appears to be negligible at low to medium level of rates, and that it rapidly increases at high rates. We choose to construct the drift functions with a series of piecewise linear functions of the underlying rate (Figure 2(a)). In our parametrization choice, for the stable and rising regimes, the drift is constant at low rates, whereas it has a mean reverting behavior at medium to high rates, consistently with Stanton (1997). For falling and deflation regimes, the higher the level of rates, the more the process is pulled toward low levels: this makes intuitive sense as, by definition, in these regimes the rate process is biased toward low interest rate levels.

Local volatility specification

Whilst the construction of the drift functions is primarily based on economically meaningful views, the local volatility specification aims to achieve consistency with the observed skew in the swaption market. We have chosen the local volatility to be of a *CEV* type (Figure 2(b)), which has received much attention in finance, and has been previously discussed by Beckers (1980); Cox (1996); Andersen and Andreasen (2000):

$$\sigma_{b_i}(L(x)) = \overline{\sigma}_{b_i} L(x)^{\beta(L(x))}. \tag{38}$$

The exponent is made dependent on the level of the underlying rates:

$$\beta(L(x)) = \frac{A + L(x)}{B + L(x)}, \tag{39}$$

where A and B are solved by imposing $\beta(0) = \beta_0$ and $\beta(L_{ref}) = \beta_{ref}$ (Figure 3). Such state dependent exponent can be interpreted as a state dependent distribution assumption: interest rates behave close to normal at very low levels and close to lognormal at high levels.

Markov generator construction

Having chosen the drift and local volatility function, conditional on each drift and volatility regime, we build the discretized Markov generator for the conditional diffusion process:

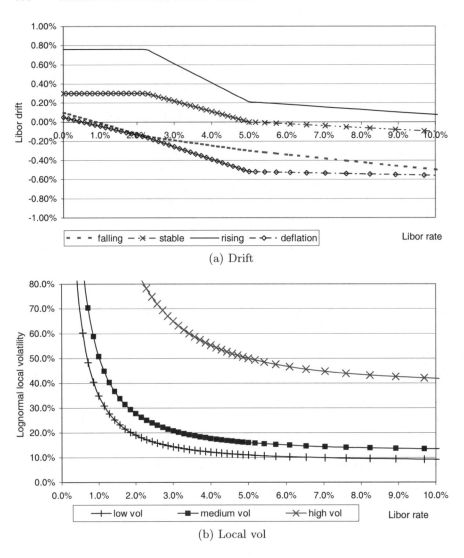

(a) Drift

(b) Local vol

Fig. 2. Specification of drift and local volatility, conditioned on the drift and volatility regimes respectively.

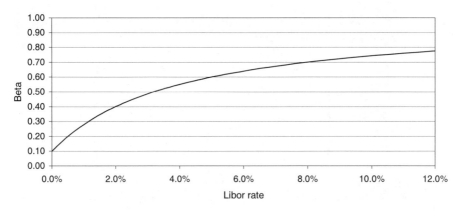

Fig. 3. Specification of beta.

$$\mathcal{L}_{\Omega_L} := \mu_{a_i} \nabla_h + \frac{\sigma_{b_i}^2}{2} \Delta_h, \tag{40}$$

for each $a_i \in \Omega_a$ and $b_i \in \Omega_b$.

Introducing Jumps

At this stage of the construction one has the option to add jumps. Although in the examples discussed in this paper we are mostly focused on long dated callable exotics, for which we find that the impact of jumps can be safely ignored, adding jumps is worth considering and implementing in other situations. To add jumps, uncorrelated to the conditional local volatility diffusion process, we use the Bochner subordinators with the Bernstein function associated to the variance gamma process previously discussed in Madan and Seneta (1990) and Madan et al. (1998):

$$\phi(\lambda) = \frac{\mu^2}{\nu} \log\left(1 + \lambda \frac{\nu}{\mu}\right), \tag{41}$$

where μ is the mean rate and ν is the variance rate of the variance gamma process. We then follow the following procedure which accounts for the need to assign different intensities to up-jumps and down-jumps. This is done in order to address the calibration of an asymmetric smile at the very short end. To produce asymmetric jumps, we specify the two parameters in (41) differently for the up and down jumps and compute separately two Markov generators (see 2.2):

$$\mathcal{L}_{\Omega_L \pm}^j = -\phi_\pm(-\mathcal{L}_{\Omega_L}) = -U_\pm \phi_\pm(-\Lambda) U^{-1}_\pm, \tag{42}$$

where:

$$\phi_\pm(\lambda) = \frac{\mu_\pm^2}{\nu_\pm} \log(1 + \lambda \frac{\nu_\pm}{\mu_\pm}). \tag{43}$$

The new generator for our process with asymmetric jumps is obtained by combining the two generators above:

$$
\mathcal{L}_{\Omega_L}^j = \begin{pmatrix}
0 & 0 & 0 & \cdots & 0 & 0 \\
l_{2,1}^- & d_{2,2} & l_{2,3}^+ & \cdots & l_{2,n-1}^+ & l_{2,n}^+ \\
l_{3,1}^- & l_{3,2}^- & d_{3,3} & \cdots & l_{3,n-1}^+ & l_{3,n}^+ \\
\vdots & \vdots & & \ddots & \vdots & \vdots \\
l_{n-1,1}^- & l_{n-1,2}^- & l_{n-1,3}^- & \cdots & d_{n-1,n-1} & l_{n-1,n}^+ \\
0 & 0 & 0 & \cdots & 0 & 0
\end{pmatrix},
\tag{44}
$$

where $l_{x,y}^{\pm}$ is the element (x, y) of the matrix $\mathcal{L}_{\Omega_L}^{j\pm}$. Here the element of the diagonal are chosen in such a way to satisfy probability conservation:

$$
d_{x,y} = -\sum_{y \neq x} \mathcal{L}_{\Omega_L}^j (x, y).
\tag{45}
$$

Also notice that we have zeroed out the elements in the matrix at the upper and lower boundary: this ensures that there is no probability leakage in the process.

At this stage of the construction, we have therefore obtained a collection of generators $\mathcal{L}_{\Omega_L}^j$ for the 3-month spot LIBOR rate process, one for each $a_i \in \Omega_a$ and $b_i \in \Omega_b$, whose dynamics are characterized by a combination of state dependent local volatility and asymmetric jumps.

Stochastic Volatility Regimes

The process for the stochastic volatility is not easily observed in interest rates. Payoffs whose primitive underlier is the volatility itself are still scarce and illiquid, although recently they have been receiving increased attention in the market. Rebonato and Joshi (2001) argue that the volatility process experiences patterns which can be effectively described by means of regimes. Also, previous econometric studies by Hamilton (1998) and Susmel and Kalimipalli (2001) on the short term interest rate give evidence for regime switching in interest rates. Consistently with this view, we model stochastic volatility by means of regime switching. The volatility is allowed to switch between three possible states: low, medium and high.

The transition probabilities between regimes are set to be independent of the rate. The values used in our model parameterisation are the following:

$$
\mathcal{L}_{\Omega_b} = \begin{pmatrix}
-2.5 & 2.5 & 0 \\
0.2 & -0.6 & 0.4 \\
0 & 0.4 & -0.4
\end{pmatrix}
\begin{matrix}
\leftarrow & high \\
\leftarrow & medium \\
\leftarrow & low
\end{matrix}
$$

The Markov generator has been constructed both by performing a fit to the volatility cube and by ensuring that transition probabilities between regimes reflect a meaningful view on the volatility process. In particular, it should be noted that the process on the regimes is typically much slower than the process on the state variable L_t. The volatility process is responsible for driving the medium to long term behaviour of the term structure of the yield curve. The rate is allowed to transition from one regime to another with low frequency and we have chosen to allow only transition between contiguous regimes. Also, the process reverts from the high to the medium volatility state faster than any other regime transition. This is in line with the qualitative behaviour observed in the market, whereby high volatile periods are typically much shorter than normal or low volatile periods.

Stochastic Monetary Policy

In our model, one has the flexibility of directly specifying the process for the stochastic drift regimes. The monetary policy represents an economically meaningful risk factor and a direct specification of its dynamics is an important benefit of this framework. This additional degree of freedom is best viewed as a powerful tool in order to provide an economically meaningful explanation of the implied volatility smile and the correlation term structure.

Evidence that the 3-month LIBOR spot rate experiences drift regimes is provided by a qualitative analysis of the historical data (Figure 1). In addition, the econometric studies by Andersen et al. (2004), and references therein, have been able to successfully explain the short term rate historical time series by incorporating a stochastic drift component in the model specification.

In our parameterisation choice, the elements of the Markov generator $\mathcal{L}_{\Omega_a}$ of the drift regime process have been chosen as shown in Figure 4. These are defined to be a function of the underlying LIBOR rate. This functional dependence introduces a correlation between rates and drift regimes. A few economically meaningful observations have driven the specification of the stochastic monetary policy Markov generator:

- the drift regime process is allowed to jump with non-zero probability only between contiguous regimes;
- apart from the deflationary regime, beyond a certain rate level the rate may switch between different regimes with equal probability;
- at very low rate levels, there is a large probability of switching from a falling to a deflation regime; whereas for increasing rate levels this probability rapidly decreases to zero; the inverse applies for deflation to falling;
- the probability of switching from a stable to a falling regime, and from a rising to a stable regime, increases up to a constant as rates increase up to a reference level; the inverse applies for falling to stable and stable to rising;

One should note that the above behaviour is qualitatively consistent with the mean reversion phenomenon observed in the interest rates market.

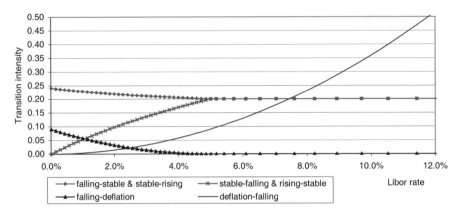

Fig. 4. Transition probability intensities between different drift regimes, as a function of the underlying LIBOR rate L.

The Combined Process for the Spot LIBOR Rate

We complete the construction of our interest rate model by combining together the conditional local volatility jump diffusion process, the volatility regime process and the drift regime process. Assuming no cross terms, the Markov generator for the combined process is defined on a finite dimensional lattice $\Omega = \Omega_L \times \Omega_a \times \Omega_b$ and is simply given by:

$$\mathcal{L}_\Omega(x_1, a_1, b_1; x_2, a_2, b_2) = \mathcal{L}^j_{\Omega_L}(x_1, x_2 | a_1, b_1) \delta_{a_1 a_2} \delta_{b_1 b_2} + \quad (46)$$
$$\mathcal{L}_{\Omega_a}(a_1, a_2 | x_1) \delta_{x_1 x_2} \delta_{b_1 b_2} + \quad (47)$$
$$\mathcal{L}_{\Omega_b}(b_1, b_2) \delta_{x_1 x_2} \delta_{a_1 a_2},$$

where δ is the Kronecker delta; (x_1, x_2), (a_1, a_2), (b_1, b_2) are two generic states for the conditional local volatility jump diffusion process, the drift regime process and the volatility regime process respectively; $\mathcal{L}^j_{\Omega_L}(x_1, x_2 | a_1, b_1)$ is the (x_1, x_2) element of the Markov generator for the subordinated local volatility diffusion process defined on the lattice Ω_L and conditional on the drift regime a_1 and volatility regime b_1; $\mathcal{L}_{\Omega_a}(a_1, a_2 | x_1)$ is the (a_1, a_2) element of the Markov generator for the stochastic drift regime process defined on the lattice Ω_a and function of the underlying LIBOR rate process; and $\mathcal{L}_{\Omega_b}(b_1, b_2)$ is the (b_1, b_2) element of the Markov generator for the stochastic volatility regime process defined on the lattice Ω_b.

In a more compact form we write $\mathcal{L}_\Omega(y_1; y_2) = \mathcal{L}_\Omega(x_1, a_1, b_1; x_2, a_2, b_2)$, where we remind that in our notation $y_i = (x_i, a_i, b_i)$. To gain an intuition on the meaning of the matrix element $\mathcal{L}_\Omega(y_1; y_2)$ we note that, for each element off the main diagonal, $\mathcal{L}_\Omega(y_1; y_2)\, dt$ represents the transition probability to move from a state $L(x_1)$ in the a_1 drift regime and b_1 volatility regime to a state $L(x_2)$ in the a_2 drift regime and b_2 volatility regime in an infinitesimal time dt.

The Markov propagator of the combined process between two consecutive times in the tenor structure (t_i, t_{i+1}) is computed by taking the exponential of the combined Markov generator:

$$u_\Omega(y_i, t_i; y_{i+1}, t_{i+1}) = e^{\mathcal{L}_\Omega(t_{i+1} - t_i)}(y_i, y_{i+1}). \tag{48}$$

Since our underlier is a 3-month rolling LIBOR rate though, we are not interested in the pricing kernel but rather in the discounted transition probability kernel, which in the 3-month spot LIBOR measure we use, is given by:

$$G_\Omega(y_i, t_i; y_j, t_j) = \sum_{y_{i+1}, y_{i+2}, \ldots, y_{j-1}} \prod_{k=i+1}^{j} u_\Omega(y_{k-1}, t_{k-1}; y_k, t_k) \frac{1}{1 + \tau L_{t_{k-1}}}. \tag{49}$$

The 3-month discounted transition probability kernel is obtained by discounting the 3-month propagator $u_\Omega(y_{k-1}, t_{k-1}; y_k, t_k)$ between times $(t_{k-1}; t_k)$ with the 3-month discount factor $\frac{1}{1+\tau L_{t_{k-1}}}$. In order to obtain the discounted transition probability kernel between two generic times $(t_i; t_j)$ belonging to the standard tenor structure, with $j > i+1$, the elementary discounted kernel is compounded $n = \frac{t_j - t_i}{\tau}$ times summing across all intermediate states.

Handling non Standard Periods

As it is the case for any model where the chosen primitive is the LIBOR rate with simple compounding, the stochastic discount factor between any two generic times that are not $j\tau$ years apart, with $j \in \mathbb{N}$, is not readily available. Whilst this is not a problem for the valuation of standard derivatives instruments with regular payment schedules, the issue arises when applying this class of models to unconventional trades. This is a matter of practical interest which should not be overlooked, as non standard payment schedules are often found in a typical trade population of exotics trading desks. Over-the-counter contracts may be tailored to match clients' unconventional cash flow schedules and this generally results in a stub period at the beginning of the trade or at exercise dates. Also, in some cases, as for example accrual or trigger swaps, one must be able to obtain the discounted transition probability matrix for non standard tenors even if the trade has a regular payment schedule. Finally, a correct evaluation of stub periods is necessary for the pricing and risk management of aging trades. In this case the tenor between the pricing date and the payment date of the first cash flow shrinks from the market standard period, at trade inception, to zero as time goes by.

Several practical solutions have been presented for the so-called LIBOR Market Model, which is also based on a simple compounding rule. In particular, a linear interpolation method on the discount factor is proposed by

Jashmidian (1997). Brigo and Mercurio (2006) suggest a drift interpolation algorithm based on Brownian bridge, whereas some practitioners choose a naive interpolation on the stochastic LIBOR rates themselves, as it is done by Morini (2005). In this work, we suggest a practical solution to this problem applicable within our modelling framework.

Without loss of generality and in order to fix ideas, we first address the issue encountered when pricing a Bermuda swaption with an unconventional exercise schedule (see Figure 5). Let t_0 be the evaluation date and, for simplicity, assume that all cash flows are τ years apart, both on the fixed and floating leg. However, assume that a given exercise date t_3^{ex} falls in between two payment dates: $t_2 < t_3^{ex} < t_3$. In order to evaluate the option at t_3^{ex} one needs to perform a backward induction between t_3 and t_3^{ex}. This requires the valuation of the discounted transition probability matrix $G_3 = G_\Omega(y_3^{ex}, t_3^{ex}; y_3, t_3)$. Also, working back on the lattice from t_3^{ex} to t_2, one also needs to know the matrix $G_2 = G_\Omega(y_2, t_2; y_3^{ex}, t_3^{ex})$.

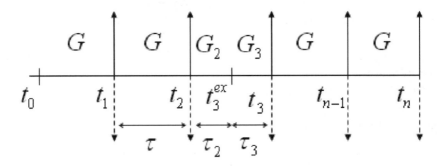

Fig. 5. Bermuda swaption with non standard exercise date t_3^{ex} falling in between payment dates t_2 and t_3.

Let $L_\tau(t)$ be the spot LIBOR rate with tenor τ and evaluated at time t. Also, let:

$$\tau_2 = t_3^{ex} - t_2, \tag{50}$$
$$\tau_3 = t_3 - t_3^{ex}, \tag{51}$$

where τ_2 and τ_3 are expressed in year fractions. If we knew the Markov generators $\mathcal{L}_{2,\Omega}$ and $\mathcal{L}_{3,\Omega}$, for L_{τ_2} and L_{τ_3} respectively, we could evaluate the elementary propagators u_2 and u_3 as follows:

$$u_2 = u_\Omega(y_2, t_2; y_3^{ex}, t_3^{ex}) = e^{\mathcal{L}_{2,\Omega}(t_3^{ex} - t_2)}(y_2, y_3^{ex}), \tag{52}$$
$$u_3 = u_\Omega(y_3^{ex}, t_3^{ex}; y_3, t_3) = e^{\mathcal{L}_{3,\Omega}(t_3 - t_3^{ex})}(y_3^{ex}, y_3), \tag{53}$$

and therefore compute G_2 and G_3 by discounting the elementary propagators with the appropriate spot LIBOR rates:

$$G_2 = \frac{u_2}{1 + \tau_2 L_{\tau_2}(t_2)}, \tag{54}$$

$$G_3 = \frac{u_3}{1 + \tau_3 L_{\tau_3}(t_3^{ex})}. \tag{55}$$

The same reasoning applies for the valuation of aging trades, as well as trigger or accrual swaps. In fact, in the most general case, the valuation of a derivative contract requires the knowledge of the discounted transition probability matrix G, associated with the standard 3 month spot LIBOR rate process, as well as some other matrices G_{τ_i} associated to the spot LIBOR rate process of tenor $\tau_i < \tau$. In the time homogeneous case, for the valuation of a given trade, one needs to compute as many G_{τ_i} as non standard periods $\tau_i < \tau$ encountered in the contract specification. For example, the trade shown in Figure 6 represents an aged Bermuda swaption with exercise dates falling in between payment dates. In this case, four time homogeneous discounted transition probability matrices G are required: one for the standard period τ, one for the initial stub τ_1, due to the aging effect, and two for the periods τ_2 and τ_3, due to the stubs at each exercise date.

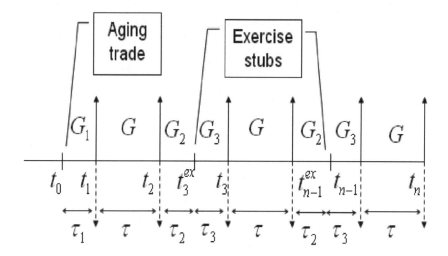

Fig. 6. Aged Bermuda swaption with exercise dates falling in between payment dates.

Therefore, solving the issue of handling non standard periods boils down to the estimation of the Markov generator for the spot LIBOR rate processes associated with tenors smaller than the standard one. For practical applications, we propose to adopt the following approximation:

$$\mathcal{L}_{\tau_i, \Omega} = \mathcal{L}_{\tau, \Omega} \quad \text{for each } \tau_i < \tau. \tag{56}$$

The above assumption is equivalent to saying that $L_{\tau_i}(t) = L_\tau(t)$, for each $\tau_i < \tau$, at any time t. Or, in other words, that the conditional yield curve dynamics for expiries less than, or equal to, τ is driven by the same stochastic process. As in this work we are mainly concerned with the valuation and risk management of long dated derivatives, the number of times when this approximation is required for the valuation of exotic trades is typically low. Also, with numerical experiments we find that the error induced by using $\tau = 0.25$ as standard lattice frequency is very small and conclude that this is an acceptable approximation for practical applications. Furthermore, we note that, for any given trade, it is possible to compute the lattice frequency $\bar{\tau}$ such that no approximation is necessary. Hence the error of this approximation can be easily quantified, and therefore controlled, by running the model with frequency $\bar{\tau}$.

As an example, we evaluate the impact of this approximation on an aged Bermuda swaption with irregular exercise dates, as described in Figure 6. Each exercise date is set to be 1m prior payment date, the first payment date is 9y and 11m from valuation date and the final maturity of the swap is 19y and 11m, with quarterly exercise and payment frequency. We evaluate the trade with a decreasing lattice frequency, from daily down to quarterly. The maximum price variation is about 0.2 basis points which, more importantly, translates in about 0.6 basis points of lognormal volatility (see Figure 7). These errors are well within calibration tolerance and certainly within bid-offer spread which, for these trade types, is in excess of 50 basis points of lognormal volatility.

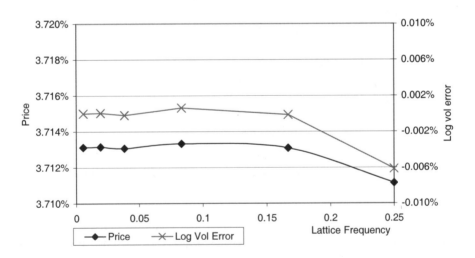

Fig. 7. Price (left axis) and lognormal volatility error (right) of an aged Bermuda swaption with irregular exercise dates, as a function of lattice frequency.

Finally we note that, having adopted the above proposed approximation, it is possible to run the model on a general time structure of this type:

$$0 < t_0 < t_1 < ... < t_n, \tag{57}$$

with t_j arbitrary.

3.5 Calibration

The Calibration Problem

In our calibration problem we focus on two main aspects:

- recover the implied market prices at one particular date;
- obtain yield curve dynamics consistent with historical data.

We perform a best fit of the model to the following calibration instruments:

- the discount curve and swap rates,
- the implied swaption volatility cube,
- the historical correlation structure.

The discount curve is evaluated at every date in the standard tenor structure we define. For the calibration of the swaption volatility cube we evaluate 6 tenors at 11 maturities for 10 strikes (see Table 1). Whereas for the calibration of the correlation structure, we evaluate the cross correlation of 6 tenors for one week horizon. For this, we consider the historical correlation of the time series of the daily differences, so that it is possible to meaningfully compare the correlation implied by the model with the historical correlation data. In fact, in this case, the correlation backed out from historical data represents a close proxy for the instantaneous correlation, which does not depend on the particular measure considered.

Table 1. Volatility cube calibration problem.

Maturity	Tenor	Relative Strike
6m	3m	-2%
1y	6m	-1%
2y	1y	-0.5%
3y	2y	-0.25%
4y	10y	0%
5y	20y	0.25%
7y		0.5%
10y		1%
15y		2%
20y		4%
30y		

Our calibration exercise aims at finding a model parameterisation that not only explains the market at one particular date but also generates yield curve shapes that, as a function of the initial condition, are consistent with historical data. We find that the model is able to obtain steep, flat and inverted yield curve shapes as a function of the initial condition. This ensures that, qualitatively, the fit is consistent with different historically realised market conditions. However we note that the model fails to recover humped yield curve shapes. In order to address this issue, the model would need to be enhanced with additional flexibility. As humped yield curve shapes can be explained with a well defined view on monetary policy in the short term, a natural and practical extension of the model would be to allow for minor time dependencies in the drift process at short maturities.

With these constraints in mind, one is able to obtain a parameterisation choice that could be used as a good first guess for a more comprehensive historical back testing, which would further enhance the calibration procedure. Whilst we leave this to future works, here we focus on the description of the calibration problem and the numerical results.

Stochastic Drift and Market Completeness

In the traditional literature, the calibration problem is formulated in such a way that, in addition to a riskless security, it is possible to associate one risky security for each risk factor introduced in the model. Broadly speaking this ensures that the market is complete. Whilst market completeness gives comfort on one side, as this implies that general contingent claims can be perfectly hedged, it often constraints model flexibility. This may lead to unsatisfactory modelling frameworks, which are not able to fit the implied market data or generate economically meaningful dynamics. The difficulty therefore lies in finding a mathematical framework that, on one side can guarantee a (quasi-)perfect hedge of contingent claims, and on the other allows for enhanced modelling flexibility. For example, for models built with geometric Lévy processes, Corcuera et al. (2005) suggest to complete the market with an additional ad-hoc risky asset. Whereas Merton (1976) assumes that jumps are uncorrelated with the marketplace and can therefore be diversified away. On a different approach, Carr et al. (2001) discuss a new framework for pricing and hedging in incomplete markets, that is somewhat in between the expected utility maximization and arbitrage free pricing theory. Finally, Schoutens (2006) states that most model are incomplete and that many practitioners believe that the market is not complete.

A market described by finite-state Markov chains provides an interesting framework in order to address this issue. In fact, on one side it provides enough modelling flexibility for practical applications and on the other side it limits the sources of randomness. The process is constrained to take a finite set of values and, in particular, the maximum number of outcomes originating from a given lattice value y_i is countable and finite. Broadly speaking, this allows to identify a replicating self-financing strategy, based on a finite number of risky securities. For example, in our case, a set of discount bonds, each with different maturity, may be used to construct a self-financing replicating portfolio. We do not engage in this discussion any further and we refer the reader to Norberg (2003) for a rigorous description of the completeness issue in a Markov chain market. The reader may notice that, within such modelling framework, market completeness conditions are reasonably easy to obtain. This contrasts with the continuous state space case for which a similar discussion can be found in Eberlein et al. (2005), where the authors consider the case of Lévy processes.

Also, we argue that, regardless of market completeness properties, the introduction of stochastic monetary policy in interest rate modelling is an interesting extension of currently known modelling frameworks and therefore it is worth exploring.

Lattice Calibration and Numerical Results

The calibration is first performed with a time homogeneous parameterisation, and subsequently refined with minor deterministic time dependent shifts for a perfect match to the term structure of the interest rates. In the time homogeneous case, discount factors conditional on an initial state y_i are given by:

$$Z(y_i; t_i, t_j) = \sum_{y_j} G_\Omega(y_i, t_i; y_j, t_j), \tag{58}$$

and swap rates can be evaluated as follows:

$$SR(y_i; t_i, t_j) = \frac{1 - Z(y_i; t_i, t_j)}{BPV(y_i; t_i, t_j)}, \tag{59}$$

where $BPV(y_i; t_i, t_j) = \sum_{k=i+1}^{j} \tau Z(y_i; t_i, t_k)$. For the calibration to the swaption volatility cube, one needs to evaluate the price of a portfolio of European swaptions, for different maturities, underlying tenors and strikes. In particular, the value of an European swaption maturing at time $t_j > t_i$, where the maturity of the underlying swap is $t_k > t_j$ and struck at k is given by:

$$SO(y_i; t_i, t_j) = \sum_{y_j} G_\Omega(y_i, t_i; y_j, t_j) \left(SR(y_j; t_j, t_k) - k \right)_+ BPV(y_j; t_j, t_k). \tag{60}$$

Time inhomogeneities can be introduced in the model in order to obtain a perfect fit to the term structure of interest rates. This can be achieved by normalising the discounted transition probability with a deterministic time dependent factor:

$$G_\Omega(y_i, t_i; y_j, t_j) \to \overline{G}_\Omega(y_i, t_i; y_j, t_j) = G_\Omega(y_i, t_i; y_j, t_j) \frac{Z_{mkt}(t_j) Z(\overline{y}_0; t_0, t_i)}{Z_{mkt}(t_i) Z(\overline{y}_0; t_0, t_j)},$$
$$(61)$$

where $Z_{mkt}(t_j)$ is the market discount factor with maturity t_j, $\overline{y}_0$ is the spot value at current time t_0 of the underlying lattice process. If the time homogeneous parameterisation is well calibrated, this time dependent change will produce shifts in the interest rates which are typically smaller than 10 basis points.

Figure 8 reports the possible shapes of yield curves obtained with the model, as a function of the initial point x_i and conditional on each drift regime. The model satisfactorily recovers steep, flat and inverted shapes, however with this parameterisation choice humped yield curve shapes can not be obtained. One can notice that the shape of the discount curve is only marginally affected by the instantaneous volatility and is largely influenced by the prevailing monetary regime. This is not surprising as the volatility process mean reverts of time scales much shorter than the time scales for the monetary policy dynamics. Hence, volatility of long term rates, spread between rates and the correlation between forward rates across the term structure are mainly controlled by the stochastic drift term.

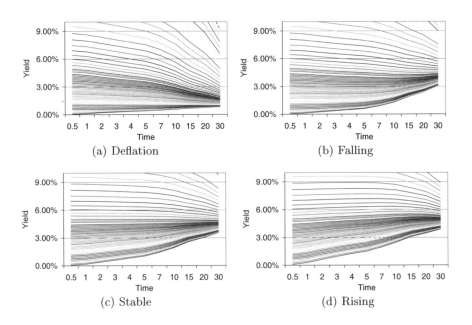

Fig. 8. Yield curve shapes as a function of the initial point and conditional on the drift regime.

The implied volatility of swaptions at various maturities is characterized by a persistent skew. The model is capable of describing this effect, and the implied volatility smiles are qualitatively in line with the market (Figure 9). This effect is primarily generated by the local volatility and stochastic volatility components at short to medium maturities. However, at longer maturities, stochastic drift becomes the predominant factor.

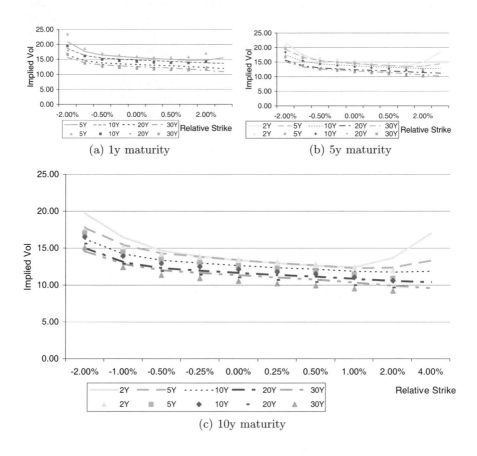

Fig. 9. Model (lines) and market (dots) swaption volatility smile at different maturities and for different tenors.

An interesting observation allows one to gain more insight on the impact of the drift regime, and in particular the deflationary regime. In fact, by removing the deflationary regime, in our experience, it has not been possible to calibrate the low strikes portion of the smile for long dated maturities without affecting the dynamics of the process at earlier maturities. Figure 10 shows that the 20y into 30y swaption smile at low strikes is too flat compared to the market.

The introduction of a deflationary regime allows to increase the probability of the rate of spanning regions of low values at long maturities, and therefore lifts up the smile at low strikes, without considerably affecting the smile at short and medium maturities. This example highlights the important fact that the monetary policy process can be used to provide direct control to the long dated behaviour of the rate process.

Finally, we note that the model implied short term correlations are also in line with historical estimates across tenors and maturities (see Table 2). The one week term correlation between two swap rates of underlying tenors τ_1 and τ_2, evaluated at t_i, is computed on the lattice as follows:

$$\rho_\delta\left(t_i; \tau_1, \tau_2\right) = \frac{\mathbb{E}_{t_i}\left[\left(S_{t_i+\delta}^{\tau_1} - S_{t_i}^{\tau_1}\right)\left(S_{t_i+\delta}^{\tau_2} - S_{t_i}^{\tau_2}\right)\right]}{\sqrt{\mathbb{E}_{t_i}\left[\left(S_{t_i+\delta}^{\tau_1} - S_{t_i}^{\tau_1}\right)^2\right]}\sqrt{\mathbb{E}_{t_i}\left[\left(S_{t_i+\delta}^{\tau_2} - S_{t_i}^{\tau_1}\right)^2\right]}}, \tag{62}$$

where $\delta = 1/52$ and we have used the notation $S_{t_i}^\tau = SR(t_i, t_i + \tau)$.

Table 2. One week term correlation structure.

	0.25	0.5	1	2	10	20	30
0.25	100.0	99.85	98.74	94.38	60.30	62.09	50.60
0.5	99.85	100.0	99.46	96.06	64.59	66.15	55.04
1	98.74	99.46	100.0	98.43	72.15	73.31	63.02
2	94.38	96.06	98.43	100.0	83.25	83.87	75.24
10	60.30	64.59	72.15	83.25	100.0	97.12	92.54
20	62.09	66.15	73.31	83.87	97.12	100.0	98.84
30	50.60	55.04	63.02	75.24	92.54	98.84	100.0

Performance Timings

For the evaluation of the calibration problem described in this section, we solve for:

- 200 discount bonds;
- 66 forward rates;
- 660 swaptions;
- 30 correlations.

The pricing of the above described calibration instruments has been executed in 1'20" on a laptop with solo Intel Pentium CPU with clock speed of 1.73GHz and 1.00 GB of RAM. We note that, in practical applications, the calibration problem is typically reduced to a much smaller subset of liquid contracts. Also, a substantial performance improvement can be obtained with currently available hardware specifications and by taking advantage of commercially available and low cost GPU cards.

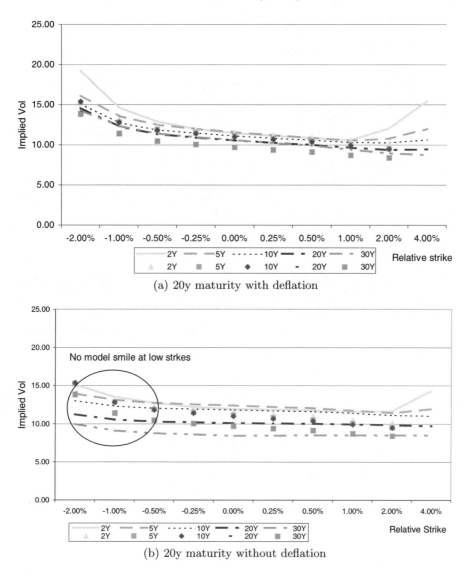

(a) 20y maturity with deflation

(b) 20y maturity without deflation

Fig. 10. Model (lines) and market (dots) swaption volatility smile at 20y maturity with and without deflation.

3.6 Applications

We have applied the stochastic monetary policy model to callable swaps and callable CMS spread range accruals. The impact of the monetary policy component built into the model is made obvious by an analysis of the price functions conditional on the drift regime, as a function of the initial condition on the 3-month LIBOR spot rate. Also, for the case of callable CMS spread range accrual, we show numerical results that highlight the impact of the drift regimes on the coupon accrual probabilities.

First we proceed with the description of the algorithm used for the pricing of callable CMS spread range accruals. Then we present numerical results. The evaluation of callable swaps is performed according to standard lattice techniques, thus its description is omitted.

Pricing Callable CMS Spread Range Accruals

We consider the case where the counterparty paying the CMS spread range accrual leg and receiving the LIBOR leg has the option to cancel the swap every year. In our example, the range accrual coupons are contingent on the spread between the 10y and 2y CMS rates. In its simplest form, the range accrual coupon is given by:

$$c_i = k\frac{n_i}{N}, \tag{63}$$

where

- n_i is the total number of days when the 10y-2y CMS spread sets between a low and a high barrier level in the coupon period i;
- N is the total number of observations per coupon period;
- k is the fixed multiplier of the structured coupon, also referred to as *fixed rate*.

In this case, the apparent path dependency in the payoff can be easily removed by taking advantage of the linearity of the payoff and the expectation operator. In fact, let t_i be the accrual start date of a range accrual flow with payment date t_j. Then the present value of the range accrual coupon can be calculated as follows:

$$PV_{t_i} = g_{t_i}\mathbb{E}_{t_i}\left[\frac{1}{g_{t_j}}k\frac{\sum_{s=1}^{N}\mathbf{1}_{\{b_l<SR(t_s,t_s+10y)-SR(t_s,t_s+2y)<b_h\}}}{N}\right] \tag{64}$$

$$= g_{t_i}\sum_{s=1}^{N}\mathbb{E}_t\left[\frac{1}{g_{t_j}}k\frac{\mathbf{1}_{\{b_l<SR(t_s,t_s+10y)-SR(t_s,t_s+2y)<b_h\}}}{N}\right], \tag{65}$$

where g_t is the numeraire associated to the chosen pricing measure, $SR(t_s,t_s+10y)$ and $SR(t_s,t_s+2y)$ are the value of the 10y and 2y swap rates respectively, evaluated at time t_s (see eq. (59)), and b_l and b_h are the low and high barrier levels respectively. Having made this observation, the pricing of callable CMS spreads range accruals may proceed with standard backward induction lattice methods, described here below.

Backward induction method

In a backward induction method, one must evaluate the 2y and 10y spot swap rates at different times. These can be evaluated by means of eq. (59), conditional on a lattice point y_s, with spot time t_s and maturity $t_s + 2y$ and $t_s + 10y$ respectively. In particular:

$$SR(y_s; t_s, t_s + 2y) = \frac{1 - Z(y_s; t_s, t_s + 2y)}{BPV(y_s; t_s, t_s + 2y)}, \tag{66}$$

$$SR(y_s; t_s, t_s + 10y) = \frac{1 - Z(y_s; t_s, t_s + 10y)}{BPV(y_s; t_s, t_s + 10y)}, \tag{67}$$

where

$$BPV(y_s; t_s, t_s + 2y) = \sum_{r=s}^{s+2y} \tau Z(y_s; t_s, t_r), \tag{68}$$

$$BPV(y_s; t_s, t_s + 10y) = \sum_{r=s}^{s+10y} \tau Z(y_s; t_s, t_r). \tag{69}$$

For the evaluation of these swap rates, the only required quantity is the stochastic discount factor with maturity τ, or integer multiples of τ. In a time homogeneous Markov framework, the stochastic discount factor, and therefore spot swap rates, depends only on time to maturity and the conditional lattice point y_s. Hence, spot swap rates may be evaluated once, cached for efficiency, and then used at any time in the lattice. In our case, we overlay a small deterministic time dependent shift to the time homogeneous discounted transition probability matrix G_Ω in order to obtain a perfect match to the term structure of interest rates. This shift must be taken into account and caching may not be possible. However, even in this case, swap rates can be computed at any time without further approximations required, by replacing the time homogeneous G_Ω with the shift-adjusted $\overline{G}_\Omega$, as described in eq. (61).

At each accrual start date t_i one needs to evaluate the discounted value of the range accrual coupon, conditional on the lattice point y_i, and payment date t_j:

$$c(y_i; t_i, t_j) = g_{t_i} \sum_{s=1}^{N} \mathbb{E}_{t_i} \left[\frac{1}{g_{t_j}} k \frac{\mathbf{1}_{\{b_l < SR(y_s; t_s, t_s + 10y) - SR(y_s; t_s, t_s + 2y) < b_h\}}}{N} \middle| y_{t_i} = y_i \right]. \tag{70}$$

Consider the generic discounted range accrual probability p_s, corresponding to the observation date t_s, with $t_i < t_s < t_j$, and payment date t_j. Its value at time t_s, conditional on the lattice point y_s is given by:

$$p_s(y_s; t_s, t_j) = g_{t_s} \mathbb{E}_{t_s} \left[\frac{1}{g_{t_j}} \left. \frac{\mathbf{1}_{\{b_l < SR(y_s; t_s, t_s + 10y) - SR(y_s; t_s, t_s + 2y) < b_h\}}}{N} \right| y_{t_s} = y_s \right]$$
(71)

$$= g_{t_s} \mathbb{E}_{t_s} \left[\frac{1}{g_{t_j}} \middle| y_{t_s} = y_s \right] \frac{\mathbf{1}_{\{b_l < SR(y_s; t_s, t_s + 10y) - SR(y_s; t_s, t_s + 2y) < b_h\}}}{N}$$
(72)

$$= Z(y_s; t_s, t_j) \frac{\mathbf{1}_{\{b_l < SR(y_s; t_s, t_s + 10y) - SR(y_s; t_s, t_s + 2y) < b_h\}}}{N},$$
(73)

where we note that the indicator function can be taken out of the expectation operator as it is a measurable function at time t_s. It follows that the discounted range accrual probability p_s at time t_i and conditional on the lattice point y_i is:

$$p_s(y_i; t_i, t_j) = g_{t_i} \mathbb{E}_{t_i} \left[\frac{1}{g_{t_s}} p_s(y_s; t_s, t_j) \middle| y_{t_i} = y_i \right].$$
(74)

This is computed on the lattice by taking the expectation of the payoff as follows:

$$p_s(y_i; t_i, t_j) = \sum_{y_s} G_\Omega(y_i, t_i; y_s, t_s) p_s(y_s; t_s, t_j).$$
(75)

From equations (73) and (75), it is clear that this algorithm requires the valuation of the discounted transition probabilities G_Ω on a non standard time structure. In particular, one must be able to evaluate the G_Ω matrix between accrual start date t_i and any observation date t_s, as well as between any observation date t_s and payment date t_j. This can be achieved by means of the approximation described in section (3.4).

Finally, the present value of the range accrual coupon at time t_i, and conditional on the lattice point y_i, can be evaluated as the sum of the range accrual probabilities, one per observation in the coupon period, times the fixed rate k:

$$c(y_i; t_i, t_j) = \sum_{s=1}^{N} k p_s(y_i; t_i, t_j).$$
(76)

Having computed the discounted value of the range accrual coupon at time t_i, the valuation of the callable CMS spread range accrual trade proceeds with standard backward induction methods.

Moments method

In a more general form, the range accrual coupon may be capped or floored. In this case it is not possible to exchange the expectation and the sum operators and we are faced with the valuation of a path dependent problem on the lattice. This can be performed by means of the moments method, which allows one to estimate the joint density of the coupon and the underlying Markov process on the lattice. Although in our application we consider CMS spread range coupons without cap and floor, we describe the moments method in Appendix D for completeness.

Numerical Results

Callable swaps

In Figure 11 we plot the price functions and exercise boundaries for a 20y maturity callable payer swap struck at 5% with annual exercise schedule. For a given initial condition, the price of a callable payer swap is always an increasing function of the monetary policy regime: the lowest price being obtained in the deflationary regime and the highest price in the rising regime. This result is not surprising and also quite obvious: the value of a callable payer swap is expected to be higher in a rising regime, when rates are biased on a rising trend, than in a falling regime, when interest rates are biased on a falling trend.

Callable CMS spread range accruals

For the case of callable CMS spread range accruals, the analysis requires more care. In fact, in this case, the monetary policy regime has an impact on both the level of the rates and the level of the CMS spread. In this example, the effect of the correlation framework based on the dynamic conditioning to monetary policy regimes becomes apparent. We plot the price functions and the exercise boundaries in Figure 12 and the range accrual probabilities in Figure 13. The latter reports the average probabilities, within a coupon period and conditional on the state at the start of the period itself, that the CMS spread sets above a specified barrier level, which in our example was set to 10 basis points. In a time-homogeneous parameterisation, these are the same for every coupon period. One can note that, for the case of the deflationary regime, the model generates nearly zero range accrual probabilities for most initial conditions, apart from very low initial condition levels. This is consistent with the fact that, conditional on a deflationary regime, the model implies a flat or inverted yield curve.

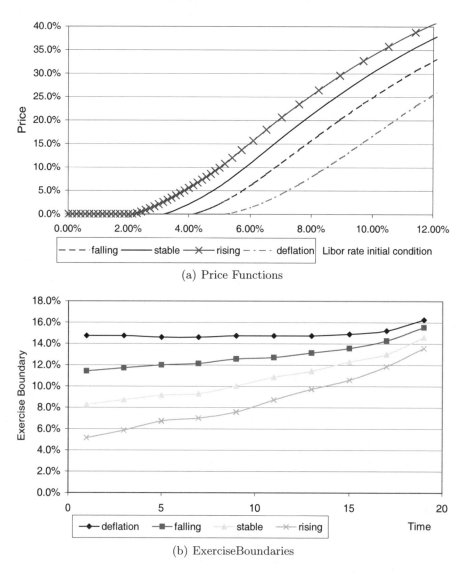

(a) Price Functions

(b) ExerciseBoundaries

Fig. 11. Price functions of a 20y maturity callable swap, as a function of the initial condition on the rate, and exercise boundaries as function of time.

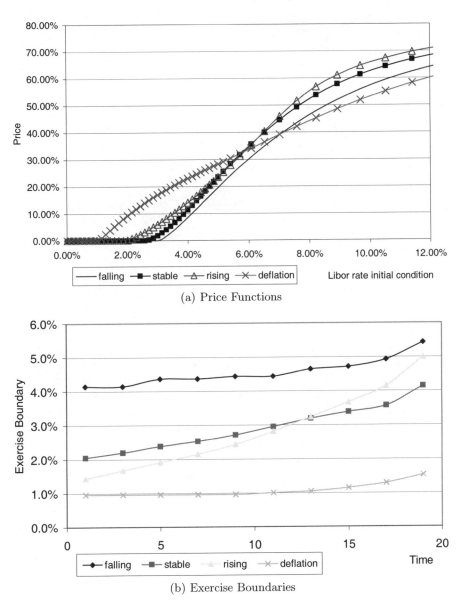

(a) Price Functions

(b) Exercise Boundaries

Fig. 12. Price functions of a 20y maturity callable CMS spread range accrual, as a function of the initial condition on the rate, and exercise boundaries as function of time.

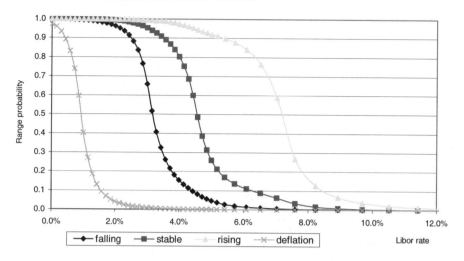

Fig. 13. Range accrual probabilities conditional on drift regimes.

4 Conclusions

We have presented an interest rate model which incorporates stochastic monetary policy in addition to a conditional local volatility component, jumps and stochastic volatility. This is a novel approach in interest rate modelling and it represents the main contribution of this paper. Time-homogeneity has been a key driver in the construction of our model and we show that, with the additional degree of freedom provided by the drift process, the model is able to describe steep, flat and inverted yield curve shapes. As a result, we find that the monetary policy component helps achieve model consistency and allows one to obtain a good qualitative fit to the swaption volatility cube and the correlation structure.

The model is built with Markov chains on finitely many states and is solved by means of continuous-time lattices. The mathematical framework and the numerical algorithm for the construction of a continuous-time lattice has been described in this work and it represents a distinctive feature of our modelling framework.

Finally, we have shown an application of the model to callable swaps and callable CMS spread range accruals. The dependence of the price functions and the range accrual probabilities to the drift regimes reveals the impact of stochastic monetary policy in interest rate derivative pricing. Perhaps the major benefit of the modelling framework proposed in this paper is that it allows one to directly incorporate economically meaningful views into the model and analyse the impact of these views on the valuation and risk management of exotic interest rate structures.

Appendix

A Continuous-time Finite-state Markov Chains

We recall the basic concepts of Markov processes and Markov generators, which play a pivotal role in our modelling framework. For a comprehensive treatment of Markov processes and the proofs of the the properties here described, we refer the interested reader to two excellent text books: Grimmett and Stirzaker (2001) and Kijima (1997).

We define a stochastic process X_t as a family of random variables $\{X(t) : t \in T\}$ indexed by some time set $T \in \mathbb{R}^+$. If the time set T is discrete, we call the process a *discrete-time* process; on the other hand, if the time set is continuous, we call the process a *continuous-time* process. The *state space* Ω of a stochastic process is the range of possible values that the random variables $\{X(t) : t \in T\}$ can take. We consider finite state spaces and, without loss of generality, we identify the *state space* with a finite set of integers $\Omega = \{0, 1, ..., N\}$, for some $N \in \mathbb{N}$. A stochastic process is characterised by its *state space* Ω, its index set T and the dependence structure between the the random variables $\{X(t) : t \in T\}$.

We are interested in stochastic processes taking values on finite sets so that they can be described on a lattice. A lattice can be thought of as a finite grid of real numbers, together with the probabilistic properties associated to them, and it represents our main numerical tool for the construction of our interest rate model.

Definition 2. *A process X_t is called a continuous-time finite-state Markov chain if it satisfies the following condition:*

$$\mathbb{P}(X(t_n) = y \mid X(t_1) = x_1, ..., X(t_{n-1}) = x_{n-1})$$
$$= \mathbb{P}(X(t_n) = y \mid X(t_{n-1}) = x_{n-1}), \tag{77}$$

for all $y, x_1, ..., x_{n-1} \in \Omega$ and any sequence of times $t_i \in T$ and such that $t_1 < t_2 < ... < t_n$.

The evolution of a Markov chain is fully determined by the specification of its transition probability matrix.

Definition 3. *The transition probability matrix function of the Markov chain is the matrix u_Ω defined to be*

$$u_\Omega(x, t_1; y, t_2) = \mathbb{P}(X(t_2) = y \mid X(t_1) = x), \tag{78}$$

for $t_1 < t_2$. Furthermore, the chain is called homogeneous if $u_\Omega(x, t_1; y, t_2) = u_\Omega(x, 0; y, t_2 - t_1)$ for all $x, y \in \Omega$ and $t_2, t_1 \in T$.

The transition probability matrix function, or *Markov propagator*, is easily interpreted as the conditional probability of moving from state x at time t_1 to state y at time t_2. We restrict our focus to a particular type of Markov propagators, referred to as *standard* in traditional literature. The following theorem summarises the key properties of a *standard* Markov transition probability matrix function.

Theorem 3. *If $u_\Omega(x, t_1; y, t_2)$ is a* standard *transition matrix function of a Markov chain, then it satisfies the following:*

$$u_\Omega(x, t_1; y, t_2) \geq 0, \tag{79}$$

$$\sum_{y \in \Omega} u_\Omega(x, t_1; y, t_2) = 1, \tag{80}$$

$$\lim_{t_2 \to t_1} u_\Omega(x, t_1; y, t_2) = \delta_{xy}, \tag{81}$$

$$\sum_{z \in \Omega} u_\Omega(x, t_1; z, t_2) u_\Omega(z, t_2; y, t_3) = u_\Omega(x, t_1; y, t_3), \tag{82}$$

for each $x, y, z \in \Omega$ and any $t_1 < t_2 < t_3 \in T$.

A matrix satisfying the first two equations is said to be *stochastic*. The third equation ensures that the transition matrix is *standard*. The last equation is the well known Chapman-Kolmogorov equation. We recall that every standard transition probability function $u_\Omega(x, t_1; y, t_2)$, as defined above, is continuous in t_2; also $u_\Omega(x, t_1; x, t_2)$ is positive and $u_\Omega(x, t_1; y, t_2)$ is either positive or identically zero, for every $x, y \in \Omega$ and any $t_1 < t_2 \in T$. This means that a standard Markov chain must visit all states with non-zero probability and that there is no periodicity.

The *Markov generator* is an operator that describes the behaviour of the Markov chain for an infinitesimal time increment. Formally, if the matrix elements of the Markov propagator $u_\Omega(x, t_1; y, t_2)$ are differentiable functions of the time parameter t_2 in a right neighborhood of t_1, then one can define:

Definition 4. *The Markov generator of the Markov chain at time t_1 is defined as the following derivative:*

$$\mathcal{L}_\Omega(x, y; t_1) = \left. \frac{d}{dt_2} \right|_{t_2 = t_1} u_\Omega(x, t_1; y, t_2). \tag{83}$$

For the case of a time homogeneous Markov chain, the Markov generator does not depend on the time variable and it is denoted by $\mathcal{L}_\Omega(x, y)$.

Theorem 4. *If $\mathcal{L}_\Omega(x, y; t_1)$ is a Markov generator, then the two following properties hold:*

$$\mathcal{L}_\Omega(x, y; t_1) \geq 0, \ \ if \ x \neq y, \tag{84}$$

$$\mathcal{L}_\Omega(x, x; t_1) = - \sum_{y \in \Omega - \{x\}} \mathcal{L}_\Omega(x, y; t_1). \tag{85}$$

Viceversa, if the matrix elements of $\mathcal{L}_\Omega(x, y; t_1)$ are differentiable functions and satisfy to the two conditions above, then there exists one unique transition matrix $u_\Omega(x, t_1; y, t_2)$ associated to it.

The above theorem establishes a strong link between the *transition matrix* and the *Markov generator*. In the construction of stochastic processes, we look for the propagator of the chain, as this contains the transition probabilities from any initial state to any final state in the lattice space. From eq. (83) it is clear that from u_Ω one can compute $\mathcal{L}_\Omega$. However in the construction of financial models the starting point is the knowledge of the local properties of the process. Hence we want to directly specify $\mathcal{L}_\Omega$ and solve for u_Ω. The following important equations allow one to perform exactly this: compute u_Ω once $\mathcal{L}_\Omega$ is known. These are reported for the case of homogeneous chains, which are of particular interest in our model construction.

Theorem 5. *If $u_\Omega(x, 0; y, t_2 - t_1)$ is the transition matrix of a time homogeneous Markov chain and $\mathcal{L}_\Omega(x, y)$ the associated Markov generator, then the following two equations hold:*

$$\frac{d}{dt_2} u_\Omega(x, 0; y, t_2 - t_1) = \sum_{z \in \Omega} u_\Omega(x, 0; z, t_2 - t_1) \mathcal{L}_\Omega(z, y), \tag{86}$$

$$\frac{d}{dt_1} u_\Omega(x, 0; y, t_2 - t_1) = - \sum_{z \in \Omega} \mathcal{L}_\Omega(z, y) u_\Omega(x, 0; z, t_2 - t_1). \tag{87}$$

The two above equations are called the forward and backward Kolmogorov equation respectively.

The solution of the forward Kolmogorov equation (86), subject to the initial condition (81), can be formally written as:

$$u_\Omega(x, 0; y, t_2 - t_1) = e^{\mathcal{L}_\Omega(t_2 - t_1)}(x, y), \tag{88}$$

where the matrix exponential is defined as:

$$e^{t\mathcal{L}_\Omega} = \sum_{n=0}^{\infty} \frac{t^n}{n!} \mathcal{L}_\Omega^n. \tag{89}$$

Therefore, for a time homogeneous Markov chain, once the *Markov generator* $\mathcal{L}_\Omega$ is known, the computation of the propagator reduces to the solution of the matrix exponentiation problem.

B Markov Generator Discretization

The Markov generator describes the local property of the process. Here we provide a sketch proof to show that in our discretization scheme equations (12) and (13) ensure that the first two instantaneous moments of the process are recovered, in addition to the zero-th moment.

We do this by considering a general finite difference scheme, with $n << N$ stencils, where N is the total number of states in the lattice Ω. We assume that we are given the first n instantaneous moments of F_t: $m_0, m_1, ..., m_{n-1}$ and that these are finite. The moments $m_i, i = 0, ..., n - 1$ may be a function of F_t and are defined as the moments of the increments. Therefore in the limit for $dt \to 0$, the following equations hold:

$$\mathbb{E}_t \left[(F_{t+dt} - F_t)^i \right] = m_i(F_t)dt, \qquad\qquad i = 0, ..., n - 1. \qquad (90)$$

We note that, as we deal with probability measures, we must have $m_0 = 1$. This condition is referred to as the *probability conservation* condition.

Given a lattice Ω, an approximating Markov chain to a stochastic process F_t is constructed through an appropriate specification of the Markov generator matrix $\mathcal{L}_\Omega$. The following theorem provides an algorithm that allows to find $\mathcal{L}_\Omega$ by solving a system of linear equations.

Theorem 6. *Let F_t a stochastic process and let $m_i, i = \{0, ..., n-1\}$ its first n instantaneous moments, with $m_0 = 1$. The moments m_i may be a function of F_t and are assumed to be known and finite. Then an approximating Markov chain that recovers these moments on the lattice Ω may be constructed by means of a Markov generator matrix which satisfies the following system of linear equations:*

$$\sum_{y \in \Omega} \mathcal{L}_\Omega(x, y) = 0, \qquad\qquad (91)$$

$$\sum_{y \in \Omega} \mathcal{L}_\Omega(x, y)(F(y) - F(x))^i = m_i(F(x)), \qquad i = 1, ..., n - 1. \qquad (92)$$

Proof. Let u_Ω be the transition probability matrix of the approximating Markov chain defined on the lattice Ω. We look for u_Ω such that the first n instantaneous moments of the Markov chain are the same as the original process F_t. In particular, u_Ω must be such that eq. (90) on the discretized system are satisfied. Solving eq. (90) on Ω we get:

$$\sum_{y \in \Omega} u_\Omega(x, t; y, t+dt)(F(y) - F(x))^i = m_i(F(x))dt, \qquad i = 0, ..., n-1. \quad (93)$$

From the definition (83) of the Markov generator operator and the probability conservation condition (81) it follows that, for a small time increment dt:

$$\mathcal{L}_\Omega(x,y)dt + o(dt) = u_\Omega(x,t;y,t+dt), \qquad x \neq y, \qquad (94)$$
$$(1 + \mathcal{L}_\Omega(x,x))dt + o(dt) = u_\Omega(x,t;x,t+dt). \qquad (95)$$

The first equation tells us that, if the chain is in state x at time t, then the chain jumps to a state $y \neq x$ at time $t+dt$ with probability $\mathcal{L}_\Omega(x,y)dt + o(dt)$; whereas the second equation states that the chain remains in the same state x with probability $(1 + \mathcal{L}_\Omega(x,x))dt + o(dt)$. Substituting (94) and (95) in (93) we have:

$$\sum_{y \in \Omega - \{x\}} \mathcal{L}_\Omega(x,y)dt + o(dt) + (1 + \mathcal{L}_\Omega(x,x))\, dt + o(dt) = dt, \qquad (96)$$

$$\sum_{y \in \Omega} (\mathcal{L}_\Omega(x,y)dt + o(dt))\, (F(y) - F(x))^i = m_i(F(x))dt, \qquad (97)$$

for $i = 1, ..., n$, where in the first equation we have used the fact that $m_0 = 1$. Rearranging (96) and taking the limit for $dt \to 0$, we arrive at (91) and (92).

Finally, for the case $n = 2$ one obtains equations (12) and (13).

C Option Pricing with Continuous-time Lattices

In this section we describe how to price call options with continuous-time lattices for diffusion processes. We consider an asset price process F_t and assume that it satisfies the following SDE in the forward measure:

$$dF_t = \sigma(F_t)dW. \qquad (98)$$

The first step in our algorithm is the construction of the discretization grid for the forward price F_t on the lattice $\Omega = \{0, 1, ..., N\}$. This can be achieved by generating an array of values for F_t, not necessarily equispaced: $\Omega = \{F(0), F(1), ..., F(N)\}$. We then compute the Markov generator on Ω, as follows:

$$\sum_{y \in \Omega} \mathcal{L}_\Omega(x,y) = 0, \qquad (99)$$

$$\sum_{y \in \Omega} \mathcal{L}_\Omega(x,y)(F(y) - F(x)) = 0, \qquad (100)$$

$$\sum_{y \in \Omega} \mathcal{L}_\Omega(x,y)(F(y) - F(x))^2 = \sigma^2(F(x)). \qquad (101)$$

The first equation is the probability conservation condition and the last two equations ensure that the discrete systems match the first and second moment of the original process. The discretized pricing kernel is recovered by computing the exponential of the Markov generator (see section 2.4):

$$u_\Omega(x,t;y,T) = e^{\mathcal{L}_\Omega(T-t)}(x,y), \tag{102}$$

for $x, y \in \Omega$. For a given initial condition x at time t, $u_\Omega(x,t;y,T)$ represents the terminal density of the process at time T as a function of y. This is all we need in order to solve our option pricing problem. In fact, the forward value of a call option, with expiry date T, strike K, evaluated in the forward measure at time t and as a function of the initial condition x, is:

$$v(x,t) = \mathbb{E}_{t,x}\left[\text{Max}\left(F_T - K, 0\right)\right], \tag{103}$$

which, on the lattice, is evaluated as follows:

$$v(x,t) \simeq \sum_{y \in \Omega} u_\Omega(x,t;y,T)\left[\text{Max}\left(F(y) - K, 0\right)\right]. \tag{104}$$

D Moments Method for Range Accruals

A CMS spread range accrual coupon, with cap and floor, is defined as follows:

$$c_i = \left[\text{Min}\left(\text{Max}\left(k\frac{n_i}{N}, k_1\right), k_2\right)\right], \tag{105}$$

where we have used the same notation as in section (3.6), and k_1 and k_2 are the coupon floor and cap level respectively.

In this case, it is not possible to exchange the expectation and the sum operators as in eq (65) and we are faced with the valuation of a path dependent problem on the lattice. This is a particular type of path dependent payoff, which falls within the category of Abelian path dependencies, introduced in Albanese (2006) and further discussed in Albanese and Vidler (2007). Abelian path dependent payoffs are characterised by a commutativity property for a certain operator algebra associated to the path dependent option. In this particular case, however, a more intuitive characterisation of the payoff is obtained by noting that the path dependent state variable, the structured coupon, is translation invariant and does not depend on its own past history. We take advantage of this property in order to solve the pricing problem, as described below.

The pricing of this type of payoffs boils down to the problem of having to find the joint transition probabilities between the Markovian process driving the dynamics of the yield curve and a path dependent process defined by an expression of the form:

$$I_t = I(y_t, t) = \int_{t_1}^{t} \varphi(y_s)ds, \tag{106}$$

for a time interval (t_1, t) and a time-homogeneous lattice function $\varphi(y_t)$. For the case of the 10y-2y CMS spread range accrual, the lattice function φ is the accrual value of the range accrual structured coupon conditional on the process being at y_s. This is given by:

$$\varphi(y_s) = k \frac{\mathbf{1}_{\{b_l < SR(y_s; t_s, t_s + 10y) - SR(y_s; t_s, t_s + 2y) < b_h\}}}{N}. \tag{107}$$

Our game plan is to evaluate the moment of I_t conditional on the initial and final states (y_1, y_2) and approximate the distribution of I_t with an analytical tractable probability distribution by means of moment matching.

We consider the problem of computing the moments of I_t. If $\varepsilon \in \mathbb{R}$, let us construct the operator:

$$\mathcal{L}_\Omega^\varepsilon(y_1, y_2) = \mathcal{L}_\Omega(y_1, y_2) + \varepsilon \varphi(y_1) \delta_{y_1, y_2}, \tag{108}$$

and consider the time interval $\tau = t_2 - t_1$. It is important to note that the operator $\mathcal{L}_\Omega^\varepsilon$ is only a function of (y_1, y_2). This is the case because the lattice function φ depends only on the Markovian process y_t, and in particular does not depend on the past history of I_t itself, and because the Markov generator $\mathcal{L}_\Omega$ is constant over the time interval τ. In fact, as our model has been specified with a time homogeneous parameterisation, $\mathcal{L}_\Omega$ is constant over the life of the trade.

Theorem 7. *If $\mathcal{L}_\Omega^\varepsilon$ is a matrix of the form (108) and I_t is of the form (106), then:*

$$\left(\frac{d}{d\varepsilon}\right)^n\Bigg|_{\varepsilon=0} e^{\mathcal{L}_\Omega^\varepsilon(t_2 - t_1)}(y_1, y_2) = \mathbb{E}_{t_1}\left[I_{t_2}^n \delta(y_{t_2} - y_2) \big| y_{t_1} = y_1\right], \tag{109}$$

for $t_1 < t_2$ and $y_1, y_2 \in \Omega$.

Proof. Consider Neper's formula for the transition probability matrix whose generator is $\mathcal{L}_\Omega^\varepsilon$:

$$e^{\mathcal{L}_\Omega^\varepsilon(t_2 - t_1)}(y_1, y_2) = \lim_{n \to \infty}\left(1 + \frac{t_2 - t_1}{n}(\mathcal{L}_\Omega + \varepsilon V)\right)^n, \tag{110}$$

where V is the matrix $V(y_1, y_2) = \varphi(y_1) \delta_{y_1, y_2}$. By collecting same powers in ε in Neper's formula, one finds Dyson's formula introduced by Dyson (1958):

$$e^{\mathcal{L}_\Omega^\varepsilon(t_2 - t_1)}(y_1, y_2) = e^{\mathcal{L}_\Omega(t_2 - t_1)}(y_1, y_2) + \tag{111}$$

$$\varepsilon \int_{t_1}^{t_2} ds_1 ds_2 \left(e^{\mathcal{L}_\Omega(s_1 - t_1)} V e^{\mathcal{L}_\Omega(t_2 - s_2)}\right)(y_1, y_2) +$$

$$\sum_{n=2}^{\infty} \varepsilon^n \int_{t_1}^{t_2} ds_1 ... \int_{s_{n-1}}^{t_2} ds_{n2} \left(e^{\mathcal{L}_\Omega(s_1 - t_1)} V e^{\mathcal{L}_\Omega(s_1 - s_1)} ...\right.$$

$$\left....V e^{\mathcal{L}_\Omega(t_2 - s_{n-1})}\right)(y_1, y_2).$$

The time-ordered integrals above are proportional to moments and in particular we have that:

$$e^{\mathcal{L}_\Omega^\varepsilon(t_2-t_1)}(y_1, y_2) = \sum_{n=0}^{\infty} \frac{\varepsilon^n}{n!} \mathbb{E}_{t_1} \left[I_{t_2}^n \delta(y_{t_2} - y_2) \big| y_{t_1} = y_1 \right]. \qquad (112)$$

Theorem (7) allows one to obtain the moments of the distribution of I_t conditional on the initial and final states (y_1, y_2). We take advantage of this and proceed with an approximation technique. We evaluate the first two moments of I_t by numerically computing the first two derivatives of eq. (109). We then approximate the distribution of I_t with an analytically tractable probability distribution such that the first two moments are assigned to the ones estimated from eq. (109). In our application, we have chosen to use the standard chi-square distribution. At this stage, the marginal distribution of I_t is fully known and the expectation of the path dependent coupon can be computed by integration across both the state space Ω and the state space for I_t. It is worth noting that, by computing further moments it is possible to assess if the approximation choice is satisfactory. In our experience, little is gained in accuracy by going further than the second moment.

References

Albanese, C.: 2006, Operator methods, Abelian path dependents and dynamic conditioning. Preprint.

Albanese, C. and Kuznetsov, A.: 2003, Discretization schemes for subordinated processes. Mathematical Finance. To appear.

Albanese, C. and Kuznetsov, A.: 2005, Affine lattice models, *International Journal of Theoretical and Applied Finance* **8**(2), 223–238.

Albanese, C. and Mijatovic, A.: 2006, Convergence rates for diffusion on continuous-time lattices. Working paper.

Albanese, C. and Trovato, M.: 2005, A stochastic volatility model for Bermuda swaption and callable CMS swaps. Working paper.

Albanese, C. and Vidler, A.: 2007, A structural model for credit-equity derivatives and bespoke CDOS. Working paper.

Andersen, L.B.G. and Andreasen, J.: 2000, Volatility skews and extensions of the LIBOR market model, *Applied Mathematical Finance* **7**, 1–32.

Andersen, T.G., Benzoni, L. and Lund, J.: 2004, Stochastic volatility, mean drift, and jumps in the short-term interest rate. Working paper.

Beckers, S.: 1980, The constant elasticity of variance model and its implications for option pricing, *Journal of Finance* **35**(3), 661–673.

Black, F. and Karasinski, P.: 1991, Bond and option pricing when short rates are lognormal, *Financial Analysts Journal.* pp. 52–59.

Bochner, S.: 1955, *Harmonic analysis and the theory of probability*, University of California Press.

Brace, A., Gatarek, D. and Musiela, M.: 1996, The market model of interest rate dynamics, *Mathematical Finance* **7**, 127–154.

Brigo, D. and Mercurio, F.: 2006, *Interest Rate Models - Theory and Practice*, Springer Verlag.

Carr, P., Geman, H. and Madan, D.: 2001, Pricing and hedging in incomplete markets, *Journal of Financial Economics* **62**(1), 131–167.

Coppersmith, D. and Winograd, S.: 1987, Matrix multiplication via arithmetic progressions, *19th Annual ACM Symp. Theory Comp* **2**(3), 1–6.

Corcuera, J.M., Nualart, D. and Schoutens, W.: 2005, Completion of a Lévy market by power-jump-assets, *Finance and Stochastics* **9**, 109–127.

Cox, J.C.: 1996, The constant elasticity of variance option pricing model, *Journal of Portfolio Management* **12**, 15–17.

Cox, J.C., Ingersoll, J.E. and Ross, S.A.: 1985, A theory of the term structure of interest rates, *Econometrica.* **53**, 385–407.

Di Graziano, G. and Rogers, L.C.G.: 2006, A dynamic approach to the modelling of correlation credit derivatives using Markov chains. Working paper.

Dyson, F.J.: 1958, Integral representation of casual commutators, *Physical Review* **110**(6), 1460–1464.

Eberlein, E., Jacod, J. and Raible, S.: 2005, Lévy term structure models: No-arbitrage and completeness, *Finance and Stochastics* **9**, 67–88.

Friedman, A.: 1975, *Stochastic differential equations and applications*, Vol. 1, Academic Press, New York.

Friedman, A.: 1976, *Stochastic differential equations and applications*, Vol. 2, Academic Press, New York.

Geman, H.: 2005, From measure changes to time changes in asset pricing, *Journal of Banking and Finance* **11**, 2701–2722.

Geman, H., Madan, D. and Yor, M.: 2001, Time changes for Lévy processes, *Mathematical Finance* **11**, 79–96.

Geman, H., Madan, D. and Yor, M.: 2003, Stochastic volatility for Lévy processes, *Mathematical Finance* **13**, 345–382.

Grimmett, G. and Stirzaker, D.: 2001, *Probability and Random Processes*, Oxford.

Hamilton, J.D.: 1998, Rational expectations econometric analysis of changes in regime, *Journal of Economic Dynamics and Control* **12**, 385–423.

Hull, J. and White, A.: 1993, One-factor interest rate models and the valuation of interest rate derivative securities, *Journal of Financial and Quantitative Analysis* **28**, 235–54.

Hunt, P.J. and Kennedy, J.E.: 2004, *Financial Derivatives in Theory and Practice*, Wiley.

Jashmidian, F.: 1997, Libor and swap market models and measures, *Finance and Stochastics* **1**(4), 293–341.

Karatzas, I. and Shreve, S.: 1991, *Brownian Motion and Stochastic Calculus*, Springer.

Kijima, M.: 1997, *Markov processes for stochastic modelling*, Champman and Hall.

Konikov, M. and Madan, D.: 2001, Option pricing using variance gamma Markov chains, *Review of Derivative Research* **5**(1), 5–115.

Kushner, H. and Dupuis, P.: 2001, *Numerical Methods for Stochastic Control Problems in Continuous Time*, Springer.

Madan, D., Carr, P. and Chang, E.: 1998, The variance gamma process and option pricing, *European Finance Review* **2**, 79–105.

Madan, D.B. and Seneta, E.: 1990, The variance gamma (v.g.) model for share market returns, *Journal of Business* **63**, 511–524.

Merton, R.C.: 1976, Option pricing when underlying stock returns are discontinuous, *Journal of Financial Economics* **3**(1-2), 125–144.

Moler, C. and Van Loan, C.: 2003, Nineteen dubious ways to compute the exponential of a matrix, twenty-five years later, *SIAM Review* **45**(1), 3–30.

Morini, M.: 2005, The implementation of the libor market model. Risk Training, New York.
URL: *www.globalriskguard.com/resources/fideriv/LMM_slides.pdf*

Norberg, R.: 2003, The Markov chain market, *ASTIN Bull* **33**, 265–287.

Phillips, R.S.: 1952, On the generation of semigroups of linear operators, *Pacific Journal of Mathematics* **2**(3), 343–369.

Rebonato, R. and Joshi, M.: 2001, A joint empirical and theoretical investigation of the modes of deformation of swaption matrices: implications for model choice. QUARC Working paper.

Schoutens, W.: 2006, Exotic options under Lévy models: an overview, *J. Comput. Appl. Math.* **189**(1), 526–538.

Stanton, R.H.: 1997, A nonparametric model of term structure dynamics and the market price of interest rate risk, *Journal of Finance* **52**, 1973–2002.

Susmel, R. and Kalimipalli, M.: 2001, Regime-switching stochastic volatility and short-term interest rates, *CEMA Working Papers 197*, Universidad del CEMA.

Vasicek, O.A.: 1977, An equilibrium characterization of the term structure, *Journal of Financial Economics* **5**, 177–88.

Whaley, R.C. and Petitet, A.: 2005, Minimizing development and maintenance costs in supporting persistently optimized BLAS, *Software: Practice and Experience* **35**(2), 101–121.

Duali: Software for Solving Stochastic Control Problems in Economics

David A. Kendrick[1], Marco P. Tucci[2], and Hans M. Amman[3]

[1] Department of Economics, University of Texas, Austin, Texas 78712, USA
   (kendrick@eco.utexas.edu).
[2] Dipartimento di Economia Politica Università di Siena, Piazza S. Francesco 7,
   53100 Siena, Italy (tucci@unisi.it).
[3] Utrecht School of Economics, Utrecht University, Heidelberlaan 8,
   3584 CS Utrecht, The Netherlands (amman@uu.nl).

Summary. Currently there is a renewed interest in the use of optimal experimentation (adaptive control) in economics. The Methods Comparison Project deals with comparing various methods for solving optimal experimentation economic models. In this context, the Beck and Wieland model and the methodology to solve this model with time-varying parameters using adaptive control is introduced. Numerical results for this model using the DualPC software are also presented.

Key words: Active learning, dual control, optimal experimentation, stochastic optimization, time-varying parameters, numerical experiments.

Preamble

When Erricos Kontoghiorghes invited us to contribute to this Festschrift for Manfred Gilli, we were delighted to be able to participate. Manfred has made many important contributions to the literature on computational economics and computational finance including recent work on methods of estimating agent based models, viz Gilli and Winker (2003). Also, Manfred has been one of those unselfish scholars who generates large externalities which facilitate the research and communication of their colleagues. For example, he has played a pivotal role in the Society of Computational Economics for many years. He was the organizer of the third meeting of this society in Geneva, a meeting which set the bar very high for future meetings in the quality of the organization. Also, he served for many years on the Advisory Board and is currently the President of the Society.

It goes without saying that we hope that Manfred will remain as productive in the coming years as he has been in the past. Thus we look forward to enjoying his cheerful company at many future scholarly meetings.

1 Introduction

This paper and its companion Amman et al. (2007) are a part of a methods comparison project initiated at the Washington D.C. meeting of the Society for Computational Economics (SCE) in 2005 by Thomas Cosimano, Volker Wieland and David Kendrick and carried into its second stage by participation in a session on the project at the 2006 meeting in Cyprus of the SCE by these three plus Hans Amman, Gunter Beck, Michael Gapen, and Marco Tucci. The methods under study are those used for *learning and control, optimal experimentation* and *adaptive* or *dual control* as the subject has been called by various authors. In essence, the methods consider dynamic stochastic models in which the control variables can be used not only to guide the system in desired directions, but also to improve the accuracy of estimates of parameters in the models. Thus there is a tradeoff in which experimentation or perturbation of the control variables early in time detracts from reaching current goals, but leads to learning or improved parameter estimates and thus improved performance of the system later in time - hence the dual nature of the control.

Cosimano (2003) and Cosimano and Gapen (2005a; 2005b; 2006) use perturbation methods, which are applied in the neighborhood of the augmented linear regulator problems discussed by Hansen and Sargent (2004). Wieland (2000a; 2000b) uses numerical approximation of the optimal decision rule in methods that are related to earlier work by Prescott (1972), Taylor (1974) and Kiefer (1989). Kendrick (1981; 2002) uses adaptive or dual control methods that draw on earlier work in the engineering literature by Tse and Bar-Shalom (1973).

Our purpose in this project is to better understand the comparative advantage of the three methods in their application to dynamic stochastic economic models. To this end we have decided to begin with the model in Beck and Wieland (2002) (BWM). A step in this direction is in Kendrick (2006), that provides a comparison of mathematical results from Kendrick (2002)) and Beck and Wieland (2002)). Also, Cosimano and Gapen (2006) have provided both mathematical and computational comparisons of their method to those of the Beck and Wieland model (BWM) and to the mathematical results in Kendrick (2006).

In this paper we present the computational results of the BWM using a number of stochastic control methods embodied in the Duali software. In the companion article, Amman et al. (2007), we provide the related results using the same and other stochastic control methods in the DualPC software. The difference between Duali and DualPC is that Duali is easier to use since it is written in C and has a Windows interface while DualPC is written in Fortran and has only a input interface. However Duali does not yet have all the features that are implemented in the DualPC software.[4]

---

[4] A copy of either *DualPC* or *Duali* can be obtained, for non commercial use, by sending a request to the corresponding author at amman@uu.nl.

Our purpose is two fold. First, we want to provide a number of models and methods which are simple enough that they can be used to check the numerical results obtained with different computer codes. Secondly, we provide a number of models and methods which provide numerical results that can be used to help in finding the comparative advantages across different methods and the implementation of those methods in various computer codes.

However, before doing this we repeat here some of the materials in Kendrick (2006) which provides the BWM first in Wieland's and then in Kendrick's notation. This is followed by a discussion of the numerical results for the BWM using four variants that depend on the stochastic specification. The first variant has no stochastic terms, the second has additive noise terms, the third has multiplicative uncertain parameters as well and the fourth has time varying stochastic parameters along with the other two kinds of uncertainty. Different variants of the model are solved with the following methods:

- Open Loop (*OL*);
- Optimal Feedback without Updating (*OFwU*);
- Optimal Feedback (*OF*);
- Expect Optimal Feedback (*EOF*);
- Optimal Feedback with Time Varying Parameters (*OFwT*);
- *EOF* with time-varying Parameters (*EOFwT*);

The names for the methods are those used in the classification system in Kendrick and Amman (2006). Then in the companion piece Amman et al. (2007) we provide solution with a subset of the methods above along with solutions with the Dual Control (*DC*) method.

2 The Beck and Wieland Model in Wieland's Notation

The BWM is a one-state, one-control model with a single time varying parameter. The system equation is

$$x_{t+1} = \gamma x_t + \beta_t u_t + \alpha + \varepsilon_t \qquad (1)$$

where

$$
\begin{aligned}
x_t &= \text{state vector} \\
u_t &= \text{control vector} \\
\beta_t &= \text{a stochastic time-varying parameter} \\
\gamma, \alpha &= \text{constant coefficients} \\
\varepsilon_t &= \text{additive noise term} ,
\end{aligned}
\qquad (2)
$$

where $t \in \{0, .., N-1\}$ is the (discrete) time subscript, $\varepsilon_t \sim N(0, \sigma_\varepsilon)$ and with an estimate $\hat{x}_0$ given as the initial condition for the state. The control variable in equation (1) in the BWM has the time subscript $t+1$ rather than t; however we follow here the convention in the engineering literature in which the control variable action at time t affects the system in time $t+1$. Also, the noise term ε has the subscript t rather than the $t+1$ used in the BWM. The time-varying parameter equation

$$\beta_{t+1} = \beta_t + \eta_t, \tag{3}$$

where η_t is an additive noise term with $\eta_t \sim N(0, \sigma_\eta)$ and with the initial condition for the time varying parameter β_0 given. Also, the noise term η_t has the subscript t rather than the $t+1$ used in the BWM.

The criterion function is a quadratic tracking equation where the goal is to find the minimum over the controls $\{u_t\}_{t=0}^{N-1}$ of

$$J = E\left[\delta^N(x_N - \tilde{x}_N)^2 + \right.$$

$$\left. \sum_{t=0}^{N-1} \delta^t\{(x_t - \tilde{x}_t)^2 + \omega(u_t - \tilde{u}_t)^2\}\right], \tag{4}$$

where

$$J = \text{criterion value}$$
$$E = \text{expectations operator}$$
$$\delta = \text{discount factor}$$
$$\tilde{x}_t = \text{desired state variable}$$
$$\tilde{u}_t = \text{desired control variable}$$
$$\omega = \text{(relative) weight}.$$

The criterion function is over a finite horizon here in contrast to the BWM where it is infinite horizon. Also, the tracking function for the last time period is separated here to indicate that the control variables are optimized only through period $N-1$.

For their numerical experiments Beck and Wieland use the following values $\alpha = 0$, $\gamma = 1$, $\sigma_\eta = 0$, $\sigma_\varepsilon = 1$, $\omega = 0$, $\forall t$ $\tilde{x}_t = 0$, $\forall t$ $\tilde{u}_t = 0$, $\delta = 0.95$, where σ_ε is the variance of ε_t in the system equation and σ_η is the variance of η_t in the parameter evolution equation. The initial conditions $x_0 = 0$, $b_0 = -0.50$, $v_0^b = 0.25$, where v_0^b is the variance at $t = 0$ of the uncertain parameter. Also, the symbol b is used to indicate the estimates of the parameter β_t, that is $\hat{\beta}_0 = b_0$ and $\hat{\sigma}_0 = v_0^b$, where the ˆ indicates an estimate.

3 The Beck and Wieland Model in Kendrick's Notation

The model in Kendrick (2002) that most closely approximates the BWM is the one in Chapter 10 since that model includes time-varying parameters. In addition, some use will be made of the notation in Amman and Kendrick (2000) because that paper includes the discounting that is used in the BWM, but is not present in the Kendrick (2002, Chapter 10) model.

The systems equations in Kendrick (2002, equation (10.7)) are

$$x_{t+1} = A_t(\theta_t)x_t + B_t(\theta_t)u_t + c_t(\theta_t) + \nu_t ,\tag{5}$$

where

$$
\begin{aligned}
t &= \text{time index}\\
x_t &= \text{state vector}\\
u_t &= \text{control vector}\\
\nu_t &= \text{additive noise vector}\\
A_t(\theta_t) &= \text{state vector coefficient matrix}\\
B_t(\theta_t) &= \text{control vector coefficient matrix}\\
c_t(\theta_t) &= \text{exogenous coefficient vector}\\
\theta_t &= \text{subset of the coefficients treated as uncertain}.
\end{aligned}
$$

The matrix A_t is a function of the subset of the uncertain coefficients in the vector θ_t. The same applies to B_t and c_t. For the BWM there is a single state variable and a single control variable so these two vectors each have a single element. Also there is a single uncertain coefficient so θ_t is

$$\theta_t = \beta_t .\tag{6}$$

Comparison of equation (5) to the BWM system equation (1) yields $A_t = \gamma = 1$, $B_t = \beta_t$, $c_t = \alpha = 0$, and $\nu_t = \varepsilon_t$. Because these notes draw on mathematics from two different sources we will occasionally encounter cases where the same symbol is used for different purposes in the two sources. When this occurs we will rely on the context to communicate the differences, for example in (5) and in the equation above ν_t is used to indicate the additive noise term for the systems equations in the Kendrick (1981) framework. In contrast, ν_t is used in Beck and Wieland (2002) to indicate the variance of the estimate of the β_t parameter. Also, we will be using equation numbers from multiple sources and, here also, we rely on context rather than special fonts to distinguish the sources.

The measurement equation in Kendrick (2002, equation (10.8)) is

$$y_t = H_t x_t + w_t ,\tag{7}$$

where

$$
\begin{aligned}
y_t &= \text{measurement vector}\\
H_t &= \text{measurement coefficient matrix}\\
w_t &= \text{measurement noise vector}.
\end{aligned}
$$

Though Coenen and Wieland (2001) have included measurement errors in one of their papers, those errors are not included in the BWM. Hence, we have $H = I$ and $w_t = 0$.

The time-varying parameter equation in Kendrick's (2002, equation (10.9)) model is

$$\theta_{t+1} = D_t \theta_t + \eta_t \, , \tag{8}$$

where

$$D_t = \text{the parameter evolution matrix}$$
$$\eta_t = \text{additive noise terms} \, .$$

For more general forms of equation (8) in the adaptive control context, including the *return to normality* model, see Tucci (2004). In the BWM there is a single time-varying parameter, also the coefficient D is one, thus

$$D_t = I = 1 \, . \tag{9}$$

Also, the additive noise term in the time-varying parameter in equation (8) is distributed

$$\eta_t \sim N(0, \sigma_\eta) \, . \tag{10}$$

The initial conditions for the systems equation (5) and the parameter evolution equations (8) in the Kendrick (2002, Chapter 10) model are

$$x_0 \sim N(E[x_0], \Sigma_0^{xx}) \tag{11}$$

and

$$\theta_0 \sim N(E[\theta_0], \Sigma_0^{\theta\theta}) \, . \tag{12}$$

Usually we do not know $E[x_0]$, Σ_0^{xx}, $E[\theta_0]$ and $\Sigma_0^{\theta\theta}$ in equations (11), (12) and we need to start our experiments with the estimates $\hat{x}_0$, $\hat{\Sigma}_0^{xx}$, $\hat{\theta}_0$ and $\hat{\Sigma}_0^{\theta\theta}$. Since there is no measurement error in the BWM and since the original state is assumed to be zero we have $\hat{x}_0 = 0$ and $\Sigma_0^{xx} = 0$. Also in the BWM the initial value of the time-varying coefficient estimate, $\hat{\beta}_0$, is set to -0.5 and the initial variance of that coefficient is set to 0.25 so we have $\hat{\theta}_0 = b_0 = -0.5$ and $\hat{\Sigma}_0^{\theta\theta} = v_t^b = 0.25$.

The additive noise terms for the systems, measurement and parameter evolution equations in the Chapter 10 model are distributed $\nu_t \sim N(0, Q_t)$, $w_t \sim N(0, R_t)$ and $\eta_t \sim N(0, \Gamma_t)$. These terms in the BWM are $Q = \sigma_\varepsilon = 1$, $R = 0$ and $\Gamma = \sigma_\eta = 0.04$.

Thus, there is a variance of one for the additive noise term in the systems equations, there is no measurement error and the additive noise term in the parameter evolution equation is set to 0.04. The criterion function in the Chapter 10 model is for a finite horizon model. That criterion with the addition of discounting as in Amman and Kendrick (2000) may be written as

$$J = E\left[\delta^N L_N(x_N) + \sum_{t=0}^{N-1} \delta^t L_t(x_t, u_t)\right], \tag{13}$$

where

$$J = \text{criterion value}$$
$$E = \text{expectations operator}$$
$$\delta = \text{discount factor}$$
$$L_N = \text{criterion function at terminal period}$$
$$x_N = \text{state vector}$$
$$L_t = \text{criterion function}$$
$$x_t = \text{state vector}$$
$$u_t = \text{control vector}.$$

Once again t is the time subscript and N is last time period of the planning horizon. The two terms on the right-hand side of equation (13) are defined as

$$L_N(x_N) = \frac{1}{2}(x_N - \tilde{x}_N)'W_N(x_N - \tilde{x}_N) \tag{14}$$

and

$$L_t(x_t, u_t) = \frac{1}{2}\left[(x_t - \tilde{x}_t)'W_t(x_t - \tilde{x}_t) + \right.$$
$$\left. (x_t - \tilde{x}_t)'F_t(u_t - \tilde{u}_t) + (u_t - \tilde{u}_t)'\Lambda_t(u_t - \tilde{u}_t)\right], \tag{15}$$

where

$$\tilde{x_N} = \text{desired state vector for terminal period}$$
$$W_N = \text{symmetric state variable penalty matrix for terminal period}$$
$$\tilde{x}_t = \text{desired state vector for period}$$
$$\tilde{u}_t = \text{desired control vector for period}$$
$$W_t = \text{symmetric state variable penalty matrix for time } t$$
$$F_t = \text{penalty matrix on state-control variable for time } t$$
$$\Lambda_t = \text{symmetric control variable penalty matrix for time } t.$$

The comparison of equations (14) and (15) to the BWM model in equation (4) above and the use of the parameter values specified on page 1367 of their article, yields $W_N = 1$, $W_t = 1$, $F_t = 0$, $\Lambda_t = \omega = 0$ and $\delta = 0.95$. Thus there is a weight of one on the state variable deviations, no weight on the cross terms, and a weight of zero on the control variable deviations. Also, the discount factor is set at 0.95 so the discount rate is ≈ 0.05. The desired paths for both the states and the controls in the BWM are set to zero i.e. $x_N = 0$, $\tilde{x}_t = 1$, $t \in \{0, ..., N-1\}$.

This completes the description of the model in Wieland's notation and in Kendrick's notation. This notation can now be used to discuss the solution of the BWM using various stochastic control methods with the Duali and DualPC software systems. We present solutions to two variants of the BW model: a simpler version without a time-varying parameter and a more complex version with a time varying-parameter. We will apply the following methods as discussed in the classification system suggested by Kendrick and Amman (2006):

- Open Loop (*OL*);
- Optimal Feedback (*OF*);
- Expected Optimal Feedback (*EOF*);
- Dual Control (*DC*) .

The *OL* method is deterministic and does not include any of the uncertain elements in the model. The *OF* method considers the additive noise terms in the system equations. The *EOF* method adds to this the uncertainty in the parameters of the system equations. The DC method includes as well adaptive control or optimal experimentation as described above.

The first three of the methods above are implemented in the Duali software, Amman and Kendrick (1999a), and we use it for these solutions. The dual control method is not yet implemented in Duali so we use the DualPC software, Amman and Kendrick (1990) for those solutions.

4 Open Loop

We begin with the simplest version of the BWM, that is one without a time varying parameter. Also, here we are dropping all stochastic elements and using a strictly deterministic version of the model. For this model the parameters of the systems equations, $\forall t$ are

$$A_t = \gamma = 1$$
$$B_t = \beta = -0.5$$
$$c_t = \alpha = 0$$
$$v_t = \varepsilon_t = 0 .$$

Also we use an initial condition for the system equation that is different than that used by BW. They use $\hat{x}_0 = 0$ but we use

$$\hat{x}_0 = 1 .$$

The reason for this is that the desired paths for the state variable x_t in BW are also zero and by using an initial value of the state that is not equal to the desired path we will have some control action in the first period. The parameters of the criterion function for the BW model are $\forall t$

$$W_N = 1$$
$$W_t = 1$$
$$F_t = 0$$
$$\Lambda_t = \omega = 0$$
$$\delta = 0.95 \, .$$

However, to avoid potential numerical problems we will not use a zero for Λ_t, but rather a small value such as $\Lambda_t = 10^{-6}$. The desired paths for both the states and the controls in the BWM are set to zero that is

$$\tilde{x}_N = 0$$
$$\tilde{x}_t = 0$$
$$\tilde{u}_t = 0$$

where $t \in \{0, ..., N-1\}$ and we use these values. This model is input to the Duali software with one state, one control and no exogenous variables. Appendix A contains notes about the specification of the model in Duali and the commands that are used for the various methods of solution of the model. The model is set up to include three time periods. This number is chosen so as to permit some dynamics in the stochastic solutions to be discussed later in the paper while keeping the model as small and uncomplicated as possible. The model is solved in Duali with the

```
Solve: QLP
```

command. The results are straight forward with the state having the value 1 in period zero and the value 0 in all other time periods. Also, the control variable has the value 2 in period zero and the value 0 in all other periods. Thus the control variable is used in period zero to move the state from its initial value of 1 back to the desired value of zero. After that no use of the control is necessary and the state remains at zero.

5 Optimal Feedback

The Optimal Feedback method is the first step into stochastic control with uncertain elements and involves the use of additive noise terms in the system equations. Thus we add the noise term ε_t to the system equation (1) in the BW model and make use of its variance σ_ε to generate the noise terms in the model with a Monte Carlo generator. The system equation noise term in the Kendrick (1981) notation is ν_t in equation (5) and its variance is Q. In the BW model this variance is $Q = \sigma_\varepsilon = 1.0$. To implement this model in the Duali software we create the file BW-OF.dui. The "Specification: Stochastic Term" dialog box is used to specify that the model is "Stochastic with Additive Noise". Also, the number of Monte Carlo runs is set to one in the "Data: Size" dialog box. See Appendix A, Section A.2, for a discussion of the details of this solution method in Duali.

The model was solved twice in Duali. In the first run the additive random terms were generated internally by the Monte Carlo routine in Duali. These random terms were printed so that they could then be used in the second run. In the second run the internal Monte Carlo generation was turned off and the additive random terms from the first run were read in. This capability of using either internally generated or "read in" stochastic terms has been used in Duali to enable us to check the results from Duali against those obtained from other codes.

The random terms for ν_t (or ε_t in BW notation) that were generated in the first run and used in the second run are displayed in Table 1.

Table 1. Random error terms ε_t.

t	0	1	2
ε_t	−0.110	0.505	−0.439

The model was then solved using the Optimal Feedback method. The optimal state and control time paths for that solution are shown below in Table 2.

Table 2. Optimal states x_t^* and controls u_t^*.

	0	1	2	3
x_t^*	1.00	−0.11	0.51	−0.44
u_t^*	2.00	−0.22	1.01	

Plots of these values for the state variable and the control variable, which are obtained with the "Results: Plot Single Run" command, are shown in Figures 1 and 2.

The solid line is the optimal path and the dashed line in the desired path in both graphs. The solution in the OF case is, of course, more interesting than that in the OL case since the additive noise term continues to bounce the state off of its desired path of zero and the control must be used to bring it back. Note that the state variable is shown for $t \in \{0, 1, 2, 3\}$ while the control variable is only shown for $t \in \{0, 1, 2\}$ since the control is not computed for the last time period because it has lagged effect on the state.

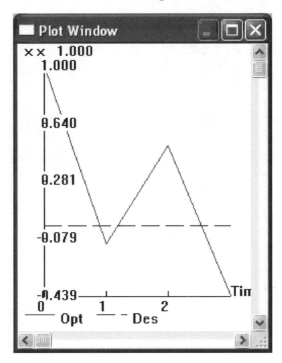

Fig. 1. Optimal state in the *OF* solution.

6 Expected Optimal Feedback

The Expected Optimal Feedback (*EOF*) method includes consideration of the uncertainty of the parameters in the model when computing the feedback rule. Also, in this section we will begin consideration of the use of the Kalman filter to update the estimate of the parameter after each observation.

Strictly speaking the method used in the previous section should be labeled *OFwU*, that is Optimal Feedback with Updating, since the parameter estimates in the model were updated after each new observation. Also, later in the paper we will be comparing the *OF* method with the *EOF* method and then we will be updating the parameter estimates in both the *OF* and the *EOF* case.

It is common when discussing *EOF* solution methods to talk about the parameters being uncertain; however, usually the parameter estimates are treated as uncertain and not the parameters themselves. Later in this paper we will use variants of the BW model that have time varying parameters. In those cases the parameters themselves, and not just their estimates, are uncertain. However, in this section the true values of the parameters will be treated as deterministic and constant and the estimates of these parameters will be treated as uncertain and time varying. Also, when discussing the estimates we will be referring to both the means and the variances of these estimates.

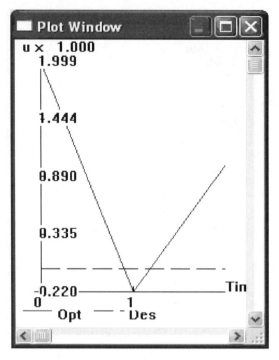

Fig. 2. Optimal control in the *OF* solution.

The BWM treats only one of the parameter estimates of the model as uncertain, namely the β_t coefficient that multiplies the control variable. So in our notation θ_t the vector of uncertain parameters includes only a single element and is the β_t. The initial estimates of the means and variance of this parameter are specified by BW as

$$b_0 = -0.5$$
$$v_t^b = 0.25 .$$

Thus the initial mean estimate of the parameter is -0.5 and the variance of that estimate is 0.25. In our notation this is specified as

$$\hat{\theta}_0 = -0.5$$
$$\hat{\Sigma}_0^{\theta\theta} = 0.25$$

where $\hat{\theta}_0$ is the estimate of the mean at time zero and $\hat{\Sigma}_0^{\theta\theta}$ is estimate of the variance of the parameter at time zero.

The solution procedure for this method is discussed in Section A.3 of Appendix A. The results of that solution including the optimal and desired states and controls as well as the estimates of the mean value of the parameter and its true value are displayed in Figure 3.

States	x opt	des	Controls	u opt	des	Parameters	beta
Time							
0	1.000	0.000		0.283	0.000		-0.072
1	0.748	0.000		0.261	0.000		-0.088
2	1.123	0.000		0.289	0.000		-0.063
3	0.539	0.000					-0.102
	Criterion OLF =		1.555				
						True Value	-0.500

Fig. 3. Optimal states and controls for the *EOF* method.

Just as in the Optimal Feedback solution the initial state is at 1.0; however in this case the state does not move quickly to the neighborhood of the desired state at 0 and varies in the neighborhood of this value as dictated by the additive noise term. Rather it varies but remains distant from the desired path of 0. The reason for this can be seen in the last column of Figure 3, which shows the path of the mean of the estimated parameter which begins in period zero at -0.072. This is some distance from the true value which is shown in the bottom right corner of Figure 3 as -0.500.

What is happening here? In Monte Carlo runs on stochastic control problems with uncertain parameters the Duali software creates a random starting value of the estimate of the mean value of the parameter. In doing so it uses the variance of the parameter, which from above is $\hat{\Sigma}_0^{\theta\theta} = v_t^b = 0.25$. Thus in this case the standard deviation of the mean of the parameter estimate is 0.5. Therefore, it is not surprising that the Monte Carlo generator in Duali would generate an initial value for β of -0.072 which is less than one standard deviation from the true value of -0.500.

Moreover, with such a large initial variance of the parameter estimate one would expect that the control would be rather timid in the *EOF* case. This mimics a situation in which the policy maker is very uncertain about the effect of the policy variable (control) on the evolution of the economy (the state) so he or she is cautious in setting the level of the control. Recall from equations (4.14) and (4.15) in Kendrick (2006) that the feedback rule for the control in the *EOF* case is

$$u_t = G_t^{\dagger} x_t + g_t^{\dagger} \tag{16}$$

with

$$G_t^{\dagger} = -\left[E\{B_t' k_{t+1} B_t\} + \Lambda_t' \right]^{-1} \left[F_t' + E\{B_t' K_{t+1} A_t\} \right]. \tag{17}$$

The expected value operator $E\{\}$ in the inverse term

$$\left[E\{B_t' k_{t+1} B_t\} + \Lambda_t' \right]^{-1} \tag{18}$$

in equation (17), is evaluated to include the covariance of the parameters in the matrix. So when (18) is large, $\|G_t^\dagger\|$ will be small and therefore the control u_t will provide a timid response to the state x_t. This can also be seen clearly in the BW notation for the feedback rule in the *EOF* case with a single state and a single control which is, from equation (4.25) in Kendrick (2006)

$$u_t = \frac{\gamma b_{t-1}}{b_{t-1}^2 + v_{t-1}^b + \sigma_\eta}(x_t - \tilde{x}_t) - \frac{\alpha b_{t-1}}{b_{t-1}^2 + v_{t-1}^b + \sigma_\eta}. \tag{19}$$

Thus when the variance of the parameter estimate v_{t-1}^b is large, the control will respond timidly to the difference between the state x_t and its desired value $\tilde{x}_t$. The upshot of all this is that when the controller begins with a parameter estimate that is distant from its true value and when the parameter estimate has a large variance, it is slow to move the system from its initial state of one toward the desired state of zero. This is shown below in Figures 4 and 5, which provide plots of the optimal states and controls for the *EOF* method.

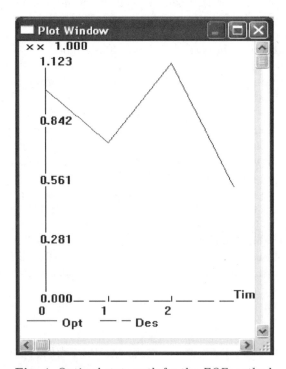

Fig. 4. Optimal state path for the *EOF* method.

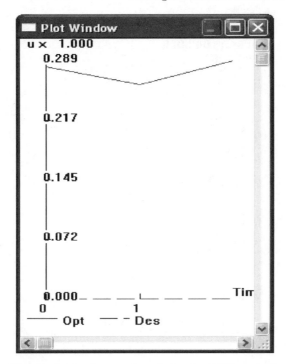

Fig. 5. Optimal control path for the *EOF* method.

Also Figure 6 provides a plot of the estimated parameter. The dashed line at the bottom of Figure 6 at -0.5000 is the true value of the parameter and the solid line at the top of the figure begins in the neighborhood of -0.0631 at -0.072 which is the initial mean value of the parameter estimate as reported in Figure 3 above. Duali automatically scales graphs, thus the values on the vertical axis for the parameter in this figure are multiplied by a factor of 1000 as is shown at the top of the figure. Recall also that the *OLF* (Open Loop Feedback) method listed in Figure 5 is equivalent to the *EOF* (Expected Optimal Feedback) method under the Kendrick and Amman (2006) classification scheme.

Also, because of passive learning the post measurement variance of the estimated parameter, $\Sigma_{t|t}^{\theta\theta}$, decreases over time. The values for $t \in \{0, 1, 2, 3\}$ are shown in Table 3.

Table 3. Post measurement variance of the estimated parameters.

t	0	1	2	3
$\Sigma_{t|t}^{\theta\theta}$	0.250	0.245	0.241	0.236

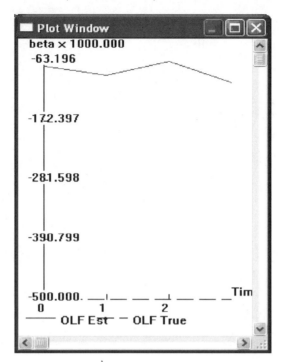

Fig. 6. Estimated parameter mean value for the *EOF* method.

Of course, with active learning as when the dual method is used the decline in this variance over time would be more rapid.

7 *OF* versus *EOF*

Next consider Monte Carlo runs that compare two methods, namely the Optimal Feedback and Expected Optimal Feedback Methods that were discussed in the two previous sections. Recall that the *OF* method used here included updating of parameter estimates and so was in fact *OFwU*, that is Optimal Feedback without Updating. So this section provides comparison of the *OF* and *EOF* methods (Optimal Feedback and Expected Optimal Feedback) - both with updating.

Also, we provide in this section two sets of numerical results. The first set is based on three Monte Carlo runs and is presented in hopes that the numerical deviates used in these runs can be used to check results across various computer codes. The second set is based on one hundred Monte Carlo runs and is a step toward providing information that can be used to help find the comparative advantage of different methods and their implementation in different computer codes. The criterion value for these three Monte Carlo runs for the two methods are in Table 4.

Table 4. Criterion values for *OF* and *EOF* for three Monte Carlo runs.

OF	EOF
18.210	1.555
2.989	2.402
16.151	10.929

The average values across the three Monte Carlo runs are 12.450 for *OF* and 4.962 for *EOF*. It is also interesting to see the average control levels by the two methods across the Monte Carlo runs for the three time periods as shown in Table 5.

Table 5. Average control levels across Monte Carlo methods.

t	0	1	2
OF	10.121	−8.723	0.950
EOF	0.514	−0.232	−0.363

For these particular runs the more cautious approach of the *EOF* method results in smaller control values on average in each time period for that method relative to the *OF* method.

Consider next 100 Monte Carlo runs to compare the two methods. The results in this case are that the *OF* method has the lowest criterion value in 44 runs and the *EOF* method in 56 runs. Also, the average criterion value for the *OF* method is 37.73 and the average criterion value for the *EOF* method is 3.56 so the *EOF* method on average does much better than the *OF* method. For results of this same type with more Monte Carlo runs, but on the MacRae model (1972; 1975), which is similar to the BWM, see Amman and Kendrick (1999b).

8 Expected Optimal Feedback with Time-Varying Parameters (*EOFwT*)

The next step is to add time varying true parameters as is done in part of the BW article. In this case the true parameter does not remain constant at −0.500 but rather varies over time according to the first order process specified in equation (8). As is shown in Figure 7, the true parameter begins in the neighborhood of −0.494 at −0.500 and progresses to −0.609 in $t = 1$, to −0.925 in $t = 2$ and finally to −0.768 in $t = 3$.

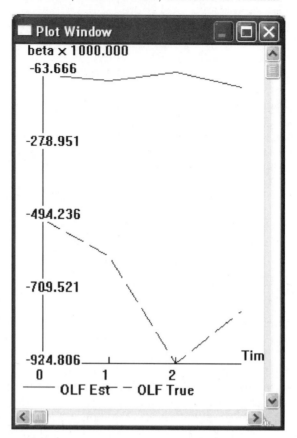

Fig. 7. True and estimated parameter beta.

In contrast, the estimate of the parameter begins in the neighborhood of -0.063 at -0.072 in $t = 0$ and progresses to -0.088 in $t = 1$, to -0.064 in $t = 2$ and finally to -0.107 in period $t = 3$.

9 *OFwT* versus *EOFwT*

Now we are in a position to compare the time-varying parameter version of the BWM and the *OF* and *EOF* methods. Thus we are comparing *OFwT* (Optimal Feedback with Time-Varying Parameters) and *EOFwT*. We will do this with 100 Monte Carlo runs. The details of this case are discussed in Appendix A, Section A.4. The results in this case are that the *OF* method has the lowest criterion value in 43 runs and the *EOF* method in 57 runs. Also, the average criterion value for the *OF* method is 138.1 and the average criterion value for the *EOF* method is 3.7 so the *EOF* method on average does much better than the OF method. These results are essentially the same as those above for 100 Monte Carlo runs on the version of the model with constant true parameters with the exception that the average value of the criterion value for the *OF* method is much higher in this case.

Thus it appears that when the parameters are time varying it is even more important to use the caution that is implicit in the Expected Optimal Feedback method rather than ignoring this uncertainty as is done in the Optimal Feedback method.

10 Conclusion

In this paper, in honor of Manfred Gilli, we have presented the capabilities of the Duali software for solving various types of stochastic control problems. For this purpose we have presented the BWM as a test bed.

References

Amman, H.M. and Kendrick, D.A.: 1990, A user's guide for DUAL: A program for quadratic-linear stochastic control problems, *Technical Report T90-4*, Center for Economic Research, University of Texas, Austin, Texas 78712, USA.

Amman, H.M. and Kendrick, D.A.: 1999a, The Duali/Dualpc software for optimal control models: User's guide, *Working paper*, Center for Applied Research in Economics, University of Texas, Austin, Texas 78712, USA.

Amman, H.M. and Kendrick, D.A.: 1999b, Should macroeconomic policy makers consider parameter covariances, *Computational Economics* **14**, 263–267.

Amman, H.M. and Kendrick, D.A.: 2000, Stochastic policy design in a learning environment with rational expectations, *Journal of Optimization Theory and its Applications* **105**, 509–520.

Amman, H.M., Kendrick, D.A. and Tucci, M.P.: 2007, Solving the Beck and Wieland model with optimal experimentation in DUALPC, *Working paper*, Department of Economics, University of Texas, Austin, Texas 78712, USA.

Beck, G. and Wieland, V.: 2002, Learning and control in a changing economic environment, *Journal of Economic Dynamics and Control* **26**, 1359–1378.

Coenen, G. and Wieland, V.: 2001, Evaluating information variables for monetary policy in a noisy economic environment. European Central Bank, presented at the Seventh International Confernce of the Society for Computational Economics Conference,Yale University, New Haven, USA.

Cosimano, T.F.: 2003, Optimal experimentation and the perturbation method in the neightborhood of the augmented linear regulator, *Working paper*, Department of Finance, University of Notre Dame, Notre Dame, Indiana, USA.

Cosimano, T.F. and Gapen, M.T.: 2005a, Program notes for optimal experimentation and the perturbation method in the neighborhood of the augumented linear regulator problem, *Working paper*, Department of Finance, University of Notre Dame, Notre Dame, Indiana, USA.

Cosimano, T.F. and Gapen, M.T.: 2005b, Recursive methods of dynamic linear economics and optimal experimentation using the perturbation method, *Working paper*, Department of Finance, University of Notre Dame, Notre Dame, Indiana, USA.

Cosimano, T.F. and Gapen, M.T.: 2006, An algorithm for approximating optimal experimentation problems using the perturbation method, *Working paper*, Department of Finance, University of Notre Dame, Notre Dame, Indiana, USA.

Gilli, M. and Winker, P.: 2003, A global optimization heuristic for estimating agent based models, *Computational Statistics and Data Analysis* **42**, 299–312.

Hansen, L.P. and Sargent, T.J.: 2004, Recursive models of dyanmic linear economies. Department of Economics, Univesity of Chicago manuscript.

Kendrick, D.A.: 1981, *Stochastic control for economic models*, 1st edn, McGraw-Hill Book Company, New York, New York, USA.

Kendrick, D.A.: 2002, *Stochastic control for economic models*. 2nd edn available at http://www.eco.utexas.edu-/faculty/Kendrick.

Kendrick, D.A.: 2006, The Beck and Wieland model in the adaptive control framework, *Working paper*, Department of Economics, University of Texas, Austin, Texas 78712, USA.

Kendrick, D.A. and Amman, H.M.: 2006, A classification system for economic stochastic control models, *Computational Economics* **27**, 453–481.

Kiefer, N.: 1989, A value function arising in the economics of information, *Journal of Economic Dynamics and Control* **13**, 201–223.

MacRae, E.C.: 1972, Linear decision with experimentation, *Annals of Economic and Social Measurement* **1**, 437–448.

MacRae, E.C.: 1975, An adaptive learning role for multiperiod decision problems, *Econometrica* **43**, 893–906.

Prescott, E.C.: 1972, The multi-period control problem under uncertainty, *Econometrica* **40**, 1043–1058.

Taylor, J.B.: 1974, Asymptotic properties of multiperiod control rules in the linear regression model, *International Economic Review* **15**, 472–482.

Tse, E. and Bar-Shalom, Y.: 1973, An actively adaptive control for linear systems with random parameters, *IEEE Transaction on Automatic Control* **18**, 109–117.

Tucci, M.P.: 2004, *The Rational Expectation Hypothesis, Time-varying Parameters and Adaptive Control*, Springer, Dordrecht, the Netherlands.

Wieland, V.: 2000a, Learning by doing and the value of optimal experimentation, *Journal of Economic Dynamics and Control* **24**, 501–543.

Wieland, V.: 2000b, Monetary policy, parameter uncertainty and optimal learning, *Journal of Monetary Economics* **46**, 199–228.

Appendix

A The Beck and Wieland Model in *Duali*

This Appendix contains a more detailed explanation about the specification of the various versions of the BWM in Duali and about the methods that are used in that software to solve the model.

A.1 Open Loop

This model is input to the Duali software with one state, one control and one exogenous variable in the file BW-OL.dui. The initial time period is specified as 0 and the terminal time period as 3 so the model has four time periods in all. This number of time periods is chosen so as to permit some dynamics in the stochastic solutions to be discussed later in the paper while keeping the model as small and uncomplicated as possible.

In fact, since the constant term in the BW model is zero one could set the number of exogenous variable to zero instead of one. However this can cause problems in Duali when we get to more complicated solution methods later so it is better to set the number of exogenous variables to one and then be sure that the C matrix and the Z exogenous vector elements are set to zero with the use of the

Data: Systems equation

dialog box. The model is solved in Duali with the command

Solve: QLP Print

and the results are displayed online and also printed in the debug file. We use the convention of naming the debug file the same as the input file, but with the extension .dbg so in this case we call it BW-OL.dbg. In this case the feedback gain matrix, G_t, is constant over time and is 2.0, while the feedback gain vector, g_t, is also constant over time and is 0.0. The results are straight forward with the state having the value 1 in period 0 and the value 0 in all other time periods. Also the control variable has the value 2 in period 0 and the value 0 in all other periods.

A.2 Optimal Feedback

This version of the model is in the file BW-OF.dui. When creating this file in the first run it was necessary to select the "Specification: Source of Random Terms" dialog box and to specify that the random terms should be generated internally. Also, in this dialog box the

Systems Equations

option was selected in the

    Noise Terms for All Periods

section of the dialog box. The model is solved with the command

    Solve: Compare Print

and then only the

    Certainty Equivalence, CE

option is selected. *Certainty Equivalence* is the name we previously used for the Optimal Feedback method. The debug file which is created when "Compare Print" is selected is named BW-OF.dbg.

In the print options the item for the generated random variables was selected. The resulting debug file contains the *xsis* which are the random variables generated for the additive noise terms. This name comes from Kendrick (2002, Chapter 5) which uses the Greek letter *xsi*, ξ, as the symbol for the additive noise in the system equations. These values are shown in Table A.1

Table A.1. Random error terms.

t	0	1	2
ξ	−0.110	0.505	−0.439

Then a second run of the model was done and the source of random variables was changed from "Generate Internally" to "Read In" and the *xsis* above were typed into Duali. Then the model was solved again and the results for the states and controls were the same (to 2 significant digits) as in the run where the *xsis* were generated internally.

The reason for first generating the random term internally and then turning off the internal generation in a second run and entering the random terms as data is to provide the random terms to others in case they wish to check the calculations performed in Duali.

In this case, as in the *OL* case, the feedback gain matrix, G_t, is constant over time and is 2.0, while the feedback gain vector g_t is also constant over time and is 0.0. The results for the optimal states and controls for this single Monte Carlo run as they are printed in the debug file BW-OF.dbg are shown below.

```
xs in CompareCe
           1.000         -0.110          0.505         -0.439
us in CompareCe
           1.999         -0.219          1.010
```

A.3 Expected Optimal Feedback

This version of the model is in the file BW-EOF.dui. This file was created by opening the BW-OF.dui file from the previous section in Duali and then using the transformations capabilities in Duali. The transformation is accomplished by selecting the "Specification: Stochastic Terms" dialog box and clicking on the "Stochastic with Parameter Uncertainty" option. Duali then prompts the user to change the number of uncertain parameters and the number of Monte Carlo runs in the "Data: Size" dialog box. For the case at hand each of these is one. Duali also suggests adding names for the uncertain parameters in the "Data: Acronyms" dialog box and reminds the user that additional data must be entered in the "Data: Parameter Uncertainty" dialog box for THO ($\hat{\theta}_0$), SITTO ($\Sigma_0^{\theta\theta}$) and ITHN (which is the mapping from i to θ and thus the mapping from the position of the uncertain parameter in the A, B or C matrices to the vector). In the case at hand the ITHN input is (1, 1, 1) i.e. the B matrix (1) and the (1,1) element in that matrix.

Also, in the "Specification: Source of Random Terms" dialog box the "Generate Internally" option is selected. In addition in the "Initial Values" section of this dialog box "Uncertain Parameters" is selected and in the "Noise Terms for All Periods" in the "System Equations" option is selected. The model is then solved with the

Solve: Compare Print

command and only the method

OLF, Open Loop Feedback

is selected in the "Method" dialog box. Open Loop Feedback is the name in our previous system for the Expected Optimal Feedback (*EOF*) method.

The debug file is given the name BW-EOF.dbg. One of the print options that are selected in the "Inputs" tab and in the "Stochastic Elements" section of this tab is

Random Variables – Final – Each Run

This causes the random variables as generated by the Monte Carlo routines to be printed in the debug file. The important ones for the method at hand are shown below, namely ξ_t, the for periods 0, 1 and 2 and the initial mean value of the estimate of the parameter beta which is in the vector theta, that is $\hat{\theta}_0$.

```
xsis in GenerateRandomTerms
        -0.110          0.505          -0.439
thetarandom in GenerateRandomTerms
        -0.072
```

In this case the post measurement variance of the estimated parameter, $\hat{\Sigma}_{t+1|t+1}^{\theta\theta}$, changes over time since there is passive learning. The values for $t \in \{0, 1, 2, 3\}$ are shown in Table A.2.

Table A.2. $\hat{\Sigma}^{\theta\theta}_{t|t}$

t	0	1	2	3
$\Sigma^{\theta\theta}_{t|t}$	0.250	0.245	0.241	0.236

Also, the feedback gain coefficients, G_t , are time varying in this case. For $t \in \{0, 1, 2\}$ these values are shown in Table A.3.

Table A.3. G_t

t	0	1	2
G_t	0.283	0.349	0.236

The feedback gain vector elements, in g_t, are zero for all time periods in this case.

A.4 *OF* versus *EOF*

The file for the models in this section is BW-OFvsEOF.dui. It is based on the BW-EOF.dui file and the first modification is to use the "Data: Size" menu to change the number of Monte Carlo runs from one to three. The model is solved with the "Solve: Compare Print" option; however this time both CE (the former name for Optimal Feedback) and *OLF* (the former name for Expected Optimal Feedback) are selected in the "Methods" dialog box.

The debug file name is BW-OFvsEOF.dbg. In the "Debug Print Options" for *OLF* dialog in the "Inputs" tab all three options in the "Model" section are selected and the "Random Variables – Final – Each Run" option is selected in the "Stochastic Elements" section. In the "Results" tab the "Estimated Values for both Means and Variances" are selected as well as the "Final options" for both "States and Controls" and for the "Criterion Value". Also, in the "Averages" tab in the "Moments Over Monte Carlo Runs" section select the "Average" option for both "State" and "Control" in the "Regular" column. Then click the "OK" button.

The "Method Count" dialog box will appear showing that the *EOF* (*OLF*) method proved to have the lower criterion value in all three Monte Carlo runs. This is followed by a dialog box that shows the average criterion values across the two methods. When this is dismissed no display of results appears. Rather it is necessary to put the debug file in an editor in order to see the results.

In order to capture the random deviates for each run it was necessary to run Duali three times and in the "Debug Print Options" for *OLF* dialog box indicate that the Monte Carlo results for the appropriate run number were to be printed in the debug file. The random deviates that are generated by Duali for each of the three runs are given in the listing below.

xsis,

	Period		
MC Run	0	1	2
1	-0.110	0.505	-0.439
2	-1.529	-0.689	-0.559
3	1.109	0.835	0.200

theta hat zero / MC Run

1	-0.072
2	-0.500
3	-0.066

The OF results for the three runs are

theta hats,

-0.072	-0.499	-0.518	-0.519
-0.500	-0.882	-0.669	-0.549
-0.066	-0.418	-0.480	-0.479

xs,

1.000	-5.911	0.516	-0.421
1.000	-1.529	-1.352	-0.902
1.000	-5.271	1.860	0.123

us,

13.602	-11.844	0.996
1.999	-1.733	-2.018
14.762	-12.592	3.872

In the above listing the rows in each section of the table are for the Monte Carlo runs $\{1, 2, 3\}$ and the columns are for $t = \{0, 1, 2, 3, 4\}$. Notice that in the top section of this table the estimates, $\hat{\theta}_t$, converge nicely to the neighborhood of the true value of -0.5 even though in runs 1 and 3 they start somewhat distant from that value. The EOF results for the three runs are given below.

theta hats,

-0.072	-0.088	-0.063	-0.102
-0.500	-0.806	-0.644	-0.512
-0.066	-0.418	-0.480	-0.479

xs,

1.000	0.748	1.123	0.539
1.000	-1.029	-1.230	-1.110
1.000	-5.271	1.860	0.123

us,

0.283	0.261	0.289
1.000	-0.976	-1.359
14.762	-12.592	3.872

The rows in each section of the table are for the Monte Carlo runs $\{1, 2, 3\}$ and the columns are for $t = \{0, 1, 2, 3, 4\}$.

For the 100 Monte Carlo runs the same input file and debug file are used. The only change is the use of the "Date: Size" option to change the number of Monte Carlo runs from 3 to 100.

A.5 Expected Optimal Feedback with Time-Varying Parameters

The file for the models in this section is created by starting with BW-EOF.dui and transforming the model to one with time-varying parameters. This is done by selecting Specification:

Specification: Stochastic Terms

and changing the selection from "Stochastic with Parameter Uncertainty" to "Stochastic with Measurement Error". We do not plan to use measurement error; however, in Duali the measurement error specification also includes time-varying parameters.

When this change is made a reminder appears asking the user to change the number of observation variables in the "Size" dialog box. In accordance with this we change the number of observation variables from blank to one. Next a reminder appears to add the acronym for the observation variable and to add the data in the

Data: Measurement Error Terms

dialog. Before doing this we use

File: Save As

to save the new file with the name BW-EOFwT.dui. Then we select the

Data: Measurement Error Terms

option and fill in the required data. We set H to one, the ZETAS to zero and R to a small number namely 0.001. Also we set D to one, ETAS to zero and GAM to 0.04 . In addition, we set XODEV and THODEV to zero and SIXXO to a small number (0.001) and SITXO to zero. The rule of thumb here is that covariance matrices that might be inverted in Duali are given small non-zero values so that they are nonsingular. Of the settings above the most important is GAM, Γ, which is the variance of the additive noise term in the parameter evolution equation (8). We set this parameter to 0.04.

Check the "Specification: Source of Random Terms" dialog box and be sure that the option for "Time Varying Parameter Eq." is selected. Otherwise the *etas*, η_t, will not be generated and the true parameter will remain constant rather than be time varying.

The model is then solved with the "Solve: Compare Print" command and only the *OLF* (*EOF*) option is selected. The debug file is given the name BW-EOFwT.dbg. In addition to the print options that have been used earlier we also select in the "Results" tab the option "Final under Uncertain Parameters: True Values".

A.6 *OFwT* versus *EOFwT*

To do these experiments of Optimal Feedback with Time-Varying Parameters versus Expected Optimal Feedback with Time-Varying Parameters we simply use the BW-EOFwT.dui file that was discussed in the previous section and change the number of Monte Carlo runs from 1 to 100 in the

Data: Size

dialog box. The model is solved with the

Solve: Compare Print

command and in the "Methods" dialog box both CE (*OF*) and *OLF* (*EOF*) are selected. The

Method Count

dialog box informs us that OF has the lowest criterion value in 43 runs and *EOF* has the lowest criterion value in 57 runs. Also the "Average Criterion Value" dialog box tells us that the OF result is 138.1 and the *EOF* results is much lower at 3.7.

Index

Printing: Krips bv, Meppel, The Netherlands
Binding: Stürtz, Würzburg, Germany